Dictionary of Energy Technology

Dictionary of Energy Technology

Alan Gilpin, BSc(Econ), PhD, CEng, FInstE, FAIE

Commissioner of Inquiry with the New South Wales Government, Australia

in collaboration with

Alan Williams, BSc, PhD, CEng, FInstE, FIGasE, CChem, FRCS

Head of Department of Fuel and Energy, University of Leeds

Butterworth Scientific

Ann Arbor Science

Ann Arbor Science is an imprint of
The Butterworth Group which has principal offices in
London Boston Sydney Wellington Durban Toronto

First published 1982
© Alan Gilpin 1982

British Library Cataloguing in Publication Data

Gilpin, Alan
    Dictionary of energy technology:
    1. Power plants—Dictionaries
    I. Title II. Williams, Alan
    621.4 TJ163.2 80–42357

    ISBN 0–408–01108–4

Typeset by CCC in Great Britain by William Clowes (Beccles)
Limited, Beccles and London

Printed in England by The Camelot Press Ltd, Southampton

Preface

This work first appeared, as a *Dictionary of Fuel Technology*, 12 years ago. Its emphasis was on the commonplace solid, liquid and gaseous fuels and fuel-burning installations. While embracing nuclear power, it paid scant attention to 'other sources of energy', and to the political, economic and environmental content of a subject which has now moved to the centre of world affairs.

This new work, while resting upon the foundations of the old, attempts to broaden the perspectives so that energy and its uses are considered in a contemporary setting. While a fair amount of technology and science pervades the book, there is no longer a tacit acceptance that energy can be treated in an isolated manner.

The work has been strengthened by my several years with the Central Electricity Generating Board in London. Furthermore, in the successful pursuit of a doctorate relating to the role of nuclear power within electricity supply systems, I have gained a deeper understanding of the economic and environmental aspects of the subject.

Much benefit has also been gained from visits to the United States of America, Canada, Scotland, Scandinavia and western European countries generally, together with the Soviet Union, to meet the principal agencies concerned with energy and the environment. Attendance at international conferences, including the UN Conference on the Human Environment in 1972, has also been valuable.

Against an enhanced background, this new book has been written. It is hoped that it will prove of interest and value to a wide range of professional people with interests in the subject, and to those concerned members of the public who follow closely key issues affecting the welfare and survival of an emerging species.

St Paul's College,   Alan Gilpin
University of Sydney,
Sydney,
Australia

Common Abbreviations

AGR	advanced gas-cooled reactor
AIE	Australian Institute of Energy
API	American Petroleum Institute
ASTM	American Society for Testing Materials
BNOC	British National Oil Corporation
BSI	British Standards Institution
BWR	boiling water reactor
CANDU	Canadian deuterium uranium reactor
CEGB	Central Electricity Generating Board
CEQ	Council of Environmental Quality (US)
ECCS	emergency core cooling system
EEC	European Economic Community
EIS	Environmental Impact Statement
EPA	Environmental Protection Agency (US)
ERDA	Energy Research and Development Administration (US)
FAO	Food and Agriculture Organisation (UN)
FBR	fast breeder reactor
FEA	Federal Energy Administration (US)
HTGR	high-temperature gas-cooled reactor
IAEA	International Atomic Energy Agency
ICRP	International Commission for Radiological Protection
IE	Institute of Energy (UK)
IEA	International Energy Agency
IMCO	Inter-governmental Maritime Consultative Organisation
IP	Institute of Petroleum (UK)
LWR	light water reactor
NCB	National Coal Board (UK)
NEAC	National Energy Advisory Committee (Australia)
NPT	Non-Proliferation Treaty for Nuclear Weapons
NRC	Nuclear Regulatory Commission (US)
OECD	Organisation for Economic Co-operation and Development
OPEC	Organization of Petroleum Exporting Countries
PHWR	pressurised heavy water reactor
PWR	pressurised water reactor
SGHWR	steam-generating heavy water reactor
VLCC	very large crude carrier
WHO	World Health Organisation
WMO	World Meteorological Organisation

Scientific Units and Abbreviations

absolute	abs	henry	H
alternating current	AC	hertz	Hz
ampere	A	horsepower	hp
ångström	Å	hour	h
atmosphere	atm		
atomic number	at. no.	inch	in
atomic weight	at. wt.	inch of mercury	inHg
boiling point	b.p.	joule	J
British thermal unit	Btu		
		Kelvin	K
calorie	cal	kilocalorie	kcal
candela	cd	kilogram	kg
centigrade heat unit	Chu	kilogram per cubic	
centimetre	cm	metre	kg/m^3
centipoise	cP	kilograms per hour	kg/h
centistokes	cSt	kilometre	km
coulomb	C	kilometres per hour	km/h
cubic inch	in^3	kilowatt	kW
cubic centimetre	cm^3	kilowatt-hour	kWh
cubic decimetre	dm^3		
cubic metre	m^3	litre	l
cubic millimetre	mm^3	lumen	lm
curie	c	lux	lx
day	d		
dyne	dyn	maximum continuous	
degree Celsius		rating	MCR
(Centigrade)	°C	megajoule	MJ
degree Fahrenheit	°F	megawatt	MW
degree Rankine	°R	melting point	m.p.
direct current	DC	metre	m
		metre per second	m/s
electromotive force	emf	metre per second	
		squared	m/s^2
farad	F	micrograms per cubic	
foot	ft	metre	μ/m^3
		micron	μm
grain	gr	milligram	mg
gram	g	milligrams per cubic	
grams per cubic metre	g/m^3	metre	mg/m^3
grams per hour	g/h	millilitre	ml
grams per kilometre	g/km	millimetre	mm
grams per square metre	g/m^2	millimetre of mercury	mmHg
		minute	min
hectare	ha	molecular weight	mol.wt.

nanometre	nm	specific gravity	sp. gr.
newton	N	specific heat	sp. ht.
newton per square		square inch	in^2
metre	N/m^2	square centimetre	cm^2
normal cubic metre	Nm^3	square decimetre	dm^2
normal temperature and		square kilometre	km^2
pressure	NTP	square metre	m^2
		square millimetre	mm^2
ohm	Ω	standard temperature	
		and pressure	STP
poise	P	stokes	St
pound	lb		
pound-force	lbf	tesla	T
poundal	pdl		
parts per hundred		volt	V
million	pphm		
parts per million	ppm	water gauge	w.g.
		watt	W
revolution per minute	rev/min	weber	Wb
		weight	wt.
second	s		
second per second	s^2	yard	yd

A

Å *See Ångström.*

Abel flash point apparatus Apparatus for determining the *flash point* of petroleum products which flash below 50° C. It consists of a test cup to hold the sample, a lid which carries a thermometer, a test flame device and a stirrer; the test cup is mounted in a water-bath with an annular air space between the two. The temperature of the water-bath is slowly raised and with it the temperature of the oil sample; at regular intervals the test flame is applied through a special aperture to the interior of the cup, the temperature at which a flash occurs in the vapour space being recorded as the 'flash point'. *See Pensky–Martens flash point apparatus.*

Abiotic Non-biological; thus an abiotic element is a physical or chemical feature of an *ecosystem* or an *environment*.

Absolute humidity The ratio of the mass of water vapour to the volume occupied by a mixture of water vapour and dry air. It is calculated:

$$d_v = \frac{m_v}{V} \text{ g/m}^3$$

in which:

d_v = density of water vapour;
m_v = mass of water vapour, g;
V = volume of water vapour and dry air, m³.

See relative humidity.

Absolute pressure A pressure expressed with zero pressure as the datum, compared with gauge pressure using atmospheric pressure as the datum. Zero pressure is -760 mmHg (approximately -14.7 lb/in²). In SI units the pascal (Pa) is the unit of pressure, with some usage of the bar (1 bar = 100 000 Pa). The bar is approximately equal to atmospheric pressure. Hence, absolute pressure is -10^5 Pa or -1 bar.

Absolute temperature Temperature measured from absolute zero; that is, the temperature at which a *perfect gas*, kept at constant

1

volume, exerts no pressure. It is equal to zero on the **Kelvin scale**; it is also equal to $0°$ R, $-273.15°$ C and $-459.67°$ F. Absolute zero is also the level at which any gas would theoretically be reduced to zero volume, since gases, at constant pressure, lose approximately $1/273$ of their volume at $0°$ C for each fall of $1°$ C in their temperature. *See temperature scales.*

Absolute thermodynamic temperature   *See Kelvin scale.*

Absolute viscosity   *See dynamic viscosity; viscosity.*

Absolute zero   *See absolute temperature.*

Absorption   The passing of a substance or force into the body of another substance; a liquid may be absorbed and held by cohesion or capillary action in the pores of a solid; or gaseous molecules may be held between the molecules of a liquid. The characteristic waves of heat or light radiations may be retained by a solid, liquid or gas, being transformed into either kinetic energy or greater molecular vibrations, when the temperature of the absorbing substance rises; or into excited atoms or molecules when the substance becomes fluorescent. *See adsorption.*

Adsorption coefficient   (1) The volume of gas, measured at normal temperature and pressure, dissolved by unit volume of a liquid under a pressure of one atmosphere. (2) The degree to which a substance will absorb radiant energy. When a parallel beam of radiation passes through a small thickness, x, of a uniform substance, the fraction absorbed is $u_a x$, where u_a is the absorption coefficient of the substance for the radiation.

Absorption control   In respect of a *nuclear reactor*, the use of a neutron absorber to absorb some of the neutrons and vary the reactivity. The absorber, usually cadmium or boron, is incorporated in control rods. The position of these rods can be varied in relation to the core of the reactor.

Absorption oil   An oil of high affinity for the light hydrocarbons. *See Absorption Plant.*

Absorption plant   A plant for recovering the condensable portion of natural or plant gas, by absorbing these hydrocarbons in an *absorption oil*, usually under pressure, followed by separation and fractionation of the absorbed material. *See natural gasoline.*

Absorptivity   The fraction of incident radiation which is absorbed by a surface on which it falls; a perfect absorber is a *black body*. The absorptivity of a material is numerically equal to its *emissivity*.

Abundance ratio   The ratio of the number of atoms of one isotope to others of the same element, in a natural or enriched material. *See isotopes.*

Acceleration   The rate of change of velocity, measured as a change in velocity in unit time—e.g. metres per second per second (m/s^2).

Accelerator   A machine for accelerating to high kinetic energy charged atomic particles such as electrons, protons, deuterons and helium ions. Common machines are the cyclotron, synchrocyclotron, synchrotron, betatron, linear accelerator and Van de Graaff accelerator.

Acetylene C_2H_2. A colourless, poisonous, gaseous fuel. It may be produced by the chemical reaction of calcium carbide and water:

$$CaC_2 + 2H_2O \rightarrow C_2H_2 + Ca(OH)_2$$

Hydrated lime is a by-product of the process. It is also manufactured from methane, heavy gas oil and naptha. In one example, oxygen and *natural gas* are preheated to 510–650° C and mixed, there being insufficient oxygen for complete combustion; the reaction achieved is:

$$2CH_4 \rightarrow C_2H_2 + 3H_2$$

Acetylene is an important industrial gas; an oxygen and acetylene mixture (used with an oxyacetylene torch) burns with the highest temperature of any common combustible gas, and, hence, has great value welding and cutting steel and other metals. The temperature of the inner brilliant part of the flame has been estimated to be of the order of 3300° C. Acetylene has a gross calorific value of about 57 MJ/m^3. *See Neutral Flame.*

Acid   A substance which, in solution in water, forms hydrogen ions. Acids neutralise alkalis, with the formation of salts. *See pH.*

Acid chimney   A chimney in which the temperature of the process waste gases is below the *dew point* of the vapours, so that acidic condensates are formed. Such chimneys require special acid-resisting linings. In cases where the problem is marginal, insulation of the chimney may be sufficient to keep the gases above acid dew point.

Acid cleaning   A common method of internal cleaning of pipework, tanks, heater shells, boilers and condensers; acid is circulated to remove dirt or scale.

Acid dew point   *See sulphur cycle; sulphur dioxide.*

Acid refractories   Refractory materials containing over 90 per cent of *silica*; they are used in open-hearth and other metallurgical furnaces to resist high temperatures and attack by acid slags.

Acid sludge   A black, viscous residue left after the treatment of petroleum oils with sulphuric acid for the removal of impurities. The sludge contains both spent acid and the impurities removed from the oil. *See acid treatment.*

Acid soot (or **acid smut**)   An agglomeration of carbon particles held together by moisture which has become acidic through combination with sulphur trioxide; soot particles range in size from about 1 mm to about 3 mm in diameter. The carbon particles are mainly

3

coke spheres produced during combustion. Acid soot emitted from chimneys leaves brown stains on materials and damages paintwork; the brown stain is caused by iron sulphate. The problem has been mainly associated with oil-fired installations equipped with metal chimneys. The potential hazard can be reduced by using fuels of relatively low sulphur content; operating plant with a minimum of excess air in order to reduce the formation of sulphur trioxide; the elimination of air in-leakage to flues; the raising of backend temperatures; the use of additives; or the insulation of chimneys and ductwork; or by a combination of such measures.

Acid treatment (1) In respect of oil wells, the use of hydrochloric acid in limestone formations to enlarge production channels and improve the movement of fluid towards the well; hydrochloric acid is used for the ease with which it will attack limestone. (2) An oil refinery process in which unfinished petroleum products, such as gasoline, kerosine, diesel fuel and lubricating stocks, are brought into contact with sulphuric acid to improve their colour, odour and other properties. *See acid sludge.*

Acre-foot A unit representing a volume of water sufficient to cover an acre of land to a depth of one foot. 1 acre-foot = 43 560 ft^3 = 1233.48 m^3.

Actinides Elements with 89 or more protons in their nuclei; they include uranium and plutonium. Many are long-lived alpha-emitters.

Activity The number of disintegrations per unit time taking place in a radioactive specimen. The unit of activity is the *curie*.

Additives *See detergent; diesel index; gasoline additives; lead susceptibility; pour point depressant; viscosity index improver.*

Adiabatic Without loss or gain of heat to a system. Thus, an adiabatic change is a change in the volume and pressure of a parcel of gas without exchange of heat between the parcel and its surroundings.

Adiabatic efficiency In respect of a *steam engine* or *steam turbine*, the ratio of the work done per pound of steam to the available energy represented by the *adiabatic heat drop*.

Adiabatic heat drop In respect of a *steam engine* or *steam turbine*, the heat energy released and theoretically capable of transformation into mechanical work during the *adiabatic* expansion of unit weight of steam.

Adiabatic lapse rate *See lapse rate.*

Administered pricing Pricing by 'price makers', not 'price takers' as under truly competitive conditions; the pricing policies of monopolistic and oligopolistic sellers. Administered prices tend to be cost-determined—i.e. they are arrived at by applying various percentages to direct labour and material costs to allow for overheads and profits. The adjustment of production to changes in demand tends to take place independently of price changes through stock adjustments; production is reduced in response to an accumulation of unsold goods and raised in the face of a depletion.

Adsorption The taking up of one substance on the surface of another; adhesion. Adsorbents in industrial use include activated carbon, activated alumina, silica gel and fuller's earth. *See absorption.*

Advanced gas-cooled reactor (AGR) An improved design of *nuclear reactor*, compared with its predecessor of Magnox design. As in the Magnox reactor, the heat of fission is removed from the reacting core by circulating carbon dioxide under pressure through it and passing this coolant gas through a boiler; graphite is also used as a moderator. The fuel elements, however, are made not of metallic uranium in a magnesium can, but of uranium oxide in a stainless steel can. This enables the heat of fission to be removed at a higher temperature in the coolant gas, and higher steam temperatures and pressures in the boilers can be achieved. The reactor shell may be protected from excessive heat by the use of a 'hot box' or heat exchanger, the gases only coming into contact with the shell after being in contact with the heat exchanger. Dungeness 'B' nuclear power station is of AGR design. *See Figure A.1.*

Adventitious ash Incombustible materials such as shale, clay, *pyrites*, *ankerite*, dirt from earthy or stony bands in coal seams, and fragments of stone from the roof or floor of a seam, which are found in *run-of-mine coal*. During mechanical cleaning, much of the adventitious material can be removed. *See ash; inherent ash.*

Advisory Council on Energy Conservation (ACEC) An advisory body responsible to the UK *Department of Energy* to assist in the formulation of plans for energy conservation. An ACEC report on energy conservation was published by the Department of Energy in 1975. The report made recommendations on ways of saving energy through prices policy, education and training, and policies relating to transport and building regulations.

Advisory Council on Research and Development for Fuel and Power (ACORD) An advisory body responsible to the UK *Department of Energy* to assist in the formulation of future plans for research and development in the energy technologies. A discussion document, *Energy R & D in the United Kingdom*, was released in 1978.

Aerated burner A *gas burner* in which the gas induces primary air immediately before the burner ports. *See bunsen burner.*

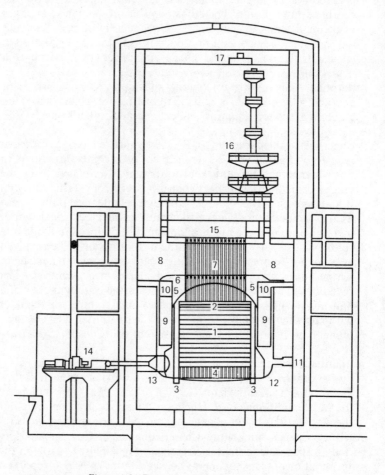

Figure A.1   Dungeness 'B' nuclear power station

1. Core and reflector
2. Top shield
3. Side shield
4. Support structure
5. Pressure cylinder
6. Thermal insulation
7. Fuelling and control standpipes
8. Concrete pressure vessel
9. Boiler
10. Steam outlet
11. Feed water inlet
12. Plenum chamber
13. Circulator
14. Circulator drive
15. Charge face
16. Fuelling machine
17. Charge face crane

(Source: Central Electricity Generating Board, London)

6

Dungeness 'B' Nuclear Power Station: operating data

Net electrical output	1200 MW
Gross generation	1320 MW
Number of reactors	2
Number of turbo-generators	2
Overall station efficiency	41.5%
Type of fuel	36 × 0.57 in pin clusters
Number of fuel elements in stringer	8
Mean fuel rating for reactor	9.5 MW/tonneU
Initial enrichment	1.47/1.76%
Feed enrichment	1.99/2.42%
Refuelling	continuous on load with axial shuffle
Lattice pitch	15.5 in
Lattice geometry	square
Active core height	27 ft
Active core diameter	31 ft
Total number of channels	465
Number of fuel channels at equilibrium	412
Channel gas inlet temperature	320° C (608° F)
Channel gas outlet temperature	675° C (1247° F)
Peak can temperature	800° C (1472° F)
Circulator outlet pressure	450 lbf/in² abs
Number of circulators per reactor	4
Type of circulator	centrifugal
Circulator speed	1500 rev/min
Speed variation	fluid drive coupling
Circulator drive	synchronous motor
Circulator installed power, each	16 500 hp
Type of boiler	once-through
Steam pressure	2315 lbf/in² abs
Steam temperature	565° C (1050° F)
Steam flow	3.7 M lb/h
Reheat pressure	556 lbf/in² abs
Reheat temperature	565° C (1050° F)
Condenser vacuum	28.9 inHg
Cooling water inlet temperature	14° C (57° F)

Aerated test burner A test burner designed to provide a rapid appreciation of gas quality; it measures the amount of air required to give a stable, well-defined, blue inner cone. The *ATB Number* is the extent to which the air inlet shutter is opened to give the standard size of inner cone.

Aerodynamics A branch of dynamics that deals with the motion of air and other gases, and the forces acting upon bodies passing through them—particularly aircraft, missiles and rockets.

Aerosol A particle of solid or liquid matter of such small size that it can remain suspended in the atmosphere for a long period of time; aerosols diffuse light, and the larger particles settle out on horizontal surfaces or cling to vertical surfaces. All air contains aerosols, the larger particles, above 5 µm in size, being filtered out in the nose or bronchia. The smaller particles, below 5 µm in size, pass into the lungs; they may be expelled immediately or retained for varying periods of time. Aerosols are classified into smoke, fumes, dust and mists.

After-burner A burner located in the exit gases from a combustion process, providing sufficient heat to destroy smoke and odours.

After-heat In respect of a *nuclear reactor*, heat produced by the decay of the fission products in the fuel elements, after the reactor has been shut down.

Agglomeration The clustering or adhering together of a number of small particles to form a larger single entity or 'agglomerate'. Agglomeration has been associated with small particles in chimney-stacks.

Air *See atmosphere; excess air; nitrogen; oxygen; primary air; secondary air; tertiary air; theoretical air.*

Air-assisted pressure jet burner A type of *pressure jet burner* in which low-pressure air is utilised to assist atomisation, provide directional stability and promote primary zone turbulence. *See oil burner.*

Air blanketing An accumulation of air in a heat exchanger or other vessel which impedes or prevents the transfer of heat.

Air count The determination of the amount of radioactivity in a standard volume of air; in one method the prescribed volume of air is drawn through a filter paper on which the radioactive solids are deposited.

Air director *See air register.*

Air-dried coal Coal exposed to the atmosphere in a dry well-ventilated place, protected against the weather, so that it loses by evaporation most of its free or surface moisture. Coal is analysed in this condition to give the percentages of *inherent moisture, volatile matter, fixed carbon* and *ash. See free moisture; proximate analysis; ultimate analysis.*

Air filter A device for removing particulate matter from an air

8

stream. *See* **electrostatic filter; roller filter; viscous oil filter; 'zig-zag' filter.**

Air heater *See* **air preheater.**

Air in-leakage The leakage of air through defective brickwork settings and joints into flues. One of the most important features of boiler maintenance is the detection and elimination of such leaks. Leaks can be readily detected by holding a lighted candle, taper or duck lamp close to cracks in the brickwork and to joints where the infiltration of cold air might occur. Cracks and joints may be caulked with rope and sealed with a mixture of tar and fireclay or cement; warped flue inspection doors should be replaced; damper slots should be sealed with a twin roller system or a special air-tight damper cover; brickwork generally should be treated with three coats of tar or limewash to seal the pores. Exclusion of unwanted cold air from the flues improves boiler efficiency.

Airlift The use of air or other gas to transfer liquids or solids from one part of a plant to another.

Air mass Formed in the *troposphere*, a vast body of air having fairly uniform meteorological characteristics. Air masses are comparable in size with continents, moving in one of the atmospheric currents of the general atmospheric circulation. Fronts are theoretical surfaces dividing one air mass from another. An air mass is designated by geographic origin as either continental or maritime, tropical or polar.

Air pollution The contamination of the *atmosphere* with undesirable solids, liquids and gases. In a strict sense, air may be considered polluted when there is added to it any substance foreign or additional to its normal composition. This definition of pollution is much too wide, however, for the purposes of practical air pollution control, and the term 'air pollution' is usually restricted to those conditions in which the general atmosphere contains substances in concentrations which are harmful, or likely to be harmful, to man or to his environment. A fuller definition of American origin is: 'substances present in the atmosphere in concentrations great enough to interfere directly or indirectly with man's comfort, safety or health, or with the full use or enjoyment of his property'. In respect of health, this reiterates the dictum that there are no such things as toxic substances, only toxic concentrations. The concept of concentration, however, cannot be divorced from time or duration of exposure; or from the acute or chronic effects likely to arise from high short-term or low long-term exposures, respectively.

While London smog has diminished under the impact of progressive measures, it is interesting to distinguish between this

9

type of smog and that of Los Angeles, which is of photochemical origin. Photochemical smog is significant in other American cities and in Tokyo, while some of the eastern state capitals of Australia, notably Sydney and Melbourne, have proved to be candidates for this affliction.

Undoubtedly, the use in technical literature of such an imprecise term as 'smog', as well as its variants such as 'smaze' and 'smust', creates confusion. The interchange of the results of air quality investigations is being made easier, however, through the adoption of a common nomenclature, consistent metric units and uniform methods; this process, however, is far from complete. *See Table A.1.*

Air Pollution Control Association (APCA) A voluntary association concerned with the control and abatement of air pollution in North America; in addition to the publication of learned papers and the holding of conferences, the Association issues guidelines in respect of pollution control in specific industries and processes.

Air preheater A device the purpose of which is to transfer heat from the flue gases to the air fed to the furnace for combustion purposes. It is usually situated between the economizer and the chimney-stack. The recovery of heat from the flue gases reduces the heat loss from the chimney and raises the overall efficiency of the plant. It is claimed that a fuel saving of 1.5 per cent is obtained for every 20° C reduction of flue gas temperature. The raising of the flame temperature in the furnace increases the rate of heat transfer by radiation and reduces the amount of excess air required. Preheated air is essential for the combustion of low-grade fuel. The combustion process is generally accelerated, which results in a short hot flame and the reduced likelihood of smoke emission. Both regenerative and recuperative air preheaters are in use in industry, the latter being the usual type for use with boilers. *See Howden-Ljungstrom air preheater; recuperative air preheater.*

Air preheater corrosion The wasting of the metal of boiler air preheaters caused by the condensation of acid from the flue gases on exposed metal surfaces.

Air register A combustion air regulator for oil burners. It permits the amount of air admitted to the combustion chamber to be controlled, i.e. kept in the right proportions with the amount of oil being consumed; it enables a flame front to be held at a satisfactory distance from the burner tip; in addition, it causes the air to mix thoroughly with the oil by giving it a rotational flow which encourages turbulence. *Figure A.2* shows details of a compound air register. Also known as an air director. *See oil burner.*

Ajax furnace A modified tilting *open-hearth furnace*, with provision for oxygen lancing; only during the charging, fettling and final

Table A.1 Air pollutants with recognised or potential long-term effects on health at usual air pollution levels (Source: World Health Organisation, *Research Into Environmental Pollution*, Technical Report Series No. 406, Geneva, 1968)

Substances with known effects on health (acute or chronic)	Substances thought to have long-term effects per se[a]	Potential long-term effects of combinations
Arsenic	Arsenic (arsenical dermatitis)	
Asbestos	Asbestos (asbestosis, mesothelioma)	
Beryllium	Beryllium (berylliosis)	B + F (fluorides potentiate pulmonary changes in berylliosis)
Carbon monoxide		Synergistic in pO_2 depression
Carcinogens		Carcinogens produce tumours in presence of promotors
Fluoride	Fluoride (fluorosis)	Fluoride (promotes or accelerates lung disease)
Hydrocarbons		$HC + O_3 \rightarrow$ tumorigen + influenza → cancer
Hydrogen sulphide (possibly with mercaptans)		Antagonizes pollutants (strictly speaking not detrimental to health)
Inorganic particulates	Inorganic particulates (pulmonary sclerosis)	
Lead		
Nitric oxide		
Nitrogen dioxide	Nitrogen dioxide (mild accelerator of lung tumours)	NO_2 + micro-organisms (pneumonia) + HNO_3 (bronchiolitis, fibrosa (obliterans) + tars (smoker's lung cancer)
Organic oxidants (peroxyacylnitrates)		
Organic particulates (asthmagenic agents)	Asthmagenic agents (asthma)	
Ozone	Ozone (chronic lung changes, accelerated aging)	O_3 + micro-organisms (lung tumour accelerator)
Sulphur dioxide, sulphur trioxide		SO_2, SO_3 + particulates aggravate lung disease

[a] Effects are given in parentheses.

phases of refining is fuel used, the necessary heat being supplied by the oxidation of the metalloids by jets of oxygen. Up to 45 m³ of oxygen per ton of steel may be used, production being much increased by this method. *See **steelmaking furnace**.*

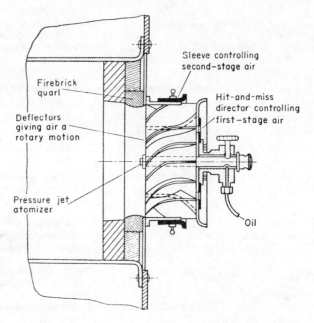

Firebrick
quarl

Deflectors
giving air a
rotary motion

Pressure jet
atomizer

Sleeve controlling
second-stage air

Hit-and-miss
director controlling
first-stage air

Oil

Figure A.2.   Compound air register

Alaskan pipeline   A 1277 kilometre (798 mile) oil pipeline from
Prudhoe Bay to the southern port of Valdez, to carry oil from the
North Slope facility developed by British Petroleum and Sohio.
Environmental and other objections imposed a 5 year delay on the
project, which did not get under way until 1975. Oil flowed in
1977; the first tanker load of crude oil from the pipeline left
Valdez in August for delivery to California. In September the first
cargo of Prudhoe Bay crude oil was delivered by tanker to a
Delaware refinery. In the same year the United States and
Canadian governments announced an agreement on the route of
a projected pipeline to carry natural gas from the Prudhoe Bay oil
field. The pipeline would run for 1200 kilometres in Alaska, and
more than 3600 kilometres in Canada, with over 1400 kilometres
of connecting lines from the United States border.

Alcohol fuel   Lower boiling point alcohols such as methyl and ethyl,
which have high octane numbers (98 and 99, respectively), and
are often blended with straight run gasolines to effect an upgrading
of *octane number*. They may be added in a proportion of 10–20 per
cent. Although these alcohols may be used alone in internal
combustion engines, they are usually more costly and have much
lower heat contents than ordinary *gasoline*.

Alga Botryococcus A minute natural bloom which appears on lakes; the bloom may occur either in the green-active growth stage or in the red-resting state. The plant has a relatively high oil content of up to 75 per cent of the dry mass in the red phase; for this reason it has been considered for harvesting to obtain the oil. Torbanite (or kerosene shale) is a Permian oil shale formed by the fossilisation of *Botryococcus* or very similar plants.

Alkali A substance which, in solution in water, forms an excess of hydroxyl over hydrogen ions. Alkalis neutralise acids, with the formation of salts. *See pH.*

Alkali, etc., Works Regulation Act British legislation initiated in 1863 to control the discharge to atmosphere from works registered under the Act; the Act was considerably widened in 1906. Until 1958 the Alkali Act was mainly concerned with processes in the heavy chemical and allied industries; in that year the scope of the Act was again extended to embrace electricity generation, oil refining, and iron and steel manufacture, in which pollution control presented problems of special technical difficulty. In 1958 the number of works supervised by the Alkali Inspectorate increased from 870 to 2200. In 1975 the Alkali Act was repealed and embodied in the Health and Safety at Work Act 1974. However, this has meant virtually no change in the actual legislation controlling emissions from the scheduled processes. The principle that industry should adopt the 'best practicable means' to control emissions has been preserved.

Alkane series Formerly called the paraffin series, the two alkane series are composed of chains of linked carbon atoms, a distinction being made between normal-alkanes containing only straight carbon-to-carbon chains, and their corresponding isomeric, branched-chain forms, called isoalkanes. Both series have the general composition C_nH_{2n+2}, and the names of the individual members end with 'ane', e.g. methane, ethane, propane. Alkanes represent a large proportion of the hydrocarbons present in many crude oils. The normal alkane series is the more abundant. *See cycloalkane series; Figure A.3.*

Alkane thiols Formerly called mercaptans, organic compounds having the general formula R—SH, meaning that the thiol group —SH is attached to an alkane radical such as ethyl, C_2H_5, or butyl, C_4H_7. The simpler alkane thiols have strong, repulsive, onion-like or garlic-like odours. These substances may be produced in oil refinery units as a result of cracking; the offensive gases may be burnt in plant heaters after removal in a scrubbing process. They occur most frequently in the distillation range to about 150° C.

Alkenes *See olefin series.*

13

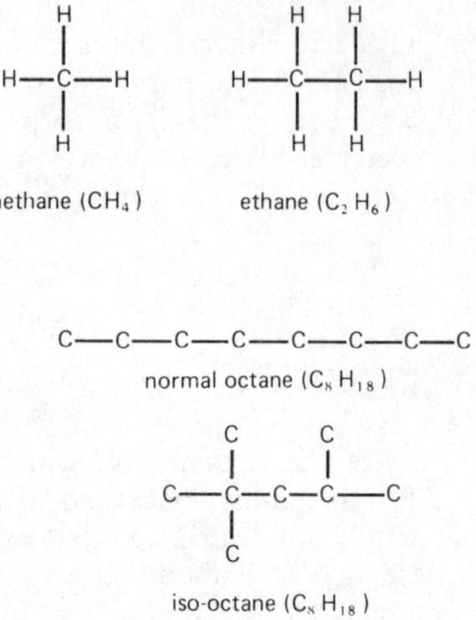

$$H-\overset{\overset{\displaystyle H}{|}}{\underset{\underset{\displaystyle H}{|}}{C}}-H \qquad H-\overset{\overset{\displaystyle H}{|}}{\underset{\underset{\displaystyle H}{|}}{C}}-\overset{\overset{\displaystyle H}{|}}{\underset{\underset{\displaystyle H}{|}}{C}}-H$$

methane (CH_4) ethane (C_2H_6)

$$C-C-C-C-C-C-C-C$$
normal octane (C_8H_{18})

$$C-\overset{\overset{\displaystyle C}{|}}{\underset{\underset{\displaystyle C}{|}}{C}}-C-\overset{\overset{\displaystyle C}{|}}{C}-C$$

iso-octane (C_8H_{18})

Figure A.3 Examples from the alkane series

Alkylation An oil refinery process developed initially to produce high-octane aviation gasolines and which is now used to produce high-octane blending components for motor fuels. In the process of alkylation, olefinic gases, e.g. butenes, are combined with isobutane to produce *isoparaffins*.

Alkyl benzene series A hydrocarbon series formed through the progressive substitution by alkanes of the hydrogen atoms on the outside of the carbon ring in the aromatic series. *See aromatic or benzene series*.

Alpha particle A positively charged particle emitted in the decay of some radioactive nuclei. It consists of two protons and two neutrons and is identical with the nucleus of the helium-4 atom. Alpha radiation has a power of penetration of less than one-tenth of a millimetre in human tissues; it is therefore, capable, of irradiating cells only in the immediate vicinity of the source of radiation.

Alternate side or wing-firing A method of firing a boiler or furnace with solid fuel. In this method, sometimes known as the side-by-side firing method, the coal is thrown to one side of the grate only. This allows the volatiles released to be consumed by the hot gases from the other side. Once the coal has coked, the other side of the grate can then be charged. As the banked coal cokes, it tends to fall into the centre. This system is excellent for avoiding the emission of smoke, provided that the little-and-often rule is strictly observed. There is less difficulty in keeping the grate properly covered with fuel, particularly at the back. This technique is more

14

suitable for large and graded coals than for coal containing a high proportion of fines, which tends to pack down, burning slowly and unevenly. See *firing by hand*.

Alternating current (AC)   An electric current the flow of which alternates in direction; the direction of flow of the electric current is reversed many times a second. The standard frequency in Britain is 50 cycles per second; in the United States, 60 cycles per second. A cycle contains two reversals—namely, from one direction to the opposite, and then back to the original direction. The time flow in one direction is known as a half-period.

Alternator or synchronous generator   An alternating current generator driven at a constant speed corresponding to the particular frequency of the electrical supply required from it. See *turbo-alternator*.

Alumina Al_2O_3. the trioxide of aluminium, an important constituent of ash; widespread in nature.

Aluminous firebrick   A *refractory*, containing from 38 to 45 per cent of *alumina*, the balance being mainly silica.

Alveoli   Innumerable minute, air-filled sacs in the human lungs; they are thin-walled and surrounded by blood vessels. It is through their surfaces that the respiratory exchange of oxygen and carbon dioxide occurs.

Ambient air quality   A general term used to describe the state of the general atmosphere respecting the presence of air impurities. In Los Angeles, California, an attempt has been made to lay down ambient air standards as a guide to what concentrations of pollutants should not be exceeded, as a basis for public alerts. Most monitoring systems attempt to break down and list the impurities measured in air, such as smoke, dust, sulphur dioxide, carbon monoxide, oxides of nitrogen and ozone. Some cities, among them New York, use their findings to construct an arbitrary index number. See *air pollution*; *Los Angeles smog; Sydney Pollution Index*.

Americium   Am. An artificial *element*, at. no. 95, at. wt. 243. It is prepared artificially from *uranium* in a cyclotron and is formed in very small quantities in nuclear reactors.

Ammeter   An instrument for measuring amperes, i.e. the current or rate of flow of electricity. See *ampere*.

Ammonia NH_3. A colourless, pungent gas with a boiling point of $-33°$ C, extremely soluble in water. It is obtained on a large scale by *ammonia synthesis*, and from the ammoniacal liquor produced in the carbonisation of coal.

Ammonia recovery   The recovery of *ammonia* from coal gas produced during *carbonisation*. Three methods of recovery are available.

(1) Direct system, in which clean hot gases, to which additional ammonia from the distillation of ammoniacal liquor has been supplied, pass into saturators containing 77 per cent sulphuric acid saturated with ammonium sulphate, crystals of ammonium sulphate falling continuously to the base of the saturators, from which they are removed and centrifuged.

(2) Semi-direct system, in which the gas is cooled to near atmospheric temperature to remove much more ammonia in liquor form, following which the gas is reheated to 60–80° C and acid extracted as in the direct system.

(3) Indirect system, in which the gas is cooled to near atmospheric temperature to remove ammonia liquor and then extracted by water in scrubbers. The solution of ammonium salts is distilled with lime to recover ammonia vapour, this vapour being condensed to give ammonia solution (up to 25 per cent ammonia) or neutralised with sulphuric acid and recovered as ammonium sulphate.

Ammonia synthesis The manufacture of *ammonia* from nitrogen in the air and hydrogen derived from *naphtha* and steam. In a typical process naphtha is vaporised and preheated in a refinery gas- or naphtha-fired heater. The naphtha vapour, together with some recycled hydrogen, is passed over zinc oxide and cobalt molybdate beds to remove sulphur compounds. Superheated steam is added to the vapour stream as it enters primary reformer catalyst-filled tubes; the tubes are heated by burners which use naphtha or refinery gas as fuel. The mixed vapour passing over the catalyst reforms into carbon dioxide (CO_2), carbon monoxide (CO), hydrogen (H_2) and methane (CH_4):

$$2C_7H_{15} + 14H_2O \rightarrow 14CO + 29H_2$$
$$CO + H_2O \rightarrow H_2 + CO_2$$
$$CO + 3H_2 \rightarrow CH_4 + H_2O$$

In the secondary reformer air is added to supply nitrogen (N_2). On passing over more catalyst, methane is converted to carbon monoxide:

$$CH_4 + H_2O \rightarrow CO + 3H_2$$

The reformed gas passes into a CO converter where carbon monoxide and steam are converted into carbon dioxide and hydrogen:

$$CO + H_2O \rightarrow CO_2 + H_2$$

The gas is scrubbed in a tower with vetrocoke solution which absorbs most of the carbon dioxide; the vetrocoke solution is heated in a regeneration tower to liberate the carbon dioxide, which is vented to atmosphere. The remaining carbon oxides react with steam in a methanator to form methane. The gas is compressed in three stages to 350 atmospheres and passed into the ammonia synthesis circuit; as gas passes through the ammonia converter, a catalyst converts hydrogen and nitrogen to ammonia:

$$3H_2 + N_2 \rightarrow 2NH_3$$

On cooling, the ammonia condenses and is separated from unconverted gas in a catchpot; the ammonia liquid passes to a storage sphere while waste heat is used to raise steam.

Amoco Cadiz incident A major oil spill which occurred in March 1978, when the supertanker *Amoco Cadiz* went aground near Portsall, France. Some 220 000 tonnes of oil was poured into the sea off Brittany from the tanker, resulting in an oil spill 16 kilometres long and 10 kilometres wide. The spill polluted 128 kilometres of beaches, destroying oyster and shellfish beds, and damaging Brittany's prosperous tourist industry. *See Torrey Canyon incident.*

Ampere The unit of electric current; the ampere is the constant current which, if maintained in two straight parallel conductors of infinite length, of negligible circular sections, and placed one metre apart in a vacuum, will produce between these conductors a force equal to 2×10^{-7} newton per metre of length. As an indication of the order of magnitude of this unit, a 2 kW electric radiator on a 240 V circuit takes a current of 8.3 amperes. Named after the famous French physicist, André M. Ampère (1775–1836).

Amyl nitrate A liquid compound, $C_5H_{11}NO_3$, used as an additive in diesel fuel to improve its ignition qualities.

Anabatic wind (or *mountain wind*) Caused by air being warmed by contact with the ground during day-time and flowing uphill. Opposite of a *katabatic wind*.

Anabolism A facet of *metabolism* relating to those changes involving the breakdown of foodstuffs and their rebuilding to form body tissues.

Anadromous fish Fish, such as the salmon, which spends most of its growing years in the ocean and, after attaining sexual maturity, ascends freshwater streams in order to spawn. The erection of power and irrigation dams, or the presence of thermal pollution, may isolate considerable numbers of fish from their traditional spawning grounds.

Anaerobic Living or active in the absence of free oxygen; an anaerobic process is one taking place in the absence of oxygen. The remaining oxygen may be combined in the form of some organic or inorganic compound, e.g. nitrate or sulphate. If

17

sulphate acts as an oxidising agent, *hydrogen sulphide* is formed, which gives rise to objectionable odour. Sulphate-reducing bacteria, though present in rivers, are normally inhibited by the presence of dissolved or free oxygen.

Anaerobic digestion A digestion process which permanently removes the offensive odour of many organic wastes so that they can be utilised on agricultural land without causing nuisance. A high proportion of the *chemical oxygen demand* can be removed with the recovery of the organic carbon as *methane*, whilst most of the lipids and other constituents which otherwise might attract flies and vermin are degraded. The wide variety of bacteria involved in the process may be classified into two broad groups:

(1) Acid-producers (non-methanogenic bacteria) which, together with their associated enzymes, degrade most types of organic material, mainly into the lower fatty acids (acetic acid accounting for about 80 per cent of the total), with much smaller amounts of lower aldehydes and ketones.

(2) Methane-producers (methanogenic bacteria) which convert the soluble products of the acid-producers into a mixture of methane and *carbon dioxide*.

Analogue computer A *computer* in which analogue representation is used, i.e. in which a variable is represented by a physical quantity, such as angular position or voltage, which is made proportional to the variable. The problems best solved by analogue computers are: dynamic problems involving fast rates of change of variables, which must be solved in real time; problems involving non-linearities, empirical relationships, etc., and requiring continuous solution. *See analogue controller.*

Analogue controller A device in which a measured signal representing a particular plant variable which has to be controlled is compared after amplification with a further signal representing the demanded value. The error signal formed is passed through a shaping network to produce the desired control equation, and then fed back to the system to actuate some device which corrects the original plant parameter. Examples include the temperature controllers on a nuclear power station and the use of analogue controllers for the run-up of large turbo/generator units. *See analogue computer.*

Andalusite An aluminium silicate used in ceramic refractories.

Anemometer An instrument for measuring wind speed.

Aneroid chamber A type of draught gauge; it consists of a sealed metal chamber with flexible sides which expand or contract with changes in the difference between the pressure inside and outside the chamber, and a mechanism to magnify and record the movement of the chamber. *See draught; draught gauge.*

Angle of inclination The natural angle which a solid fuel takes when piled; the angle will be increased as the moisture content increases. Examples are: coke, 45°; graded coal, 40°; washed small coal, 55°.

Ångström A unit of length, symbol Å, equivalent to 10^{-8} cm (one-hundred-millionth of a centimetre). It is employed for measuring wavelengths of light.

Aniline point The temperature at which an oil first becomes completely miscible with an equal volume of aniline. A convenient laboratory test for assessing the proportion of aromatic hydrocarbons in hydrocarbon mixtures such as petroleum fuels; a low aniline point indicates a high aromatic content and a low *cetane number*.

Anion A negatively charged ion of an electrolyte which migrates towards the anode under the influence of a potential gradient.

Ankerite A carbonate of calcium, magnesium and iron; it may be found as white partings within a coal seam, although it is frequently associated with iron ores.

Annealing A *heat treatment* process; it is used to eliminate the effects of the cold working of metals by removing internal stress. It may also be used to improve electrical, magnetic or other properties.

Anode The electrode of an electrolytic cell at which oxidation occurs. Usually, in corrosion processes, it is the electrode that has the greater tendency to go into solution.

Anomaly A deviation from uniformity; a geological feature considered capable of being associated with commercially valuable *hydrocarbons* or minerals.

Anoxia A deficiency of oxygen in the blood.

Antagonistic effect The tendency of some chemicals and processes to react together to form combinations which may have a less powerful effect than the substances or processes taken separately; opposite of a *synergistic effect*. The term is also applied where the growth of one organism is inhibited by another through the creation of unfavourable circumstances, e.g. by exhaustion of the food supply.

Anthracene oil A coal tar fraction, boiling above 270° C; in addition to anthracene, it contains phenanthrene, chrysene, carbazole and other hydrocarbon oils.

Anthracite The highest-ranking coal to be produced by the physicochemical alteration of peat; it is hard, dense and lustrous, does not break easily and is clean to handle. Difficult to ignite, it burns with a short intense flame and with the virtual absence of smoke. Anthracite contains 5–9 per cent *volatile matter*; the

19

calorific value is of the order of 34 kJ/g. The best anthracites in the United Kingdom are mined in South Wales.

Anthracitisation The progressive conversion into *anthracite* of plant remains which might have produced *bituminous coal* under less severe geological influences.

Anthraxylon A glossy constituent in coal formed from woody parts of plants, trunks, branches and twigs. *Vitrain* and *clarain* are predominantly anthraxylon. A term used in the United States of America.

Anticipatory pricing Or hedge pricing, or inflated premium pricing, a pricing strategy by companies that allows for future anticipated cost increases in both operating and capital costs. It is an attempt, in inflationary conditions, to maintain profit margins. If companies believe that government policies will lead to a greater rate of inflation, then this belief will be reflected in price structures—particularly in respect of goods and services to be delivered at some future time.

Anticyclone In meteorology, a high-pressure area with winds rotating clockwise around the centre in the northern hemisphere and anticlockwise in the southern hemisphere. There is a general slow descent of air over a wide area. Normally this subsiding air brings welcome fair weather because the cold upper atmosphere carries little water vapour and the air warms as it slowly flows groundward, superheating any vapour entrained. In summer, in the British Isles, an anticyclone means fine, warm, sunny weather. In winter, however, anticyclones bring dense sheets of stratocumulus cloud giving typical winter gloom; as the air descends, it is compressed and heated so that a deep inversion layer is formed which very often results in fog. *See cyclone; inversion; subsidence inversion*.

Antifoams Usually polyoxide or polyamide compounds which, when added to boiler water, reduce or prevent the formation of foam. In consequence, a higher level of dissolved solids content can be tolerated.

Antiknock additive An additive in petrol which permits the use of engines of higher compression ratio without the phenomenon of 'knock', i.e. the detonation of part of an air/fuel mixture before the flame front reaches it. *See tetra ethyl lead*.

Antioxidant A substance which when added in small amounts to *petroleum products* will delay or inhibit undesirable changes such as the formation of gum, sludge and acidity which are brought about by oxidation.

Anti-Trust Laws A series of Acts introduced by the United States Federal Government to check and control the formation of trusts, i.e. the combination or amalgamation of large firms exercising monopoly control over the commodity produced. The first was the Sherman Anti-Trust Act of 1890. The Clayton Act of 1914 also sought to check the development of monopolies by prohibiting the amalgamation of firms producing a large proportion of the total output of a commodity. In 1979 the United States courts ruled that four Australian companies had violated the US anti-trust laws through the arrangements made for marketing uranium. The court proceedings were brought by the Westinghouse Electric Corporation.

API gravity *See degrees API.*

Apparent specific gravity The *specific gravity* of coal or coke inclusive of any voids within the pieces included in the sample subjected to test. *See porosity; true specific gravity; voidage.*

Aramco An American consortium comprising Exxon, Mobil, Texaco and Standard Oil of California; they produce and sell a large part of Saudi Arabian oil.

Aromatic or benzene series A hydrocarbon series, the simplest member of which is benzene, composed of a ring of six carbon atoms, six associated hydrogen atoms and three pairs of unsaturated carbon bonds between members of the ring (*see Figure A.4*). The series has the general composition C_6H_6. Other members include toluene, $C_6H_5CH_3$, and xylene, $C_6H_4(CH_3)_2$.

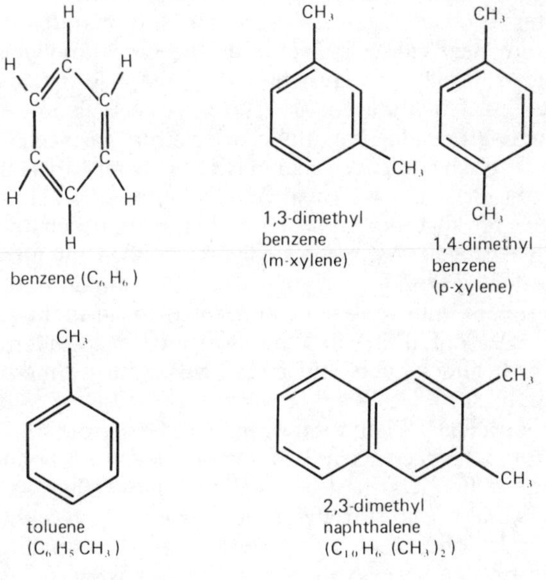

Figure A.4 Examples from the aromatic or benzene series

21

Aromatic rings can also possess side chains, and two or more rings may be linked together through two common carbon atoms. Naphthalene, $C_{10}H_8$, consists of two benzene rings linked in this way; benzpyrene has five rings. Aromatic hydrocarbons such as benzene, toluene and the xylenes are used as solvents, and are required in large quantities as 'intermediates' for conversion by chemical means to many familiar plastics and synthetic fibres, e.g. nylon and terylene. The heavier aromatic concentrates such as those extracted from lubricating oil base-stocks are used as plasticisers and extenders in PVC and synthetic rubber manufacture.

Arsenic As. An *element* in coal, at. no. 33, at. wt. 74.9216. The arsenic content of a coal seam may vary within wide limits over a relatively small area of a coalfield; the result of a survey of coals in the East Midlands of England suggested a range of 15–40 ppm As_2O_3.

Asbestos A broad term embracing several fibrous minerals, with chrysotile, a hydrated magnesium silicate, being the most common form. It has been widely used as an insulating material for temperatures up to about 600° C. However, concern has been strongly expressed in recent years regarding asbestos fibres in the atmosphere; these may enter the general environment from the activities of asbestos cement industries, from the handling of asbestos products, and from repair and maintenance work. Exposure may cause a disease of the lungs known simply as asbestosis; it may also cause mesothelioma, a form of cancer. For this reason, substitutes for asbestos have been vigorously sought.

Ash The inert residue remaining when a fuel has been completely burnt. The ash content of coal may range from 1 to 30 per cent or more; in Britain 12–15 per cent is typical. If ash melts, the particles run together to form a molten mass which on cooling is known as *clinker*. Ash largely consists of silica and alumina. The ash content of liquid fuels is of considerable importance, as it is an indication of impurities which may cause wear in the fuel pumps and nozzles and, if they find their way into the cylinder, increased wear of cylinder liners and piston rings and pitting of exhaust valves and seats.

Ash handling plant Plant for the removal of ash from a boiler house. Ash from a furnace accumulates in the boiler ash hoppers; high-pressure water jets may be used to remove the ash from the hoppers and wash it through sluice ways into a receiving hopper or sump.In some plants the ash is removed by a grab for loading into lorries or barges; in others the ash is passed through a comminutor and pumped to lagoons where it settles out, Ash from grit arresters may be removed by a suction-and-water system.

Ash quench   The process of cooling hot boiler ash with water.

Asphalt   Mixtures in which bitumen is associated with inert mineral matter. The term is normally qualified by an indication of type or origin, e.g. lake asphalt, natural asphalt. Known in the United Kingdom as asphaltic bitumen.

As received   A basis for reporting an analysis of coal; it includes the total moisture made up of *free moisture* and *inherent moisture*, *volatile matter*, *fixed carbon* and *ash*. *See proximate analysis; ultimate analysis.*

Associated gas   *Natural gas*, dissolved in crude oil, or in contact with crude oil, in an underground petroleum reservoir. The former is known as 'solution gas', the latter as 'gas-cap gas'.

ASTM   The American Society for Testing Materials. The organisation is reponsible for the issue of many of the standard methods used in the energy industries.

ASTM Coal Classification   A system of coal classification devised by the American Society for Testing Materials. Under this system, coals containing less than 31 per cent *volatile matter* on a mineral-matter-free basis are classified on the basis of *fixed carbon* only; these are divided into five groups:

$$\left.\begin{array}{l}\text{above 98 per cent fixed carbon} \\ \text{98–92 per cent fixed carbon} \\ \text{92–86 per cent fixed carbon}\end{array}\right\} \text{anthracites}$$

$$\left.\begin{array}{l}\text{86–78 per cent fixed carbon} \\ \text{78–69 per cent fixed carbon}\end{array}\right\} \text{bituminous coals}$$

The remaining bituminous coals, sub-bituminous coals and lignites are divided into groups as determined by the *energy value* of the coals containing their 'natural bed moisture', there being eight groups in all, with calorific values ranging from about 34 kJ/g to below 21 kJ/g. The classification also differentiates between consolidated and unconsolidated lignites, and between the weathering characteristics of sub-bituminous and lignitic coals. *See coal classification systems.*

ATB number   *See aerated test burner.*

ATK   Aviation turbine kerosine.

Atmosphere   (1) A unit of pressure; the pressure which will support a column of mercury 760 mm high (29.92 in) at $0°$ C, sea level and latitude $45°$. One normal atmosphere = 101 325 pascals or newtons per square metre = 14.72 lb/in^2 approximately. Atmospheric pressure tends to fluctuate about this norm. One bar = 10^5 Pa; hence, the normal atmosphere is very approximately one bar. *See absolute pressure.* (2) The gaseous envelope of air surrounding the earth, the principle constituents of which are nitrogen and oxygen in proportions by volume of about 79.1 per cent and 20.9 per cent, respectively. *Carbon dioxide* is also present to the extent of about 0.03 per cent, together with very small amounts of inert gases such as argon, krypton, xenon, neon and helium. Also present are water vapour, traces of ammonia, organic matter, ozone, salts and suspended solid particles.

Atmosphere, standard The atmosphere at 'normal' pressure; one standard atmosphere, at $0°$ C, is equal to the following:

$$14.695 \text{ lb/in}^2$$
$$2116 \text{ lb/ft}^2$$
$$1.033 \text{ kg/cm}^2$$
$$760 \text{ mmHg}$$
$$29.92 \text{ inHg}$$
$$33.9 \text{ ftH}_2\text{O}$$
$$407 \text{ inH}_2\text{O}$$

The zero mark on a steam *pressure gauge* means a pressure of one standard atmosphere and must be distinguished from *absolute zero*.

Atmospheric temperature inversion. *See inversion.*

Atom The smallest quantity of an *element* which can take part in a chemical reaction; it has a diameter of the order of 10^{-8} cm and consists of a *nucleus*, of diameter 10^{-12} to 10^{-13} cm, around which electrons move in orbits.

Atomic number The number allocated to an *element* when arranged with other elements in order of increasing atomic weight; it is equal to the number of protons in the *nucleus* and, in the neutral atom, to the number of extra nuclear electrons.

Atomic Time (AT) Time measured by atomic characteristics; in the SI system of measurement the unit of time interval is the second, defined in terms of the length of time for a certain frequency of vibration of caesium-133 atom.

Atomic weight, relative A number that gives the mass of an atom of an element relative to that of the isotope of carbon, ^{12}C, with an assigned atomic weight of 12. An early scale took hydrogen, with atomic weight 1, as the reference standard. From 1900 to 1961 oxygen was used as a standard of reference, with an assigned atomic weight of 16. In 1961 the International Committee on Atomic Weights decided that the isotope of carbon, ^{12}C, be taken as the standard of reference, and that it be given the atomic weight 12. The replacement of oxygen as the reference changed the established atomic weights by about 0.004 per cent; the relative atomic weight of oxygen became 15.9994. The International Committee also decided that the term 'atomic weight' should be changed to 'relative atomic weight'.

Atomisation The breaking into fine particles of a liquid fuel to ensure intimate mixing with combustion air, a prerequisite for efficient combustion. Some burners utilise burner tips of special design to atomise fuel oil supplied under pressure. Other burners utilise air or steam as the atomising medium. The normal size range of particles for good atomisation is from 300 μm down to 5 μm. *See oil burner.*

Atomiser *See oil burner.*

Attemperator Or de-superheater; a device to control any excessive superheat in a boiler. There are two main types. In the spray-type attemperator there is direct contact between the superheated steam and the cooling steam; in the surface-type attemperator the superheated steam passes through tubes around which the cooling steam circulates.

Attritus A dull constituent of coal formed from finely divided materials, leaves, pollens, spores, seeds, resin, etc. A term used in the United States of America, synonymous with *durain.*

Audibert–Arnu test A test for determining the coking properties of coal.

Auger mining A technique of mining employed in many strip mines where the overburden is too thick to permit economical continuous stripping; augers are used to bore horizontal holes into the exposed coal seams. The coal falls from the augers into a conveyor.

Austenitic steel Characteristically, steel which is non-magnetic, corrosion-resistant, and with poor thermal conductivity, a high coefficient of expansion and good creep-resistance qualities. These characteristics are brought about by the presence of substantial amounts of alloying elements, notably nickel.

Australian Institute of Energy A body established in 1977 'to advance the scientific, technological and social study and professional practice of energy disciplines and enhance the contribution of energy disciplines to the promotion of the public welfare . . . and to consider, initiate and promote improvements in and alterations to the law relating to energy disciplines . . . and to oppose or support any law relating to energy disciplines in the Commonwealth of Australia'. The formation of the new institute followed a decision to dissolve the Institute of Fuel (Australian Membership), a branch of the London Institute. Membership includes fuel technologists and chartered engineers, and others involved with the economic, legal, policy-making and management aspects of the subject. The Institute fulfils an important role in providing a forum where discussion and debate can occur, with participants from a wide variety of backgrounds; it brings together many disciplines previously restricted to disparate approaches in Australia. Initially, there were 500 members. *See Institute of Energy.*

Australian Minerals and Energy Council An Australian ministerial council, comprising the mines and energy ministers from the federal government and from the six state governments. A consultative committee of the Council makes recommendations on training, advisory services, energy conservation, pricing and taxation policies. In 1979 the Council agreed to establish a national petroleum advisory committee to advise the government on allocation arrangements and priorities should the need for emergency measures arise.

Autofining A fixed-bed catalytic process for de-sulphurising distillates; the pelleted catalyst is cobalt molybdate on alumina and may be regenerated in place.

Autoignition Pinking, knocking or detonation, denoting the ignition of gas due to adiabatic compression; usually associated with the spark-ignition engine when operating on an unsuitable fuel. Knock resistance is usually measured by the *octane number* of the fuel.

Automatic boiler control A system which automatically controls the fuel and air supplies to a boiler, adjusting for the required steam output while simultaneously maintaining the best fuel/air ratio for high combustion efficiency.

Automatic voltage control In respect of the *electrostatic precipitator*, a system designed to maintain optimum electrical conditions and high efficiency of collection under all boiler-operating conditions. The system generally works by slowly increasing the voltage until the current, because of flash-over, exceeds a predetermined value; the voltage is then reduced a little and the cycle repeated.

Autunite A uranium-bearing mineral, being hydrated calcium uranium phosphate.

Auxiliary-fuel firing equipment Fuel-burning equipment for supplying additional heat to a boiler or furnace to achieve higher temperatures than would otherwise be achieved for raising additional steam, increasing output or completing the combustion of combustible solids, vapours and gases.

Availability Plant capacity available for use, usually expressed as a percentage of the maximum capacity. The percentage available may refer to a particular time, such as a time of peak demand for steam or electricity, or may be the average taken over a period—for example, a year. The availability factor is calculated:

$$\text{availability factor} = \frac{\text{hours operated or operable}}{\text{total hours in period}}$$

See *capacity*.

Avgas Abbreviation for *aviation gasoline*, used in piston-driven aircraft.

Aviation gasoline Or avgas, a gasoline suitable for piston-driven aircraft; these gasolines have rigid controls imposed on their distillation temperatures to ensure adequate volatility for normal cold starting, and safeguards against vapour lock, fuel system vent losses and *carburettor icing*. The final boiling point of 170° C maximum prevents fuel maldistribution and crankcase oil dilution. This gasoline must be distinguished from aviation turbine gasoline, which is intended for use in some types of turbine.

Aviation mixture methods Tests to determine the octane numbers of high octane aviation fuels. In the Aviation Lean Mixture Method an engine speed of 1200 rev/min is employed; in the Aviation Rich Mixture Method an engine speed of 1800 rev/min is employed. *See octane number; performance number*.

Avogadro hypothesis A statement that equal volumes of different gases at the same temperature and pressure contain the same number of molecules.

Avogadro number Or constant; the number of molecules in a gram-molecule. It is equal to $6.022\,52 \times 10^{23}$ mole^{-1}.

Avtag Aviation turbine gasoline.

Avtur Abbreviation for *aviation turbine fuel*, notably *kerosine*.

Axial-flow fan A type of fan in which the direction of flow of the air from inlet to outlet remains unchanged. *See centrifugal fan*.

Axial-flow turbine A turbine in which the flow of steam or gas is essentially parallel to the rotor axis.

B

Bacharach Smoke Scale A scale of ten shades from white to black used for the assessment of smoke concentrations in flue gases; a smoke stain is obtained by drawing a sample of flue gases through a filter paper in a prescribed manner, the resultant stain being compared with the scale. *See Ringelmann Chart*.

Backing wind The anti-clockwise change of direction of a wind, e.g. from N. through NW.; an opposite change of direction to veering. The same definition applies whether in the northern or southern hemisphere.

Back-pressure turbine A *steam turbine* in which exhaust steam is utilised for process work or for heating. Thus, the steam is not condensed at the turbine, being utilised at the temperature and pressure at which it leaves the turbine. *See pass-out turbine*.

Backward-curved fan blading A design of blading for *centrifugal fans* which is considered one of the most suitable for supplying forced draught to a boiler; it has reserves of pressure with which to overcome dirty boiler conditions—that is, it permits a wide variation in delivery pressure for the same quantity of air delivered. Blading of this design offers a high efficiency with a higher initial cost and lower running costs compared with *forward-curved fan blading. See Figure B.1; fan.*

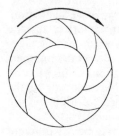

Figure B.1 Backward-curved fan blading

Bacteria A large group of unicellular or filamentous microscopic organisms, lacking *chlorophyll* and multiplying rapidly by simple fissure. Bacteria occur in air, water, decaying organic material, animals and plants.

Baffle A refractory structure whose function is to change the direction of flow of air, fuel or the products of combustion.

Bagasse The fibrous portion of sugar-cane remaining after the juice has been extracted. Mills crushing sugar-cane commonly use bagasse as fuel for steam production. In mills not used for the refining of sugar, the supply of bagasse equals and often exceeds plant steam demand; there has been little incentive to burn bagasse efficiently, and steam boilers serve also as incinerators. A typical percentage analysis of bagasse as fired is: moisture, 52; volatile matter, 40; fixed carbon, 6; ash, 2. The bulk density is a low $25-50$ kg/m^3, and the gross energy value about 9.0 kJ/g. Work on the control of particulate emission to atmosphere from the chimneys of bagasse-fired boilers has made much progress in recent years. Low-cost scrubbers have been developed; the ash filtrate from the scrubbers is mixed with mill mud and returned to the cane fields as a soil conditioner.

Bag filter A device for removing particulate matter from the waste gases of industrial processes. The filter medium is a woven or felted fabric usually in the form of a tube. The bags may be up to 10 metres in length and up to 1 metre in diameter. The upper ends are closed, while their lower ends are connected to a gas inlet which also serves as a hopper to catch the dust which is shaken down from the bags by mechanical or air reversal methods. The collecting efficiencies of bag filters are high, between 99 and 99.9 per cent; low gas velocities of the order of $1-3$ m/min are required. The choice of materials for bag filters is limited by temperature

28

limitations, being: (a) wool or felt, 90° C; (b) nylon, 200° C; (d) glass fibre, siliconised and graphite impregnated for longer life, 260° C. Plants have been constructed at British and American power stations to appraise the suitability of bag filters for power station emissions.

Bahco centrifugal dust classifier Apparatus for dividing samples of dust into size fractions by means of centrifugal force acting against an inward air current. By using different inward air velocities while the centrifugal force remains constant, the sample can be divided into a number of size grades.

Bailey wall A wall of plain water tubes in the combustion chamber of a *water-tube boiler*, faced with metal blocks.

Balanced draught A combination of *forced draught* and natural or *induced draught*, which ensures that the pressure of the atmosphere outside a furnace and the pressure of the bases inside (over the fire grate) are virtually the same; this implies zero on a water gauge. In practice, to avoid accidents, a slight suction is kept inside the furnace. *See* **Draught**.

Ball mill A low speed *pulveriser* for the production of *pulverised fuel*. It consists of a large-diameter shell, tubular or conical in shape, containing a charge of steel balls of about $1\frac{1}{4}$ in diameter. The shell is mounted on hollow trunnions which allow entry and exit for the dust carrying air. The mill rotates at a speed of about 15–25 rev/min. The ground product passes through a classifier which separates the overlarge particles and returns them to the feed end of the mill. The outlet temperature aimed at is usually 50° C.

Ball-mill method A method for determining the *grindability index* of a coal. A sample of coal is placed in a ball mill and the number of revolutions required to grind it so that 80 per cent of the sample passes a No. 200 US sieve (74 μm) is ascertained.

Ball-race mill A *pulveriser* which grinds by crushing; a lower ball-race is turned by a motor causing the balls to rotate. Coal enters the centre of the mill and is thrown out against the moving balls, which crush it against the races. The fine coal spills over the lower race, where it is picked up by air and carried around the upper ring to the centrifugal classifier section. Particles of sufficient fineness are carried with the air stream to the burner; the oversize particles are returned for additional pulverisation.

Ballast gas An inert gas introduced into a mixture of combustible gases to adjust its specific gravity.

Balloons, 'zero weight' Special hydrogen-filled balloons, adjusted to 'zero weight', used in air pollution research into the rise of chimney plumes under varying meteorological conditions. The balloons are introduced into a chimney and carried aloft into the general atmosphere by the chimney gases, their trajectory tending to follow the path of the chimney plume. A specially mounted range finder can be used to plot the distance and the height of each balloon from time to time.

Band screens Rectangular perforated screen plates, connected together as a continuous band. They are used for screening circulating water in power stations.

Banking The retention of a fire on a grate but at a very slow rate of combustion; this enables the boiler to be brought up to full steam quickly when steam is demanded. The banking of fires overnight or at the weekend is a common practice. The amount of fuel burned during the banked period should be only just sufficient to make up for natural cooling and to produce any small amount of steam that may be required.

Bar A unit of pressure in the centimetre-gram-second (CGS) system; the following relationships apply:

$$1 \text{ bar (b)} = 10^5 \text{ pascals (Pa)}$$
$$= 10^5 \text{ newtons per square metre}$$
$$= 10^6 \text{ dynes per square centimetre}$$
$$= 750 \text{ mm of mercury (approximately)}$$

One bar is just below normal atmospheric pressure. It is a non-SI unit but remains in use because of its exact relationship with the SI unit, the pascal. *See pascal*.

Barn An area of 10^{-24} cm^2; a unit used to measure the cross-sections of a *nucleus*.

Barometric condenser A device which, by condensing steam, produces a partial vacuum in a piece of equipment; it is used in conjunction with oil refinery vacuum distillation towers. The barometric seal legs end in a hot well which contains both hydrocarbon and aqueous condensates. The hot well should be sealed and fitted with a vent whereby released vapours can be passed to a furnace and burned. The barometric water flows to an oil interceptor; the problem of odorous water may be largely solved by the application of odour-masking compounds.

Barometric damper A pivoted, balanced plate, normally situated in the flue between furnace and stack, and actuated by the draught. *See Damper*.

Barrel (United States) Normal measure used throughout the oil industry, the US barrel equals 42 US gallons or 35 Imperial gallons. Since the density of different crude oils varies and the density of refined petroleum products varies widely, there is no universally applicable conversion factor from barrels to tons. However, for crude oils of a gravity of about 33–36 *degrees API*, corresponding to the averages for Iran, Iraq and Kuwait, there are

30

roughly 7½ barrels to a long ton (2240 pounds). For refined products the conversion factors vary from about 6¾ barrels of heavy fuel oil per ton to about 9 barrels of aviation gasoline per ton. A useful rule of thumb in expressing oil supplies (e.g. rate of flow of wells or the rate of exports of a given country) is that one barrel of crude oil a day represents 50 tons a year.

Barrels per stream day   The quantity of crude oil processed by a refinery in a continuous 24 hour period, expressed in barrels.

Barring gear   Equipment for slowly rotating a turbine shaft when the turbine is off-load.

Basal energy requirement   The energy required by a person (or any creature) when resting quietly, producing just enough energy to maintain basic body processes such as heart beat and breathing. For an average man weighing 65 kilograms, the basal energy requirement is 7.4 megajoules each day; for an average woman weighing 55 kilograms, the basal energy requirement is 5.4 megajoules each day. The total amount of energy needed by people whose activities are not very strenuous is about one-third higher than the basal energy requirement: about 11.1 megajoules per day for a 65 kilogram man and 8.2 megajoules per day for a 55 kilogram woman. Male office workers need about 10.5 megajoules each day, female office workers about 8.0 megajoules. Male university students need about 12, steelworkers about 14 and farmers 10.5–20 megajoules a day. *See **Food and Agriculture Organisation (FAO)***.

Base exchange process   *See **Ion exchange process***.

Base load.   A steady demand equal to a high percentage of the installed capacity of a plant. *See **base load station***.

Base load station   A power station whose generating costs are low and which therefore, stands high in the 'order of merit'; in consequence, it is under load by day and by night for long periods. Nuclear power stations are by their nature base load stations, because of low generating costs and of a technical need for continuous operation at high load factors. *See **merit order***.

Batching oils   Heavy gas oil or light lubricating oil used in the jute industry to facilitate spinning by lubricating the fibres

Batching pigs   Rubber and/or metal plugs used to separate out products in a pipeline and minimise contamination between succeeding batches of different products, although their use is often unnecessary with correct operating conditions.

Batch process   A process for carrying out a reaction in which the reactants are fed in discrete and successive charges; not a continuous process. Examples are the *batch still, coke oven* and *horizontal retort*.

Batch still   A still in which distillation is carried out in 'batches', the entire charge being introduced before heating begins and the distillation being completed without the introduction of an additional charge; an intermittent or non-continuous process.

Battersea gas washing process A method of wet scrubbing the flue gases from the coal-fired Battersea power station, located in London on the Thames, to remove large quantities of sulphur dioxide. The washing medium comprises water from the Thames with the addition of chalk. About 35 tons of water is required for each ton of coal burnt. The efficiency of removal of sulphur dioxide is of the order of 95 per cent. The effluent is discharged into the river as an almost saturated solution of calcium sulphate. A similar system is used at Bankside power station, also situated in London on the Thames, where oil containing 3–4 per cent sulphur is consumed. The process has several drawbacks: (a) the plume is cooled, and the residual sulphur dioxide and moisture fall readily on to the surrounding district; (b) the sulphur is not recovered in a useful form; (c) enormous quantities of water are required and water pollution occurs; (d) it promotes corrosion problems and adds significantly to generating costs.

Baum coal washer A wet cleaning plant for separating 'dirt' in the form of sandstone, shale, clay, pyrites, calcite, and so on, from *run-of-mine coal*. Coal is passed over an immersed perforated grid; alternating pressure and suction cause the water to move up and down through the perforated bed on which the coal is travelling. The dirt settles while the clean coal rises to the upper layers.

Baumann exhaust A special design of the exhaust end of a turbine to give additional effective area for the exhaust steam to pass to the condenser.

Bearing chatter Vibration of a shaft journal, arising from excessive clearance at the bearing surfaces.

Beaufort scale A scale for estimating and reporting wind force ranging from 0 for calm up to 12 for a hurricane.

Bed moisture The total moisture in a coal seam prior to working. *See moisture*.

Beijer Institute *See International Institute for Energy and Human Ecology*.

Belgian grate-fired kiln An annular longitudinal arch kiln used in the brick-making industry; coal is fired on to grates situated near the bottom of the kiln. In some cases oil firing through the wickets is practised. *See brick kiln*.

Beneficial use A use of the environment or any element or segment of the environment that is conducive to public benefit, welfare, safety or health, and that requires protection from the effects of waste discharges, emissions, deposits and despoilation. Beneficial uses include: (a) potable water supply for drinking, domestic and

32

municipal use; (b) agricultural and industrial water supply; (c) habitats for the support and propagation of fish and other aquatic life; (d) recreational activities such as bathing, fishing and boating; (e) scenic and aesthetic enjoyment; (f) navigation; (g) wildlife habitats. A residual use is a use other than a beneficial use in respect of the environment, e.g. the disposal of liquid effluents.

Beneficiation In relation to ores, a concentration process which may include drying to reduce the water content, roasting to reduce the sulphur content, and washing to remove some of the gangue or undesirable impurities.

Benefit–Cost analysis *See cost–benefit analysis.*

Bentonite The clay mineral montmorillonite, a magnesium aluminium silicate; it is used in oil refineries as a treating agent, in drilling as a mud component and in greases.

Benzene An aromatic liquid hydrocarbon obtained from coal and petroleum; it is used as a solvent, as a fuel and in the manufacture of plastics. A benzene ring is the arrangement of six carbon atoms in the benzene molecule.

Benzine A mixture of liquid hydrocarbons obtained from petroleum and used as a solvent, particularly in dry cleaning.

Benzole A mixture, predominantly composed of aromatic hydrocarbons, obtained as a by-product of the carbonisation of coal, either from coal gas by absorption or from coal tar by distillation. This mixture, after washing and rectification to concentrate the *benzene* and eliminate undesirable constituents to any desired extent, is classified in the United Kingdom according to purity as 'motor benzole', 'industrial benzole', '90s benzole', 'pure benzole' and 'pure benzole for nitration'. Specifications for these grades are issued by the National Benzole Association.

Benzpyrene A hydrocarbon present in coal smoke and cigarette smoke; it is known to be carcinogenic to man. *See* **carcinogenic compounds**.

Bergius process A process developed in 1924 for the manufacture of oil from coal. Made into a paste with heavy oil, coal is heated in the presence of a catalyst to a temperature of 450–470 C; the carbon of the coal reacts with hydrogen to yield a variety of hydrocarbons. The process made an important contribution to German war capabilities; in 1944 twelve factories were in operation, producing 2.25 million tonnes of petrol and 1.25 million tonnes of fuel oil per annum. Named after F. Bergius (1884–1949). *See* **coal hydrogenation; Fischer–Tropsch synthesis.**

Beryllium Be. An *element* and light metal, at. no. 4, at. wt. 9.0122. It is used in crystalline form for transistors and in fluorescent lighting, and in fabricated form for small aircraft parts. Its low atomic weight, low neutron-capture cross-section and high thermal conductivity render it suitable for use as a moderator and reflector in nuclear reactors; its considerable strength and high melting point make it possible canning material for fuel elements. It is expensive, toxic and difficult to fabricate. *See cross-section, total.*

Bessemer converter   A *steelmaking furnace*, consisting of a cylindrical body of steel plates, brick-lined, with a base provided with a series of air holes; the capacity of a converter varies from 10 to about 60 tonnes of metal. Molten pig iron is poured into the converter; air is forced through the liquid metal, producing a flame at the mouth of the converter; no additional heat source is necessary. The tropena is a small side-blown converter. The scope of steelmaking by the Bessemer process has been widened by the adoption of an oxygen/steam blast. The process is then known as the VLN (very low nitrogen) steelmaking process; by this process low-nitrogen steels may be produced.

Beta particle   An *electron*, or a *positron*, emitted in the decay of some radioactive nuclei. In beta radiation the rays travel distances of up to a few millimetres in human soft tissues before they are fully absorbed.

'Big oil'   An American description for the large domestic oil companies in the United States, especially when appearing to act collectively; and a recognition of the commanding role of oil in the national economy.

Bimetallic strip thermometer   A device for measuring temperature. It comprises two strips of metal having different coefficients of expansion welded side by side; a temperature change causes the strips to curl or deflect slightly. The use of this device is usually limited to simpler types of thermostat; it is rarely used in industry for the straightforward measurement of temperature. *See pyrometer; thermometer.*

Bin and feeder system   A system for supplying *pulverised fuel* to boilers or furnaces. It consists of a pulveriser of the slow-, medium- or high-speed type, from which pulverised coal is extracted and delivered by an exhauster fan to a cyclone situated above a bin; this cyclone extracts the coal dust, which is deposited in the bin, at the base of which are a number of feeders, usually of the screw or rotary drum type, which distribute the pulverised fuel to the boilers or furnaces. The fuel falls from the feeder into an air stream from a primary air fan which conveys it through ducts to the burners.

Biochemical oxidation   The process by which microorganisms within an aerobic treatment process transform organic pollutants into settleable organic or inert mineral substances.

Biochemical Oxygen Demand (BOD)   The weight of oxygen taken up mainly as a result of the oxidation of the constituents of a sample of water by biological action. The result is expressed as the number of parts per million (or grams per litre) of oxygen taken up by the sample from water originally saturated with air, usually over a period of 5 days at a temperature of 20° C. The result gives some measure of the amount of biologically degradable organic material in polluted waters, although when samples contain substances such as sulphites or sulphides, which are oxidised by a purely chemical process, the oxygen so absorbed may form part of the BOD result. The BOD test is no longer an adequate criterion for judging the presence or absence of pollution, for many relatively new pollutants must be considered, e.g. pesticides, industrial organic compounds, fertilising nutrients, dissolved salts, soluble iron and manganese, and heat.

Bioconversion Or biological conversion; the conversion of the energy stored in plant materials such as trees, crops, algae and water plants, to produce ethanol, methanol and methane. Crop wastes, urban wastes, sewage and animal excreta can also be converted to liquid and gaseous fuels such as methane and hydrogen.

Biological shield A thick wall or shield, usually consisting of 3 metres thickness of concrete, surrounding the core of a *nuclear reactor*, to absorb neutrons and gamma radiation for the protection of operating personnel.

Biomass The total weight of all living matter in a particular habitat or area. Biomass is often expressed as grams of organic matter per square metre. Biomass differs from productivity, which is the rate at which organic matter is created by *photosynthesis*.

Biomass energy The concept of using vegetation as a continuous source of future energy, relying on crops which offer a high yield in energy terms. In Sweden experiments have been undertaken at the College of Forestry using specially selected strains of willow and poplar; high yields have been achieved using a system called short-rotation forestry. It has been argued, though yet to be demonstrated, that biomass could contribute a significant proportion of Sweden's energy before the turn of the century.

Biome A community of plants or animals extending over a large natural area; a major regional ecological community such as a tropical rain forest.

Biometeorology Studies involving pollutants and infectious agents, and the interaction of pollutants and weather factors.

Biosphere The sphere of living organisms (plants and animals, including micro-organisms); it comprises parts of the atmosphere, the hydrosphere (oceans, inland waters and subterranean water) and the lithosphere. The biosphere includes the human habitat or *environment* in the widest sense of these terms.

35

Biostimulants Substances which stimulate the growth of aquatic plants. For example, the addition of large amounts of nitrogen and phosphorus compounds to lakes may stimulate massive growth of microscopic plants such as blue–green algae or the larger water-weeds. The process is called cultural eutrophication. Sewage is a major source of biostimulants, particularly nitrogen and phosphorus. A large part of the phosphorus found in sewage is derived from domestic and industrial detergents which contain phosphorus compounds to enhance their cleaning properties.

Biosystematics The study of the biology of populations, particularly in relation to their breeding systems, reproductive behaviour, variation and evolution.

Biota The animal and plant life found within an *environment* or a geographical region.

Biotic Relating to life and living systems, rather than the physical and chemical characteristics of an *environment*. Biotic factors are influences in the environment that emanate from the activities of living organisms.

Birdnesting The adhesion of fused ash to boiler tubes, cementing together unburned carbon.

Biscuit firing A stage in the firing of pottery ware in which the clay is transformed into fired ware at a temperature of about 1150° C. *See enamel firing; glost firing*.

Bismuth Bi. A grey-white metallic *element* at. no. 83, at. wt. 208.98. The metal is used as a constituent of fusible alloys. The naturally occurring isotope is the heaviest stable nuclide. The metal has a low neutron-capture cross-section and a low melting point of 271° C; it has been considered as a solvent for uranium in a liquid-metal-fuelled *nuclear reactor*, and as a liquid carrier for fuel slurries.

Bitumen oxidising plant Plant in which air is blown through bitumen in order to convert it to a rubbery consistency which is required for special uses such as roofing felt and battery tops. The spent air from the blowing is malodorous because of incipient cracking and oxidation, and is piped to process furnaces where the odour is destroyed by combustion.

Bituminous coal The best-known of solid-fuels; the description 'bituminous' was originally applied because of the tendnecy to burn with a smoky flame and melt when heated in the absence of air. The volatile content varies from about 20 to 35 per cent; the *fixed carbon* between 45 and 65 per cent; the *inherent moisture*, between 3 and 10 per cent; and the calorific value, between 26 kJ/g and 32 kJ/g. Most bituminous coals have a banded or laminated structure, and a shiny black appearance. *See coal*.

Black body Surfaces with an *absorptivity* and *emissivity* of unity.

The term 'black body' does not necessarily mean that the body appears black to the eye. The radiation emitted by a black body depends upon its surface area and temperature, and follows a relationship known as the *Stefan–Boltzmann law*.

Black centre burning The condition of a fuel bed in which an area or areas of unignited coke appear at the surface. It may occur in underfeed stoking if the air rate is too great. Black centre burning may also occur due to the formation of 'coke trees' from strongly coking coals. Black centre burning is accompanied by smoke emission. *See red top burning*.

Black oils A general term applied to the heavier and darker-coloured petroleum products such as heavy diesel fuel, fuel oil and some *cylinder stocks*. It is used mainly in connection with shipping and storage; a black oil tanker is one used for carrying black oils, and would require cleaning before being used for *white oils*.

Black smoke Smoke produced when particles of carbon are derived from the cracking of hydrocarbon gases, following sudden cooling. Smoke is visible evidence of incomplete combustion. Black smoke is frequently considered to be smoke as dark as, or darker than, shade 4 on the *Ringlemann Chart*. *See brown smoke*.

Blast atomiser A type of *oil burner* in which the atomising medium may be air or steam. Blast atomisers are classified into three types according to the pressure of air or steam: (a) low-pressure, operating with an air blast with about 25–40 per cent of the necessary combustion air used for atomisation; (b) medium-pressure, operating with an air blast with about 3–5 per cent of the necessary combustion air used for atomisation; (c) high-pressure, operating with a steam or air blast.

Blast furnace A shaft furnace used for the reduction of ore to metal; used in the smelting of iron ore, copper, lead, antimony, tin, nickel and cobalt. In the steel industry the furnace shaft is a vertical steel cylinder lined with firebrick standing up to 30 m or more in height; the hearth diameter varies up to over 10 m. Immediately above the hearth section the diameter of the furnace increases to form a 'bosh' which accommodates the *tuyeres*; above the bosh the furnace diameter decreases gradually. The conical shape of the stack allows the charge of ore, coke and flux to swell; the bosh assists in checking descent. The charge is admitted through a double bell arrangement at the top. *See blast-furnace gas; blast.furnace stove*.

Blast-furnace gas A gas used extensively in the metallurgical industries, being produced in blast-furnaces during the reduction of iron ore. The calorific value of the gas varies, but a figure of 4 MJ/m^3 is general. A typical analysis of the gas, the constituents being expressed in percentages, would be: carbon monoxide, 28;

hydrogen, 2; carbon dioxide, 12; nitrogen, 58. The gas as it leaves the blast furnace has a high dust burden; cleaning in two stages is commonly practised.

Blast-furnace stove A stove for supplying preheated air to a *blast-furnace*. A typical stove consists of a vertical steel cylinder with a spherical dome top; the stove contains a combustion chamber with a lens-shaped horizontal cross-section, while the major part of the volume of the stove is filled with *chequer-brickwork*. A stove passes through a two-stage cycle lasting perhaps 4 hours. In the first, 'on-gas', stage the chequer-brickwork is heated by the combustion of blast-furnace gas for about 3 hours; in the second, 'on-wind', stage the chequer-brickwork is cooled by the passage of cold air for about 1 hour. At first only part of the air to be heated passes through the stove, emerging at perhaps 1100° C, when it is mixed with cold air to supply a blast at about 590° C. As the chequer-brickwork cools, an increasing proportion of the blast air passes through it; in this way the final temperature of the air supplied to the blast-furnace is maintained. With a battery of four stoves a blast-furnace receives a constant supply of preheated air. Also known as a Cowper stove. *See Howden–Ljungstrom air preheater; pebble stove.*

Blast saturation temperature (BST) The criterion of the proportion of water vapour in the air-steam blast of a *gas producer;* the maximum thermal efficiency of a gas producer is usually attained when the BST is about 60° C.

Blending The mixing of two or more fuels of different properties and characteristics to produce a fuel with certain qualities; similarly, the mixing of various refined fractions to obtain lubricating oils having particular properties.

Block heating *See district heating.*

Block tariff A method of charging for gas or electricity in which the price per therm or unit is highest for a specified initial number, or block, of therms or units consumed within a prescribed period, and lower for additional quantities of therms or units consumed within the same period. *See tariff.*

Bloom In respect of an oil, its colour by reflected light when this differs from its colour by transmitted light; many petroleum oils which appear red or yellow by transmitted light exhibit a blue or green bloom.

Blow-by Gases consisting of carburetted gasoline together with some exhaust products which blow past the piston rings into the crankcase of an internal combustion engine. For many years American-made automobiles utilised a system of crankcase ventilation employing a road draft tube. With this system, air flowing past the moving vehicle aspirated crankcase fumes through the tube from the crankcase. Replacement air was drawn through a combination oil filter cap and air inlet. Devices to control crankcase emissions became mandatory in California from 1961 onwards. Control has consisted in returning crankcase ventilation air and gases to the engine intake manifold for consumption during operations. The negative pressure of the engine induction system is used to establish a positive flow of the crankcase ventilation air through the crankcase.

Blow-down   The release of water from the lowest part of a boiler to free it of sludge and reduce the dissolved solids content of the water. The amount of blow-down required per day depends upon the maximum total dissolved solids (TDS) which can be tolerated without risk of *priming*.

Blow-down valve.   A valve connected to the lowest part of a boiler to fulfil two functions: (1) the boiler can be emptied for inspection and maintenance purposes; (2) the sediment can be removed and the salt concentration in the boiler water reduced. In horizontal boilers the valve is connected to the base of the boiler front plate. In water-tube boilers it is connected to the mud drum by a substantial steel pipe. Many boilers are equipped with continuous blow-down valves through which a constant stream of water leaves the boiler. Without blow-down, sediment would simply accumulate and the dissolved solids concentration increase until foaming and priming occurred, resulting in wet steam. Blow-down valves should be provided with a safety device to prevent the removal of the key until the valve is fully closed. The parallel slide valve is the best type that can be obtained.

Blow-holes   Holes which appear in a fuel bed through localised acceleration of combustion rates.

Blown bitumen   A semi-solid or solid oxidised product obtained primarily by bubbling air through hot liquid bitumen, with a resultant increase in the melting point and a modification of other physical properties.

Blown oil   Fatty oil, of which the viscosity has been increased by blowing with air at an elevated temperature; a term used in the petroleum industry.

Blow-out preventer   A special hydraulically operated gland-like device, employing synthetic rubber, designed for use when drilling for oil, to maintain pressure control of the drilling fluid.

Blue water gas   *See water gas*.

Boghead coal   A coal derived from *sapropel*, similar to *cannel coal*, but characterised by the presence of algal remains and substantial enrichment in hydrocarbon compounds. A variety of boghead coal known as torbanite is largely a compound of minute algal bodies, and yields over 90 per cent of volatile matter.

39

Boiler A device for making steam from water, although the term is also applied to many hot water appliances in which the water is not meant to boil. A more modern term for those in the former category is that of 'steam generator'. *See Cornish boiler; economic boiler; horizontal boiler; La Mont boiler; Lancashire boiler; locomotive boiler; Manchester steam generator; magazine boiler; marine boiler; package boiler; sectional boiler; supercritical once-through boiler; vertical boiler; waste-heat boiler; water-tube boiler; WIF boiler.*

Boiler availability The number of days per year that a boiler remains in service without shutdown for cleaning or overhaul, expressed as a percentage.

Boiler cleaning methods Methods for the removal of internal scale in boilers and external deposits which impede heat transfer and may lead to dangerous overheating of the metal. The tendency to scale formation is reduced by proper boiler water conditioning. Sometimes substances may simply be added to the water, causing the small amount of residual hardness to precipitate in a flocculent form, which allows it to be removed by blow-down. In many instances, however, external softening is necessary before the treated water goes forward to the boiler. Conditioning may be by a *lime soda process* or an *ion exchange process*. Scale is generally so hard as to resist a wire brush and chipping hammers, and specially shaped chisels and scrapers must be used. For water tube boilers, rotary chippers are used in which hammers are thrown against the tube. Inhibited acid cleaners may be used to assist with the removal of scale. Sometimes compressed-air guns may be employed which blast out air in short bursts at high pressure. These bursts cause shock waves which break deposits from heating surfaces. External cleaning consists chiefly in the removal of soot from heating surfaces. *See soot blower; soot blowing; water lancing.*

Boiler drum A vessel usually solid forged or fabricated by fusion welding which provides a reserve of water for the tubes of a *water-tube boiler* and as a collecting space for the steam that is generated.

Boiler efficiency The amount of heat in useful form produced by a boiler expressed as a percentage of the potentially useful heat put into it, i.e.:

$$\text{Boiler efficiency }\% = \frac{\text{heat output}}{\text{heat input}} \times 100$$

where   heat input = weight of fuel × energy value per unit weight;
heat output = weight of steam × (total heat/kg steam − heat/kg feed-water)

40

For hot water boilers
> heat output = weight of water leaving boiler × (outlet temperature − inlet temperature)

Boiler feed pumps and injectors Devices for feeding water into a boiler which is under pressure: the main methods are the use of (a) the *injector*, (b) the *displacement pump*, (c) the *centrifugal pump*. The *injector* is suitable for the smaller boiler only.

Boiler and turbine unit plant A boiler and turbine having no interconnections with adjacent boilers and turbines; self-contained.

Boiling water reactor A *nuclear reactor* using water as the *moderator* and *coolant*. The water boils under pressure in the reactor and the steam may be used to drive a *steam turbine*; the steam being radioactive, the turbine must be shielded. *See biological shield*.

Bomb calorimeter Apparatus used to determine accurately the gross calorific value of oil or coal. One gram of oil, or finely powdered coal, is placed in a small platinum or silica crucible and inserted in a strong stainless steel 'calorimetric bomb' with an air-tight cover. The bomb is filled with oxygen to a pressure of about 25 atmospheres and immersed in a calorimeter vessel containing water. The oil or coal is ignited by passing a momentary current of electricity through a thin wire within the bomb. The heat evolved by the combustion of the oil or coal is measured by the rise in temperature produced, proper corrections being applied for the water equivalent of the bomb and calorimeter vessel and for temperature changes due to radiation. The gross calorific or energy value is then calculated.

Bonded deposits Deposits formed on the heating surfaces of furnaces and auxiliary plant consisting of the bonding of *fly ash* with fusible materials produced during *combustion*. See *high-temperature deposits; low-temperature deposits*.

Bond energy The energy characteristic of a chemical bond between two atoms; it is measured by the energy required to separate the two atoms.

Bone and Wheeler apparatus Apparatus used in the analysis of fuel gases.

Bone seeker A radioactive element which tends to be deposited preferentially in the bones of a body, e.g. strontium, radium or plutonium.

Bord and pillar mining A method of coal extraction used in underground coal mines; a series of straight, parallel roadways (bords) are driven into the coal seam, and connected at intervals by cut-throughs usually driven at right angles to the bord. Large blocks of coal (pillars) are left to support the overlying strata while

41

mine development, or 'first working', proceeds. Development is divided into panels on two or more parallel roadways; the main development roadways within panels are called 'headings', which are interconnected by cut-throughs. Once an area is exhausted, the pillars may be systematically removed, allowing a controlled collapse of the mine roof to take place. Bord and pillar mining is generally fully mechanised, a large electrically operated continuous mining machine being allocated to a panel. This cuts coal from the face in both first working and pillar extraction, and loads into shuttle cars of about 10 tonnes capacity. The shuttle cars provide intermediate coal transportation from the face to the main coal haulage system; this is usually a high-capacity conveyor belt travelling to a surface storage facility.

Boron B. A non-metallic *element*, at. no. 5, at. wt. 10.82. It has a high neutron-capture cross-section, and boron steel is often used for control rods in nuclear reactors.

Bottled gas *Butane* or *propane*, or butane–propane mixtures, liquefied and bottled under pressure for industrial, commercial or domestic use.

Bottoms Liquid which collects in the bottom of a vessel, during a distillation process or while in storage; hence 'tower bottoms', 'tank bottoms'.

Bottom sediment and water (BSW) Description of contaminants of fuel oil; the greatest amounts of sediment and water are found in residual fuel oils.

Boundary layer That part of a fluid extending from a physical boundary into the bulk of the fluid in which the motion of the fluid is significantly affected by the frictional drag caused by the boundary.

Boundary lubrication A state of partial lubrication existing between two surfaces in relative motion in the absence of full film or fluid lubrication, due to the existence of an adsorbed monomolecular layer of lubricant on the surfaces; thus, a static load may squeeze out all the lubricants except that held by adsorption on the metal surfaces. *See lubrication.*

Bourdon gauge A *pressure gauge*, commonly fitted to steam boilers. It comprises a tube, one end of which is open for connection to the boiler, the other end being sealed. The tube is coiled and the sealed end connected to the pointer mechanism of the gauge. When steam is admitted, the pressure inside the tube tends to straighten it; the effect of this is a movement of the sealed end and pointer mechanism. The gauge is connected to the boiler by means of a siphon containing condensate which prevents live steam from coming into contact with the gauge.

Boyle's law A gas law which states that the volume of a given mass of a perfect gas varies inversely as its absolute pressure, provided that the temperature remains constant. Thus, at constant temperature:

$$P_1 V_1 = P_2 V_2$$

where P_1 and P_2 represent two different pressures, and V_1 and V_2 the volumes corresponding to those pressures. *See Charles's law; gas laws; perfect gas.*

Boys calorimeter A *gas calorimeter*, in use in Britain in which the *energy value* of a known volume of gas is calculated from the increase in temperature of a measured volume of water heated by the gas under test. The hot gases from the burner in the instrument pass over nine turns of copper tubing; the cooled gases leave the instrument through a number of holes in the lid. The temperature of the water is measured at the inlet to, and outlet from, the instrument. Both the gross and net energy value can be calculated.

Bradford breaker A machine for breaking coal to a predetermined size while rejecting large refuse and tramp iron. it comprises a large cylinder rotating at about 20 rev/min; the cylinder consists of steel screen plates, the size of the screen openings determining the size of the crushed coal. The coal is broken by dropping; as it enters at one end of the cylinder, it is picked up by lifting shelves and carried up until the angle of the shelf permits the coal to drop. *See pick breaker.*

Brannerite A uranium-bearing mineral, being a mixture of uranium, calcium, titanium and other metal oxides.

Brass An alloy of copper and zinc, though other elements such as aluminium, iron, lead, manganese, nickel and tin are often added; there are many varieties.

Brasses Mineral impurities in coal which have a yellow metallic appearance; the impurities consist mainly of iron sulphides.

Breathing The movement of oil vapours or air in and out of the relief valves or vent lines of storage tanks due to the alternate heating and cooling of the contents.

Breeches-flued boiler *See Galloway boiler.*

Breeching The connection between a furnace or incinerator and its stack.

Breeding The process of producing fissile material from a fertile material such as uranium-238 or thorium-232.

Brick kiln *See Belgian grate-fired kiln; Hoffmann kiln; intermittent kiln.*

Bridge-wall A partition wall between chambers or sections of a flue over which pass the products of combustion.

Bright   Applied to lubricating oils, a term meaning clear or free from moisture. In 'blowing bright' air is used to carry off moisture.

Bright stock   A lubricating oil of high viscosity prepared from a cylinder stock by further refining, e.g. by solvent de-asphalting, de-waxing, acid treatment and/or earth treatment.

Bringing in   In relation to an oil well, the stage following the completion of drilling and withdrawal of the drill pipe when the *mud* is withdrawn or 'bailed down', or displaced with a fluid of lower specific gravity, and the oil is able to overcome the static head and flow into the well.

British Gas Corporation   A central body responsible for finance and general policy in respect of the gas industry in the United Kingdom; it superseded the Gas Council and the twelve regional gas boards in 1972. The UK gas industry was nationalised in 1949.

British National Oil Corporation (BNOC)   A state oil company created in 1976 under the Petroleum and Submarine Pipelines Act to exercise control over the North Sea oil industry and engage in other initiatives. The company has powers to: (a) search for and exploit petroleum anywhere in the world; (b) take over the British Government's participation interests in North Sea oil; (c) buy and sell petroleum and trade in its derivatives; and (d) build, hire or operate refineries, pipelines and tankers. Profits are paid into a national oil account. BNOC has options to lift 51 per cent of North Sea oil; it buys and sells back the oil to British companies with refineries at market prices. *See North Sea bonanza*.

British Nuclear Fuels Ltd   A specially created company to which the nuclear fuel business of the United Kingdom Atomic Energy Authority was transferred under the Atomic Energy Authority Act 1971; the shares of the company are held by the Authority.

British system   An older system of measurement still in fairly wide use, despite the strong trend towards the *International System of Units* (SI). Length is measured in inches, feet and yards; weight in ounces, pounds and (long) tons; and time in seconds, minutes and hours.

British thermal unit (Btu)   The amount of heat needed to raise the temperature of 1 pound of water through 1 degree Fahrenheit at or near 39.1 degrees Fahrenheit. The following relationships apply:

$$
\begin{aligned}
1 \text{ Btu} &= 1055.06 \text{ joules} \\
&= 1.05506 \text{ kilojoules} \\
&= 251.996 \text{ calories} \\
1.8 \text{ Btu} &= 1 \text{ Chu} \\
1.8 \text{ Btu} &= 453.59 \text{ calories} \\
10^5 \text{ Btu} &= 1 \text{ therm}
\end{aligned}
$$

One Btu/lb × 2.326 = joules/gram (J/g). Obsolescent types of British thermal unit are the 60° F Btu, which was the heat needed to raise the temperature of 1 pound of water through 1 degree Fahrenheit, from 60 to 61° F; and the mean Btu, which was 1/180 of the heat needed to raise the temperature of 1 pound of liquid water from 32 to 212° F. *See calories; Centrigrade (Celsius) heat unit; joule.*

Bronchitis Inflammation of the mucous membrane of the bronchial tubes existing in either an acute or chronic form. Epidemiological studies have indicated that there is a correlation between high concentrations of air pollution and morbidity and mortality among chronic bronchitics, although the ultimate progress of this disease is sensitive, particularly in winter, to many stimuli, of which air pollution is but one. Despite the complexity of the problem, the British Medical Research Council has concluded that cigarette smoking and air pollution are two important factors in the causation of the disease.

Brown coal *Coal* representing an early stage in the 'coalification' of *peat*; it is brown to black in colour, is devoid of lustre, contains appreciable quantities of *volatile matter* and has a moisture content ranging from 30 to 50 per cent. Textures vary in appearance from that of fibrous and woody to that of true coal. On a dry basis the brown coals of Victoria, Australia, show the following percentage variations in ultimate analysis: carbon, 64–71; hydrogen, 4.4–6.6; oxygen, 19–28; sulphur, 0.1–5.0; chlorine, 0.02–0.6; and nitrogen at a fairly constant 0.6. The gross dry energy value of most Victorian brown coals ranges between 23.5 and 28.0 kJ/g.

Brown smoke Smoke produced by tarry *volatile matter*, q.v., given off at relatively low temperatures. Unless burnt it appears at the chimney top as brown smoke. *See Black smoke.*

BSW *See bottom sediment and water.*

Btu meter A device, varying in design, for measuring the heat consumption between the flow and return pipes of a heating system. A typical example comprises the following parts: (a) a flowmeter transmitter connected to an *orifice plate* or *venturi tube*; (b) two electrical resistance thermometers in pockets and in common with the flowmeter transmitter connected to a Btu integrator; (c) a Btu integrator incorporating an electrical bridge network and a motor-driven chopper-bar integrating mechanism with reading scale and counting dial; (d) a flow indicator measuring water flow.

Bubble cap An inverted cup with a notched or slotted periphery to disperse the vapour in small bubbles beneath the surface of the liquid on the bubble plate of a distillation column. *See distillation tower.*

45

Bulk supply tariff A preferential *tariff* for the purchase of supplies of a commodity or service in bulk. The *Central Electricity Generating Board* sells electricity in bulk to the Area Electricity Boards, who then resell it to the consumer. The tariff is based upon the aggregate costs of generation incurred by the Generating Board, and the surplus or 'balance of revenue' to be earned. It is a *two-part tariff*, with a kilowatt demand charge and a running charge per kilowatt-hour.

Bund An earthwork or wall surrounding a tank or tanks to retain the contents in the event of a fracture of the tank.

Bunker 'C' fuel oil A heavy residual fuel oil supplied to ships and industry.

Bunker coal Coal supplied as fuel to ships.

Bunsen burner A *gas burner*, in which air is inspirated by the gas and mixed in the bunsen tube; the fuel and oxygen are heated together, the result being a short blue flame with little or no luminescence. The hydrocarbons and oxygen form hydroxylated compounds before ignition. If combustion is incomplete, aldehydes can be detected by their faint acrid odour. *See fishtail burner*.

Bureau of Mines Correlation Index A United States system of classifying crude oils into types, based on the predominant hydrocarbon series which is present. As many crude oils exhibit changes in dominant type in different regions of their distillation ranges, the Correlation Index quotes the dominant types in two nominated fractions in the distillation range, giving crude oil classifications such as naphthenic–naphthenic, naphthenic–paraffinic.

Burner A device which produces a flame. A burner must mix the fuel and oxidizing agent in proportions suitable for *ignition* and steady combustion; and it must supply the mixture at rates which ensure the stability of the flame. *See gas burner; oil burner*.

Burner management system The equipment used to supervise the light-up, firing and shut-down of a burner under all conditions; the concept does not include combustion control equipment for regulating the firing rate according to demand.

Burning area The total area of a grate or hearth, or combination thereof, on which combustion takes place.

Burning oil Paraffin oil or *kerosine*.

Burning rate The amount of fuel or waste consumed, usually expressed as pounds per square foot of *burning area*, or *grate*.

Burn-up The amount of the fissile material in a *nuclear reactor* which is destroyed by fission or neutron capture, expressed as a percentage of the original quantity of fissile material present; alternatively, the heat obtained per unit mass of fuel, expressed generally as MW/day tonne. In nuclear reactors using metallic uranium fuel elements, an average burn-up of 3000/4000 MW/day tonne is obtained; with uranium dioxide elements in advanced gas-cooled reactors, an average burn-up of 18 000 MW/day tonne is expected.

Burst can (cartridge or slug)   A fuel element can in which a leak has developed, permitting the escape of radioactive fission products.

Bursting discs   Carefully machined discs designed to burst at a predetermined pressure; they are inserted in the shell of a pressure vessel, protecting it from overpressurisation. Equivalent to a fuse in an electric circuit.

Busbars   A continuous set of separate copper conductors extending the full length of a switchboard, to which all the circuits on the switchboard are connected individually by some form of circuit-breaker or switch.

Bussed   A term describing the connecting together of a number of independent plant items through a common main to provide flexibility in operation.

Butadiene   A co-product from the cracking of *naphtha* or *propane*, containing four carbon atoms; it is used for making synthetic rubber.

Butane   A hydrocarbon gas, C_4H_{10}, being a member of the *alkane series* it can be stored under pressure as a liquid at atmospheric temperatures, and as 'bottled gas' it is used widely for cooking and domestic heating. When blended into gasoline in small quantities, it improves volatility and *octane number*. Commercial butane is a mixture of gaseous paraffins, mainly normal-butane and isobutane (both C_4H_{10}). Butane burns in air to carbon dioxide and water; the sulphur content is negligible, as with *propane*. *See liquefied petroleum gases*.

Butanol   Or butyl alcohol; an alcohol containing four carbon atoms, used for making plastics.

Butterfly damper   An adjustable, pivoted, plate normally installed in the flue between furnace and stack. *See damper*.

By-product fuels   Fuels produced incidentally in the manufacture of a product. Examples are bark, black liquor, bagasse, sawdust, blast furnace and coke oven gas; plutonium is a by-product of nuclear power generation.

C

Cadmium   Cd. A white metallic *element*; used as an alloy, it gives strength and ability to withstand high temperatures. Alloys include cadmium–nickel alloy, cadmium–silver alloy and cadmium–copper alloy. It is used in bearing metals, in electroplating and for some types of nuclear reactor control rods. Cadmium has toxic qualities; it occurs in minerals used for zinc production and is a potential air pollutant around zinc smelters. Human intake of cadmium can be from food and drink, as well as air.

Caesium An alkali metal; the isotope Cs^{137} is a product of fission in a *nuclear reactor*, being a radioactive gamma-ray emitter with a *half-life* of 30 years.

Caisson vessel system (CVS) A system of extracting and despatching offshore oil. Wells are drilled through a sea-floor template, the wells being linked back to the platform through vertical riser pipes; the oil is stored and then off-loaded to waiting tankers. The system obviates the need for pipelines connecting the rig to the shore.

Caking coal A coal which liberates, at temperatures of the order of 350–450° C, substances which cause the main mass of the coal to become more or less fluid or 'plastic'. As the coal becomes plastic, gases are evolved within the material, setting up a pressure; as the gases escape, the material becomes cellular. At higher temperatures, of the order of 450–500° C, decomposition proceeds rapidly and the coal hardens into coke. *See **non-caking coal**.*

Calcium cycle The circulation of calcium atoms brought about mainly by living things. Thus, calcium is taken up from the soil by trees and other plants, and deposited in roots, trunks, stems and leaves. Rain may leach some calcium from the leaves and return it to the soil. Creatures such as insects, rabbits and other herbivores obtain their share of calcium from the plants and leaves; birds acquire it by eating the insects. Animals and birds die, leaves and branches fall and decay, and thus the calcium component returns to the soil. Some calcium may be lost from an *ecosystem* by leaching and surface run-off carried to bodies of water; it is recycled through phytoplankton, zooplankton, fish, and lake and ocean water. Sea spray taken up to the atmosphere and airborne dust return calcium to the land. In a balanced ecosystem gains equal losses.

Calcium silicate An insulating material, light in weight, moisture resistant, and possessing a low thermal conductivity; it is not suitable for surfaces much above 750° C.

Calder Hall The location in Cumbria, England, of the first nuclear power station to supply electricity to the national grid; generation began in 1956.

Calorie Originally, the quantity of heat needed to raise 1 gram of water through 1 degree Celsius, although more precisely defined calories have come into use:

(a) 15° C calorie (cal_{15}), being the heat needed to raise the temperature of 1 gram of water through 1 degree Celsius from 14.5° C to 15.5° C = 4.1855 J.

(b) thermochemical or 'defined' calorie (cal_{th}) = 4.184 J.

(c) International Table calorie (cal_{IT}) = 4.1868 J.

In the energy field the 15° C calorie and the International Table calorie have been the main measures in use. The calorie is sometimes called a small or gram calorie, the kilocalorie a large or kilogram calorie. The calorie is being replaced by the SI unit for measuring energy in all its forms, the joule (J).

Calorific value *See energy value.*

Calorifier A vessel for supplying hot water; it is essentially a heat exchanger, the source of heat being a steam or hot water pipe coil running through a storage vessel.

Calorimeter Any vessel used for containing a liquid during heat experiments; it is generally essential to calculate its *water equivalent*.

Can A container in which the fuel rods of a *nuclear reactor* are sealed to: (a) prevent the escape of fission products into the *coolant*; and (b) protect the fuel rods from corrosion by the coolant.

Canadian pressurised heavy water reactor (CANDU) A Canadian-designed heavy water moderated and cooled natural uranium reactor; the moderator in being contained in a calandria vessel is kept separate from the coolant. The reactor is designed for on-load refuelling, but the heat release per tonne of fuel is relatively low and frequent fuel element replacement is necessary. Plants of this kind require substantial initial quantities of heavy water, although the subsequent make-up requirements to compensate for running losses are relatively small. A modification of CANDU, aimed at reducing both capacity and operating costs, is the Boiling Light Water-cooled Pressure-tube Reactor (CANDU-BLWR). *See nuclear reactor.*

Candela Unit of luminous intensity; the magnitude of the candela is such that the luminance of the total radiator, at the temperature of solidification of platinum, is 60 candelas per square centimetre (60 cd/cm^2).

Cannel coal A coal derived from *sapropel*, typically dark-brown to black in colour, with a dull, greasy lustre. Cannel is composed predominantly of minute *vitrinite* particles, with little or no admixture of *micrinite* or *fusinite*. It is highly volatile, up to 56 per cent, and usually non-caking. The calorific value may be as high as 34 kJ/g on a dry ash-free basis. *See boghead coal.*

Canvey Island terminal A terminal on the Thames Estuary near London, England, equipped to receive liquefied natural gas

(LNG) imports. In 1979 storage capacity was 110 000 tonnes. Algerian LNG is delivered throughout the year, one tanker arriving on average every 5 days. *See methane tankers.*

Capacitor   An arrangement of conductors in the form of either sheets or foil separated by a thin dielectric. It provides what is known as 'capacity', and is capable of storing a comparatively small electric charge.

Capacity   The estimated maximum level of production from a plant on a sustained basis, allowing for all necessary shut-down, holidays, etc. The capacity factor is calculated:

$$\text{capacity factor} = \frac{\text{output for period}}{\text{rated capacity} \times \text{hours in period}}$$

See availability; installed capacity; load factor, plant; refinery capacity.

Capacity costs   Or fixed or overhead costs or charges; costs and derived charges which do not vary with the total amount of good or service produced, but only with additions to the total capacity of the system. In respect of electricity generation, capacity costs are those incurred in the initial construction of power stations (including the cost of land and, in the case of nuclear power stations, the initial charges of fuel and heavy water), transmission and distribution networks, and control centres.

Capital, cost of   The cost of obtaining the total *capital employed* by a business, expressed as a rate of interest.

Capital charges   Charges which include interest on the amount of capital employed, and provision for depreciation or repayment of principal.

Capital cost component   In relation to the total production cost of each unit of output, the capital cost component is given by:

$$\frac{\text{capital charges}}{\text{units of output}}$$

where the capital charges are the sum of the annual interest charges on *capital employed* and the annual depreciation charges, and the units of output relate to the year under consideration.

Capital employed   The capital in use in a business; it consists of the total assets *minus* the current liabilities. Current liabilities are those liabilities providing resources for less than a year; their exclusion helps avoid fluctuations in total assets which might otherwise impair comparisons based on several years' trading. *See capital employed, return on.*

Capital employed, return on   The relating of profit to the estimate of average *capital employed* to give a ratio, commonly called the primary ratio, as follows:

$$\frac{\text{profit}}{\text{capital}}$$

It is essential that the long-run return on capital employed in a business should be sufficient to ensure a fair return to shareholders, provide for the normal expansion of the business, attract new capital when required and retain the confidence of creditors and employees.

Capital-intensive   Description of forms of production in which there is a considerable use of capital equipment per person employed. In a capital-intensive industry *capital charges* may account for 50 per cent or more of production costs, excluding the value of raw materials or fuel. In chemical processes the value of capital equipment per worker is very high. In the British coalmining industry, which is comparatively highly mechanised, wages still account for over half the costs of production.

Cap rock   An impervious layer, e.g. clay, which overlies a reservoir and prevents wholesale leakage of petroleum to the surface.

Capture   The process in which an atom or nucleus acquires an additional particle; for example, the capture of neutrons by nuclei, which often results in the production of a radioactive isotope of the capturing element.

Carbon   C. A non-metallic *element* at. no. 6, whose allotropic modifications include diamond, graphite, charcoal and coke. When a sufficient amount of oxygen is available, carbon may be burned completely to *carbon dioxide*, the reaction being expressed $C + O_2 \rightarrow CO_2$. The calorific value of carbon is about 34 kJ/g. Its low neutron-capture cross-section and low atomic mass render it suitable, in the form of graphite, for use as a moderator in a thermal nuclear reactor to slow down the fast neutrons produced in fission. It is also used as a neutron reflector.

Carbon bisulphide   A volatile solvent manufactured by reacting wood charcoal with sulphur vapour at a high temperature. It is used in the manufacture of a family of chemicals known as xanthates; these are used in mineral ore separation and also in the manufacture of viscose rayon.

Carbon black   Fine, fluffy carbon particles resulting from the incomplete combustion of either gaseous or liquid hydrocarbons. It is used as a reinforcing agent in rubber, and as a pigment and filler in plastics, paint, varnish, ink and paper. The particles are essentially pure carbon. Carbon black is produced in a reducing atmosphere. In the oil furnace process either a part of the feedstock or an auxiliary fuel is burned in the furnace to generate the heat required to decompose the oil into carbon black and gas.

Carbon cycle The circulation of carbon atoms brought about mainly by living things. Organic carbonaceous matter in rivers and other bodies of water is derived from domestic and industrial effluents, animals and plants, and soil erosion. Such carbonaceous matter is oxidised to carbon dioxide by aerobic bacteria thriving in the presence of dissolved oxygen. Carbon dioxide released may be neutralised in whole or in part by basic elements present (e.g. sodium, calcium or magnesium) to bicarbonates or carbonates. The process is analogous to respiration in animals, in which oxygen is absorbed and carbon dioxide released. The reverse process, the production of oxygen from carbon dioxide, is carried out by chlorophyll-containing (green) plants only in the presence of bright sunlight (photosynthesis). The carbon is utilised for the synthesis of complex organic compounds (e.g. fats and carbohydrates) and the oxygen released. In the absence of dissolved or free oxygen, anaerobic bacteria metabolise organic compounds (putrefaction), which results in the production of methane. This process occurs in septic tanks and sewage sludge digestion tanks, as well as in deposits of sludge, mud and vegetation. *See photosynthesis*.

Carbon dioxide CO_2. A colourless gas produced when carbon is burned in a sufficient supply of oxygen to complete the reaction $C + O_2 \rightarrow CO_2$. It is present in flue gases to a varying extent, but in typical cases amounts to 12 per cent of the gases by volume. If the supply of combustion air is insufficient, *carbon monoxide* is formed. Carbon dioxide is a normal constituent of the atmosphere to the extent of about 0.03 per cent by volume.

Carbon dioxide recorder An instrument of the thermal conductivity type. *Carbon dioxide* and water vapour possess similar thermal conductivities, which are about half those of oxygen and nitrogen; if, therefore, a sample of flue gas is passed over one arm of a heated Wheatstone Bridge circuit, the 'out-of-balance' set up in the circuit is a measure of the CO_2 percentage in the flue gases. *See 'Fyrite' CO_2 apparatus; Orsat apparatus*.

Carbon disulphide CS_2. A colourless liquid, when pure, turning yellow on exposure to air; it is extremely inflammable. The unpleasant characteristic smell of commercial carbon disulphide is due to its impurities. It is used extensively in many industries as a solvent for sulphur and rubber. *See Carpenter–Evans process*.

Carbon monoxide CO. An invisible, tasteless, odourless and highly poisonous gas. It burns with a pale blue flame to form *carbon dioxide*. It results from the incomplete combustion of carbonaceous materials, being encountered in many industrial works and occurring in the exhaust gases from internal combustion engines and in smoke from chimneys. Carbon monoxide has an affinity for the haemoglobin of the blood some 300 times that of oxygen and readily forms carboxyhaemoglobin in the red blood corpuscles. As the red blood corpuscles cannot carry their full quota of oxygen to the tissues of the body, the tissues suffer from oxygen starvation and show the symptoms of 'carbon monoxide poisoning'. Significant concentrations of carbon monoxide, expressed in parts per million by volume, are: maximum allowable concentration in an atmosphere in which a man works for 8 hours at a time, 100; headache after 3–4 hours, 200; severe headache, palpitation, nausea and dizziness, after 1–1½ hours, 500; prolonged exposure may be fatal, 1000. *See combustion*.

Carbon residue The quantity of solid deposits obtained when medium and heavy fuel oils are subjected to evaporation and *pyrolysis* at elevated temperatures. While the bulk of the sample evaporates, the residue decomposes and forms carbonaceous deposits. This property is of importance for oils used in burners, gas-making, compression ignition engines and certain lubricating oils. *See Conradson test; Ramsbottom test*.

Carboniferous period A period, between 200 and 250 million years ago, in which conditions suitable for the accumulation and preservation of coal-forming deposits were most widely established over the earth's surface.

Carbonisation The *destructive distillation* of coal to produce coke, gas and liquid products. Low-temperature carbonisation means carbonisation at temperatures ranging up to 800° C; high-temperature carbonisation means carbonisation at temperatures above 800° C. Coking coals with more than 33 per cent *volatile matter* are used for the manufacture of *town gas* and *coke*; coal is carbonised in horizontal, intermittent vertical and continuous vertical retorts. *See continuous vertical retort; horizontal retort; intermittent vertical retort*.

Carbonised briquettes Briquetted solid fuels *carbonised* in the process of manufacture. *See 'Phurnacite'*.

Carboxyhaemoglobin (COHb) A combination of *carbon monoxide* and the respiratory pigment occurring in the blood plasma, haemoglobin. The haemoglobin is deprived of its oxygen-exchanging properties, with resultant poisoning and suffocation of the body. Haemoglobin, while having a marked affinity for oxygen to form oxyhaemoglobin, has an even greater affinity for carbon monoxide. Individuals absorb carbon monoxide as smokers, from the emissions of petrol-driven vehicles and sometimes from occupational exposure. A normal and non-smoking person, breathing air devoid of carbon monoxide, will have blood containing about 0.4 per cent COHb; this is called the 'background COHb level'.

Carboxylic resin A cation exchange resin; it has the property of accepting calcium and magnesium ions in exchange for hydrogen. Used as an additional stage in water softening if water is too alkaline for base exchange softening; it removes the bicarbonates. The hydrogen combines with the carbonate ions to produce carbonic acid. Water leaving this de-alkalisation unit contains virtually no alkaline hardness. The carbonic acid is subsequently removed.

Carburation The mixing of air with a volatile fuel to provide the combustible mixture for use in an internal combustion engine; the device in which this occurs is the *carburettor*.

Carburetted water gas Water gas enriched with gasified hydrocarbon oil; about 10 litres of gas oil may be used per 30 m^3 of gas. The calorific value is about 20 MJ/m^3, similar to that of *town gas*. The gas burns with a yellowish luminous flame. This type of plant has often been used to meet winter peak loads. A typical analysis, the constituents being expressed in percentages, is: hydrogen, 36; carbon monoxide, 30; carbon dioxide, 6; nitrogen, 5; methane, 15; hydrocarbons, 8.

Carburetting In *town gas* manufacture, the enrichment of gas by the addition of gaseous hydrocarbons, e.g. from gas oil cracked in a carburettor.

Carburettor A device for producing a suitable fuel–air mixture for a *gasoline engine* under a variety of operating conditions. Small quantities of fuel are drawn from a jet by a current of air passing through a venturi in which the jet is situated; the gasoline is then atomised and partly vaporised. By the use of a *choke* and compensating jets the fuel composition can be varied to meet engine requirements. Also a device for cracking *gas oil*.

Carburettor icing Ice formation when, at high humidity and low ambient air temperature, water vapour entering a *carburettor* freezes on the throttle blade and venturi; this is partly the result of the withdrawal of heat from a carburettor in vaporising the fuel passing through it, when the throttle blade will be several degrees below the surrounding air temperature. In severe climatic conditions, therefore, a heated induction system is necessary.

Carburising A case hardening or cementation *heat treatment* process for raising the carbon content of steel by heating in a carbonaceous medium. *Carbon monoxide* carburises and is essential for the heat treatment of high carbon steel. *See decarburisation*.

Carcinogenic compounds Complex chemical compounds producing cancer in experimental animals and strongly suspected of contributing to lung cancer in man. One of the best-known carcinogens is 3,4-benzpyrene. In 1933 Cook and co-workers, with the assistance of fluorescence spectroscopy, isolated from 2 tons of pitch a substance characterized by synthesis as 3,4-benzpyrene; it is now known that coal tar contains about 1.5 per cent of the carcinogen. In recent years *benzpyrene* and other carcinogenic hydrocarbons have been identified in soot, carbon black, processed rubber, the exhaust gases of gasoline and diesel engines, coal gas, coffee soots, smoked food, and cigarette and tobacco smoke. Relatively large concentrations of some carcinogenic hydrocarbons have also been found in the atmospheres of some cities. Some of the complex polycyclic hydrocarbons formed in smoke have been shown to be carcinogenic when applied to the skin of animals. Coal tar and certain petroleums become carcinogenic only after being heated.

Carless days A New Zealand scheme, introduced in 1979, to save petrol; each motorist must elect on which day of the week a car will not be used. Carless day stickers are displayed on cars, unless exempt.

Carnot cycle An ideal cycle for a *heat engine*, giving maximum theoretical efficiency; no practical cycle is able to equal the Carnot cycle. In the ideal cycle, heat is added at a constant temperature T_1 and rejected at a lower temperature T_2:

$$\text{thermal efficiency} = \frac{\text{work done}}{\text{heat added}}$$
$$= \frac{\text{heat added} - \text{heat rejected}}{\text{heat added}}$$
$$= \frac{T_1 - T_2}{T_1}$$

Carnotite A uranium-bearing mineral, being hydrated uranium potassium vanadate.

Carpenter–Evans process A process for the removal of *carbon disulphide* from coal gas or petroleum oil gases; the gas stream passes over a nickel catalyst at about 450° C, carbon disulphide being converted to *hydrogen sulphide*. The hydrogen sulphide is then removed in a separate process. *See hydrogen sulphide removal.*

Carpet losses Losses which occur when solid fuel is stored outside directly on the earth; these can amount to 3 per cent of the initial amount stored.

Carry-over Water droplets and impurities carried by steam from a boiler to the superheater.

Cartel An arrangement among formerly competing producers to charge the same or similar prices. It may or may not include an agreement to set up a central selling organisation, or to share out the market either geographically or by production quota. The members of a cartel agree, in effect, to operate as a monopoly. The **Organisation of Petroleum Exporting Countries** (OPEC) is an outstanding example of an oil cartel. The Queensland Sugar Industry is organised as a cartel by Queensland State Government legislation. Air fares for international travel are fixed by agreement through the International Air Transport Association, an agreement which includes sharing the market between national carriers. The **European Economic Community** has set up an iron and steel cartel, Eurofer. The attitude of governments to cartels may vary from disapproval and positive measures to restore competition, to positive approval with active support. See **administered pricing; anti-trust laws.**

Carter energy policy   A national energy policy for the United States of America enunciated by President Carter in 1977; the aim was to shift the energy consumption pattern of the nation from a major reliance on oil and gas to coal, nuclear power, and solar and geothermal energy, and to attain specific goals by 1985. The President outlined a programme with strong emphasis on energy conservation, tax and price increases to discourage the use of oil products and natural gas and encourage the use of coal, a speed-up in the licensing of new nuclear power plants, the reform of electricity rates, and incentives for the development and use of solar and geothermal energy; a particularly controversial item was the proposal to impose a graduated excise tax on 'gas guzzling' automobiles. Congress turned back this programme in 1977, but after the new OPEC price increases in 1979 became willing to adopt parts of it. In July 1979 President Carter enunciated a new energy policy, comprising six points:

(1) Foreign oil imports to be cut in half by 1990, or by 4.5 million barrels a day.

(2) Import quotas to be used to ensure that imports never exceed those of 1977.

(3) An Energy Security Corporation to be set up to develop alternative sources of liquid fuels and energy, the Corporation to be allowed to issue $5 billion in 'energy bonds' and to be allowed to draw upon the proposed windfall profits tax to be imposed on the gains from the phased decontrol of indigenous oil prices.

(4) Congress to be asked to approve a 50 per cent reduction in the use of oil by electricity utilities.

(5) An Energy Mobilisation Board to be set up, along the lines of the War Production Board of World War II.

(6) More money to be provided for public transport and energy conservation measures generally.

In 1979 the United States was consuming 18.5 million barrels a day, of which 8.2 million barrels were imported.

Carter nuclear policy   A policy enunciated by President Carter in 1977, whereby the United States of America would avoid the development of nuclear power systems heavily dependent on plutonium, in the interests of the non-proliferation of nuclear weapons. He announced that he had decided to halt research aimed at perfecting an experimental fast breeder reactor, and the construction of plant for reprocessing irradiated fuels. In an expanding nuclear power system this would give priority to the older light water reactors. It would defer what has been described as the *plutonium economy*.

Cartridge   The fuel element of a *nuclear reactor*, consisting of a rod of fuel, such as uranium, hermetically sealed in a container or can. *See burst can*.

Car tunnel kiln   A continuous ceramic kiln comprising a tunnel some 30–100 m in length, a furnace being situated at about the centre of its length. Rail tracks are laid through the kiln and on these special cars carry the ware to be fired. The ware passes slowly through preheating, firing and cooling zones. Car tunnel kilns are fired by town gas, electricity, fuel oil or producer gas; the conversion from coal-fired intermittent kilns to continuous car tunnel kilns has been a major factor in increasing productivity and reducing smoke emission in the pottery industry.

Cascade heat exchanger   A *heat exchanger* which uses small refractory particles as a heat carrier. The particles are heated by a counter-current flow of hot gases in one chamber, and then pass to a second chamber, where they give up their heat to air or any other gas that is to be preheated.

Cascade impactor   An instrument in which the particulates in a gas sample are separated into a number of fractions in different size ranges.

Cascading   The interconnection of feed-heater drains, so that the flash steam from one heater can be used as heating steam in another heater at lower pressure.

Casing   The steel lining of an oil well, the main purposes of which are to prevent caving of the sides of the well, to exclude water or gas from the well, and to provide means for the control of well pressures and oil production. To complete an oil well it is usually necessary to perforate the last string of casing cemented in the borehole, opposite the oil-bearing formation, in order to allow the oil to flow into the well.

Casinghead gasoline   Gasoline extracted from natural gas; also known as *natural gasoline*.

Cassava   A tropical plant of genus *Manihot* with tuberous roots; it yields starch and flour, and may be used to produce alcohol from the fibrous residue of both the leafy tops and the tuberous roots.

Cast-iron An alloy of iron and carbon; a typical analysis of rainwater goods, the constituents being expressed in percentages, is: carbon, 3.4; silicon, 2.5; manganese, 0.7; phosphorous, 1.0; sulphur, 0.12; balance, iron. *See grey cast-iron; malleable cast-iron; white cast-iron.*

Catalyst A substance which accelerates or retards the rate of a chemical reaction, without itself undergoing any permanent change in its composition, and which is normally recoverable when the reaction is completed.

Catalytic cracking *See Fluid Catalytic Cracking Unit (FCCU).*

Catalytic hydrogenation A process involving the reaction of hydrogen with a slurry of coal in process-derived oil at high pressure and temperature, in the presence of a catalyst. The crude liquid product is separated from the solid residue of mineral matter and unconverted coal, and upgraded to liquid fuels by one or more secondary hydrogenation steps. The technology was developed in Germany from research by Bergius.

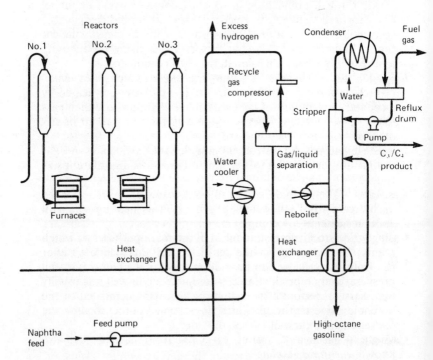

Figure C.1 Catalytic reforming process. (Source: British Petroleum)

Catalytic reforming   A process for upgrading straight-run gasoline and naphtha fractions in an oil refinery; de-sulphurised stocks are passed with hydrogen over a catalyst, usually platinum supported on alumina. The catalyst converts the normal- and cyclo-paraffinic hydrocarbons into higher octane number isoparaffins and aromatics. A by-product of the reforming process is a hydrogen-rich gas stream suitable for use in other refinery catalytic processes. *See Figure C.1.*

Catalytic rich gas (CRG) process   A *Gas Council* pressure process for manufacturing rich gas from *naphtha*; it has a high calorific value. The gas may be supplied direct to the consumer after blending.

Cathodic protection   A method of protecting metal from corrosion, whereby the metal at risk is made the cathode of an electro-chemical cell.

Caustic embrittlement   A cause of metal failure occurring in steam boilers at riveted joints and at tube ends, believed to result from the action of certain constituents of concentrated boiler water upon steel under stress.

Cavitation   The production of localised cavities in a liquid as a result of reduction of pressure in pumps and other equipment; the effects of cavitation may include erratic pumping, noise and erosive damage.

Cell   The unit of a battery. A 'primary' cell consists of two dissimilar elements (usually carbon and zinc), known as 'electrodes', immersed in a suitable liquid or paste known as the 'electrolyte'. Such a cell will produce a direct current of 1–1.5 volts. A 'secondary' cell or accumulator is of somewhat similar design, but is made suitable for use by passing a direct current of correct strength through it in a certain direction. When 'charged' in this way, current may be obtained from it for a number of hours. Each cell of a lead–acid accumulator produces approximately 2 volts; a 12 volt car battery contains six cells. *See fuel cell.*

Cellular respiration   A complex biochemical process which breaks down the carbohydrates produced by *photosynthesis* (either in the plants themselves or in the animals that eat them). Cellular respiration is a slow combustion or oxidation process and the chemical energy thus liberated is used by the plants or animals for their living functions. The end products of cellular respiration are carbon dioxide and water which are returned to the atmosphere, completing the cycle of carbon dioxide.

Celsius heat unit (Chu)   The amount of heat necessary to raise the temperature of 1 pound of water through 1 degree Celsius, or, more exactly, 1/100 part of the amount of heat required to raise the temperature of 1 pound of water from 0 to 100° C. The following relationships apply:

$$1 \text{ Chu} = 1.8 \text{ Btu}$$
$$= 453.592 \text{ cal}$$
$$= 1899.10 \text{ J}$$

See British Thermal Unit; calorie; Celsius temperature scale; joule.

Celsius temperature scale A temperature scale based on two fixed points, (a) the lowest (0) being the freezing point of water and (b) the highest (100) being the boiling point of water, at normal atmospheric pressure. The scale was originally devised by Anders Celsius (1701–44). Formerly called Centigrade, the use of that name was abandoned by the General Conference on Weights and Measures in 1948. However, the term Centigrade remains in fairly common use. The relation between Kelvin temperature (K) and the Celsius temperature (C) is given by: $K = C + 273.15$. The International Practical Temperature Scale of 1968 is expressed in both kelvins and degrees Celsius. One Celsius degree = 1.8 Fahrenheit degrees; 0° C = 32° F and 100° C = 212° F.

Cement kiln A rotating steel cylinder up to 250 m in length and to 4 m in diameter, lined with alumina bricks. The cylinder is slightly inclined to the horizontal so that the contents may gradually descend as the cylinder rotates. The kiln is heated by pulverised coal or oil, temperatures of the order of 1300° C being attained. Chalk or limestone, and clay or shale, mixed with water in the required proportions, are fired in the kiln to form cement clinker. The clinker is ground, with the addition of about 5 per cent gypsum, to form cement. The slurry of lime and clay enters the upper end of the kiln, passing through three zones. In the first zone it is dried; in the second zone the chalk loses its *carbon dioxide* and becomes quicklime; in the third or lower zone the oxides of calcium, alumina and silicon combine to form cement clinker.

Cenospheres Small hollow spherical particles arising during the combustion of atomised liquid or pulverised solid fuels.

Centistokes An international unit for the measurement and expression of all kinematic viscosity values. The conversions are:

$$100 \text{ centistokes} = \begin{cases} 1 & \text{stokes} \\ 407 \text{ seconds} & \text{Redwood No. 1} \\ 460 \text{ seconds} & \text{Saybolt Universal} \\ 46 \text{ seconds} & \text{Saybolt Furol} \\ 13.3 & \text{degrees Engler} \end{cases}$$

Central Electricity Generating Board (CEGB) A central body established under the Electricity Act 1957 'to develop and maintain an efficient, co-ordinated and economical system of supply of electricity in bulk for all parts of England and Wales'. The Board owns and operates most of the power stations, both conventional and nuclear, and most of the transmission lines in England and Wales. Over half the Board's total output is from high-efficiency 500 and 660 MW generating units. Electricity is sold through a bulk supply tariff to twelve area electricity boards for distribution to consumers. In 1978, 70 per cent of the energy generated was from coal-fired power stations (using 75 million tonnes of coal); 17 per cent came from oil and gas-fired stations; and 13 per cent from nuclear stations. Following the completion of a series of magnox nuclear power stations, the Board introduced advanced gas-cooled reactors. Hinkley Point 'B' became the first of the Board's AGR nuclear power stations to raise power. Another station has been commissioned at Hunterston, Scotland, by the electricity board of Scotland. In 1978 the Board was operating 137 power stations with a workforce of 61 000. The thermal efficiency of electricity generation averaged 31.51 per cent. The installed capacity of the system was 56 000 MW. *See advanced gas-cooled reactor; Table C.1.*

Centrifugal fan   Or radial flow fan; a *fan* in which air flows into the 'eye' and is discharged from the periphery at right angles to the direction of entry. *See Figure C.2.*

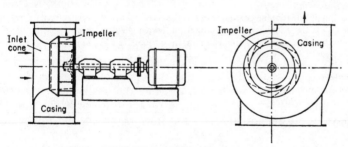

Figure C.2   A centrifugal fan.

Centrifugal pump   A pump using the principle of centrifugal force; the liquid enters the 'eye' or centre of a rotating impeller through which it flows radially under centrifugal force, leaving through an exit at the periphery; in this way the kinetic energy of the liquid is converted into pressure energy. This type of pump may be used as a boiler feed pump; it is most suitable for higher pressures and outputs. These pumps are usually electrically driven; some of the larger sizes are turbine-driven, the exhaust steam being utilised for heating the feed-water.

Centrifuge   A device for separating substances of different densities from the fluid carrying them, by centrifugal motion.

61

Table C.1 CENTRAL ELECTRICITY GENERATING BOARD FOR ENGLAND AND WALES DATA FOR 1977–78

Sales of electricity

Total sales (to area electricity boards and direct consumers)	TW h[a]	209.828
Sales to area electricity boards only	TW h	203.700
Average charge per kW h sold (total sales)	pence	1.6848

Finance

Total income	£m	3 546.4
Total expenditure on revenue account (excluding interest)	£m	3 230.6
Operating profit	£m	315.8
Interest	£m	297.1
Profit after interest	£m	18.7
Capital expenditure on property, plant and equipment	£m	500.4
Return on average net assets: net	per cent	8.0
Return on average net assets: gross	per cent	17.8

Generation

Number of power stations		137
Declared net capability of power stations (end of year)	MW s.o.	56 326
New generating plant brought into operation	MW s.o.	57
Electricity supplied from CEGB power stations	TW h	211.914
Maximum demand met	MW	42 803
Thermal efficiency of generation: coal, gas and oil-fired steam stations	per cent	31.51
Works costs per kW h supplied	pence	1.1933

Transmission

Length of line (double or single circuit) in service: total	km	7 740
Length of line (double or single circuit) in service: 400 kV	km	5 051
Length of line (double or single circuit) in service: 275 kV	km	2 089
Number of 400 kV and 275 kV substations		199

Employees

Total number at end of year		60 691
Sales per employee	mn.kW h	3.457

[a] One TW h equals one thousand kW h.

Cerenkov radiation Visible radiation emitted when charged particles move through a transparent medium with a velocity greater than the velocity of light in that medium; Cerenkov radiation is observable as a faint blue glow in the vicinity of the fuel elements of a *swimming pool reactors*.

Ceresin A hard, brittle wax obtained by purifying ozokerite; the commercial product is nearly always adulterated with paraffin waxes.

Cetane number A rating for diesel fuels comparable to the octane rating for gasolines. It is the percentage of cetane, $C_{16}H_{34}$, which must be mixed with alpha-methyl naphthalene to give the same ignition performance, under test conditions, as the fuel under examination. Thus, it is a measure of the time required for a liquid fuel to ignite after injection into a compression ignition engine. On an empirical scale 0–100, a short delay period is indicated by a cetane number from 40 to 70; a long delay by a cetane number below 40. Most diesel engines operate satisfactorily on fuels in the 40–50 cetane number range. *See octane number.*

CFR engines Cooperative Fuel Research engines used to measure the anti-knock quality of production and experimental fuels, including motor, aviation and automotive diesel fuels; engines developed by the Cooperative Fuel Research Council in America and used by all the oil companies for this purpose.

CGS system *See metric system.*

Chain grate An endless chain of short bars linked together to carry a fuel bed through a furnace.

Chain grate stoker A *mechanical stoker* consisting of an endless chain of cast-iron links driven through sprockets and moving from front to rear; the fuel is fed by gravity, being spread over the grate to a thickness regulated by the fire door. The fuel burns as the grate moves along. At the rear end, ash plates remove the ashes and clinker from the grate. The first metre or so of the grate is shrouded by a refractory ignition arch which extends down to the level of the *chain grate*, on either side. The arch is heated by the burning fuel beneath it and so acts as a source of radiation to stabilise ignition. The fire door or guillotine is adjustable so that the thickness of the fuel bed may be varied to suit the conditions, e.g. the larger the fuel the thicker should be the fuel bed. The stoker may be operated with natural, induced or forced draught.

Chain reaction A self-sustaining process; in a *nuclear reactor* the neutron-induced fission of uranium-235 results in the liberation of neutrons which cause the fission of further uranium-253 atoms.

Char The by-product solid fuel from low-temperature *carbonisation* processes; it is the residue after extracting from coal valuable tars, bitumen and gases for use as raw materials in the chemical industry.

Characterisation factor, *K* For crudes and petroleum fractions, the ratio of the cube root of the normal boiling point in degrees Rankine to the specific gravity at $15.6°$ C $(60°$ F). United States crude appear to lie in the range 11.5–12.5.

Charcoal The solid residue, black in colour, of the *destructive distillation* of wood; it is a brittle porous material retaining the original shape of the wood while its micro-structure preserves the vegetable cell structure. It burns without smoke, and is used for heating and cooking; in the United States, and elsewhere, it is used for grilling and barbecuing. It is also used in the manufacture of gunpowder, as an absorbent and decolorising agent, in hop drying and in work where a very pure fuel is required. Activated charcoal (produced by blowing steam through charcoal) absorbs organic compounds; it is used in sugar refining and solvent recovery.

Charles's law A gas law stating that the volume of a given mass of a perfect gas varies directly as the absolute temperature, provided that the pressure remains constant. Thus, at constant pressure:

$$\frac{V_1}{T_1} = \frac{V_2}{T_2}$$

where T_1 and T_2 represents the two different temperatures, and V_1 and V_2 the volumes corresponding to those temperatures. *See Boyle's law; gas laws; perfect gas.*

Chelating agents Substances suitable for removing the residual hardness from soft or softened waters; they take polyvalent metal ions, such as calcium and magnesium, into their molecular structure and prevent them from precipitating; the effect is the same as if the hardness had been removed, and there is consequently no precipitation or scale formation.

Chemical energy The energy liberated in a chemical reaction, as in the combustion of solid, liquid and gaseous fuels.

Chequer-brickwork Multiple openings in a baffle wall to permit combustion gases to pass yet promote turbulent mixing of the products of combustion. In a *regenerator* serving a steel or glass furnace, chequer-brickwork is utilised as a means of storing heat.

Cheval vapeur (ch or CV) The French term for metric *horsepower*.

China syndrome, the The concept that a core melt-down in a nuclear reactor could burn a hole through the Earth 'to China'; the basis of a film distributed by Columbia Pictures around the time of the *Harrisburg nuclear power plant incident. See melt-down.*

Chlorinated dioxins A family of toxic chemicals formed in minute quantities in normal combustion processes. Dioxins include compounds of a most toxic nature. In addition to combustion processes, dioxins also occur during the manufacture of pesticides and herbicides. The findings in relation to combustion arose from research by the Dow Chemical Corporation.

Chlorination The addition of chlorine in gaseous form to power plant circulating water to control slime formation in condensers, and mussel growth in culverts.

Chlorine (Cl) An *element* in coal, at. no. 17, at. wt. 35.453; it occurs in varying amounts, chiefly as sodium chloride. The average amount present in British coals is about 0.23–0.24 per cent, with an upper limit of 0.75 per cent. In all combustion equipment at least 90 per cent of the chlorine in the coal appears to be emitted as hydrochloric acid in the flue gases; in certain conditions this acid can give rise to serious corrosion problems and high-temperature bonded deposits on heating surfaces, particularly in *water-tube boilers*. A heavy residual fuel oil may contain about 0.004 per cent of chlorine.

Christmas tree The complete assembly of valves and connections at the head of an oil well.

Chromatrography A technique for separating and analysing mixtures of chemical substances; a flow of solvent or gas causes the components of a mixture to migrate differentially from a narrow starting zone through a special medium. Separations are usually made in columns of selected sorptive powders or liquids, or in strips of fibrous media such as paper. In gas–liquid partition chromatography (GLC) the column packing is a liquid solvent distributed on an inert solid support; in gas–solid chromatography (GSC) the column packing is a surface-active sorbent such as charcoal, silica gel or activated alumina. Gas chromatography is used principally as an analytical technique for the determination of volatile compounds. The composition of the emerging gas is monitored by a suitable detecting device capable of indicating the presence and amount of the individual components; the most popular detectors are the thermal conductivity detector (katharometer) and the flame-ionisation detector. *See spectrophotometry*.

Chrome brick A *refractory* containing 35–45 per cent of chromic oxide, 15–35 per cent of magnesium oxide, 10–25 per cent of alumina and 12–20 per cent of iron oxide.

Chrome-magnesite brick A *refractory* containing 40–60 per cent of magnesium oxide and 20–35 per cent of chromic oxide.

Circuit-breaker A switch for carrying normal electric currents, but which is able to 'break' safely the faulty currents caused by short-circuits; the contacts may work in oil or air.

City and Guilds Boiler Operators Certificate A certificate issued by the City and Guilds of London Institute to persons who complete an approved course, with written and oral examination, in the subject of plant operation in relation to energy conservation and clean air requirements. By 1979, over 12 000 operators had qualified.

Clarain Soft humic coal which presents a shiny black, well-laminated appearance. It consists largely of *vitrinite* and some *micrinite*, with traces of *fusinite*.

Classifier In relation to *pulverised fuel*, a device to permit the passage of material of desired fineness while returning oversize material for further grinding.

Claus kiln An oil refinery unit for the recovery of sulphur from hydrogen sulphide rich gases; hydrogen sulphide is burned in an insufficient supply of oxygen for complete combustion, the principle of *preferential combustion* giving the reaction:

$$2H_2S + O_2 \rightarrow 2H_2O + 2S$$

Clay treating The removal of oxygen, nitrogen and sulphur compounds from certain oils following refining; removal is effected by adsorption on attapulgus clay, fuller's earth and other suitable materials.

Clean Air Council (UK) A consultative council appointed by the Secretary of State for the Environment under the Clean Air Act 1956 (Section 23); the purpose of the Council is to keep under review progress made in abating pollution of the air in England and Wales. The Council has 30 members. In 1974 it published a summary of achievements in this area entitled *Clean Air Today*.

Clean air legislation National, state or city legislation introduced to control *air pollution* in the ambient atmosphere; most industrial countries have now introduced legislation.

The United Kingdom has had two Acts of Parliament relating solely to air pollution: (1) the Alkali, etc., Works Regulation Act 1906 (now embodied in the Health and Safety at Work Act 1974); and (2) the Clean Air Act 1956 and 1968. The Alkali Act was designed to control emissions from a wide range of scheduled industrial processes, including power stations, oil refineries and petrochemical processes, the chemical industry, and the iron and steel industry. The Clean Air Act deals with emissions of smoke, grit and dust, from a very wide range of non-scheduled industry, and smoke from domestic dwellings. Under this Act, smoke control areas have been established. By 1977, some 5000 smoke control orders had been confirmed covering more than 7 million premises. The Control of Pollution Act 1974 augmented the Clean Air Acts in providing new powers to deal with pollution from motor vehicles; the Secretary of State may make regulations as to the composition and content of fuel used in motor vehicles, and also to control the sulphur content of fuel oil. Through these provisions the emission of lead can be controlled.

In the United States of America the primary responsibility for air pollution control rests with state and local governments, with support from the federal government. The Clean Air Act 1963 (amended 1965) made available federal technical and financial assistance to air pollution control agencies throughout the nation, generally seeking to encourage co-operative action by the states and local authorities. Under the 1965 amendment, the federal government acquired the authority to control air pollution from new motor vehicles. The federal Air Quality Act of 1967 provided for the development and issue of air quality criteria and the establishment of air quality control regions. Under the Clean Air Amendment Act 1970 the Environmental Protection Agency (EPA) sets national ambient air quality standards for oxides of sulphur, oxides of nitrogen, carbon monoxide, photochemical oxidants, hydrocarbons, particulates, toxic metals, other hazardous substances, odours and noise. The aim has been to reduce motor vehicle pollution by 90 per cent, while requiring the adoption of the best available control technology for all new pollution sources. Under the National Environmental Policy Act 1970 an environmental impact study is required for all proposed projects. Under the Clean Air Act 1977 companies planning new plant or major extensions to old plant in polluted areas are required 'to take out as much pollution as is put in'; this means that a company may well need to finance the control of other sources in the area, whatever the ownership of those sources, or close down plant of its own to make way for new plant.

Other countries with effective air pollution control legislation include Canada, Australia, New Zealand, South Africa, France, West Germany, Italy, Scandinavia, the Benelux countries, the Soviet Union, East Germany, Yugoslavia and Japan. *See Alkali, etc., Works Regulation Act.*

Clean Air Society of Australia and New Zealand Established in 1965, a voluntary society to promote the control of air pollution in Australia and New Zealand. Clean air conferences are held, often international in character. The Society publishes a journal, *Clean Air.*

'Cleanglow' A *gas coke* of a highly reactive nature produced in continuous vertical retorts from specially selected coals; it is a smokeless fuel, very free-burning, and suitable for any domestic appliance. *See authorised fuels; smoke control area.*

Cleaning fires The removal of ash and clinker from a fuel bed; two methods are adopted with fixed grates: (1) the side method, in which the live coal is pushed to one side of the grate and then to the other, enabling the ash and clinker on the grate to be removed; (2) the front-and-back method, in which the fire is pushed back against the bridge while the nearest parts of the grate are cleaned. As the whole grate cannot be cleaned by this method, it is only justified when the demand for steam is high; the side method should be employed as soon as circumstances permit.

Climate The average weather conditions of a place or region throughout the season. It is governed by latitude, position in relation to continents or oceans, and local geographical conditions. Near the equator, climate is almost synonymous with *weather*, owing to the more stable conditions. It is described in terms of atmospheric pressure, temperature, solar radiation, wind speed and direction, cloudiness, humidity, rainfall, evaporation, incidence of fog and temperature inversions, lighting and thunderstorms. Concern has been expressed in recent years that a gradual increase in carbon dioxide in the atmosphere may warm the earth through the retention of heat energy (the greenhouse effect), while an increase of aerosols may cool the earth through the reflection of light energy, either effect producing possible profound climatic changes. Both the carbon dioxide and the aerosols are attributable, in the main, to the increasing combustion of fossil fuels. The arguments either way are at present inconclusive.

Clinker Hard material formed from the cooling of fused *ash*. Clinker is generally undesirable in a furnace. It obstructs the flow of primary air through the fuel bed and increases draught requirements; it leads to *blow-holes* forming in other parts of the fuel bed; and its removal necessitates the use of the slice. Clinkering depends on the fusion temperature of the ash and the temperatures to which ash is exposed. Steam or water jets may be used to help reduce the formation of clinker. *See clinker prevention.*

Clinker prevention Measures adopted to prevent *clinker* formation: these are to (a) avoid thick fires; (b) use tools sparingly; (c) avoid mixing ash with burning fuel; (d) keep the fire level by careful firing; (e) operate as long as possible without disturbing the fires and then clean out thoroughly; (f) in last resort to use water sprays under the grate.

Closed-loop control system Or feedback control system; a control system in which a controlled variable is measured and compared with a standard representing the desired performance. Any deviation from the standard is 'fed back' into the control system so that the deviation of the controlled variable from the standard can be corrected. *See feedback controller; open-loop control system.*

Cloud point The temperature at which a cloud or haze appears when a sample of oil is cooled under prescribed test conditions. Clouding may be due to separated waxes or to water coming out of solution in the oil. The main use of the cloud point is to give an indication of the lowest temperature at which an oil may be used without causing the blockage of filters.

CO_2-acceptor process A possible technique for the gasification of coal. Coal is gasified in steam in a fluid bed in the presence of lime; the carbon dioxide formed by the reaction between steam and carbon reacts exothermically with the lime, producing sufficient heat to support the endothermic gasification reactions; and the hydrogen-rich gas that is produced hydrogenates part of the coal substance to form methane. Lime carbonated in the process is calcined for recycling. The process has been developed by the Conoco Coal Development Company.

Coal   A general name for a firm, brittle, sedimentary, combustible rock derived from vegetable debris which has undergone a complex series of chemical and physical changes during the course of many millions of years. *See* **anthracite; bituminous coal; brown coal; lignite; peat; sub-bituminous coal; welsh dry steam coal; wood.**

Coal classification systems   *See ASTM Coal Classification; International Coal Classification (ECE); Mott's Classification; National Coal Board Classification; Parr Classification: Seyler's Classification.*

Coal cleaning   *See Baum coal washer; cyclone washer; dense medium process; dry cleaning process; froth flotation.*

Coal equivalent   A common measure customarily used in energy statistics where the consumptions of different fuels in any economy are being compared. The 'coal equivalent' of any fuel, other than coal, is the heat content of the fuel expressed as a quantity of coal with the same heat content. In British practice one tonne of oil is taken as equivalent to 1.7 tonnes of coal, and 280 therms of natural gas as equivalent to one tonne of coal, these having (on average at any rate) the same heat content. Nuclear power and hydroelectricity are equated to the amount of coal needed to produce electricity at the current efficiency of conventional power stations.

International conversion factors differ in that when oil is expressed in 'coal equivalent', one tonne of crude oil is taken as equal to 1.3 tonnes of coal and 1 tonne of refined products as equalling 1.5 tonnes. For crude and refined products together, the Organisation for Economic Co-operation and Development uses a factor of 1.43 tonnes of coal per tonne of oil. For estimating 'tonnes of coal equivalent', the United Nations assumes a net energy value for coal of 29.3 kJ/g or 7000 kcal/kg.

Coal gas   Gas obtained from the *destructive distillation* of suitable bituminous coals in closed retorts at high temperatures.

Coal hydrogenation   The addition of hydrogen to coal at elevated temperatures and pressures to increase the hydrogen/carbon ratio towards that prevailing in petroleum, thus yielding liquid and/or gaseous hydrocarbons, notably motor spirit. Clean small coal of selected rank at low ash content is ground and mixed with recycled heavy oils to give a paste which is fed with recirculated hydrogen at about 260 atmospheres pressure to coal hydrogenation units. The paste and hydrogen are heated to about 410° C before entering converters. Catalyst is added partly to the coal paste and partly directly to the converters. The total product of the converters is separated in a hot catchpot which collects a heavy asphaltic fraction, and a cold catchpot which collects distillate oil. After this 'liquid phase' hydrogenation process some of the products pass through a 'vapour phase' hydrogenation process. Petrol is washed to extract phenols, and refined. The yield of regular grade motor spirit from a suitable coal of about 84 per cent carbon and 2.5 per cent ash is 46.1 per cent, i.e. 1 tonne of petrol requires 2.17 tonnes of clean coal. If the coal required for hydrogen production is taken into account, a total of 4.50 tonnes of coal is required for 1 tonne of petrol.

'Coalite' A reactive coke of the Coalite and Chemical Co Ltd, produced under 'low-temperature' carbonisation conditions when coal is heated in retorts to a temperature of the order of 600° C. It ignites readily, burns with very little smoke emission and is well-suited for all domestic appliances, including old-fashioned stool-bottom grates. The volatile content is about 7 per cent. It is an authorised fuel in a *smoke control area. See authorised fuels; coke; 'Rexco'*.

Coal Research Establishment A research establishment maintained at Stoke Orchard, near Cheltenham, England, by the *National Coal Board*, the main areas of research being: improvements in industrial coal-burning equipment; fluidised bed combustion; coal gasification; coal liquefaction; the development of domestic appliances for burning ordinary coal smokelessly; improved methods of making metallurgical coke; and the use of mineral material produced during mining.

Coal segregation The tendency for lumps of coal to separate out from the fine coal between, say, the coal bunkers and mechanical stoker hoppers.

Coal tanker A road tanker for delivering coal, using pneumatic methods for discharging coal via a pipeline into storage facilities. The *National Coal Board* for example, operates a fleet of coal tankers. Pneumatic discharge can reach storage heights of 20 metres.

Coal tar A by-product of the manufacture of coke and coal gas; a viscous mixture consisting mainly of aromatic compounds. It has a calorific value of from 38 to 40 kJ/g. *See coal tar fuels*.

Coal tar fuels Fuels produced by the distillation of *coal tar*. Following distillation six standard types of coal tar fuel are produced; each grade is identified by the degree of preheat (° F) required for optimum atomisation. The six grades are: CTF 50, grade 'A' creosote; CTF 100, grade 'B' creosote; CTF 200, pitch–creosote mixture; CTF 250, heavy pitch-creosote mixture; CTF 300, heavy pitch–creosote mixture: CTF 400, pitch. The carbon/hydrogen ratio is high in all grades. The sulphur content is low, rarely above 1 per cent; ash content is very small. The principles governing the combustion of coal tar fuels are similar to those for fuel oils of comparable viscosity; oil-firing equipment for petroleum oils is generally suitable for coal tar fuels. *See Table C.2.*

Table C.2 SOME PROPERTIES OF THE MORE COMMON COAL TAR FUELS

Designation	CTF 50	CTF100	CTF 200	CTF 250
Approx. equiv. Redwood seconds at 100° F	60	100	1200	4000
Gross calorific value, kJ/g	39.6	39.0	38.3	38.0
Flash point, ° C	82	93	99	105
Sulphur, %	about 0.75	⟶		about 1.0
Ash, %	about 0.05	⟶		about 0.25
Specific gravity	1.0	⟶		1.12
Air reqts, kg/kg		around 12.5		

Note: For coal tar fuels, the viscosity is measured as the temperature required to give the fuel a viscosity of 100 Redwood 1 seconds, reported in degrees Farenheit; hence CTF 200, etc.

Coal Utilisation Research Laboratories (UK)   Formerly the British Coal Utilisation Research Association (BCURA), research laboratories located at Leatherhead, near London, England. A key centre of research into *fluidised-bed combustion*, the laboratories won US contracts in 1978 for research associated with a new technology for fume-free combustion of coal.

Coal washery reject material   The discarded material after recovery of good-quality coal from run-of-mine coal. Coarse coal reject material, or chitter, consists of rock fragments, carbonaceous shales and poor-quality coal; this waste may contain up to 30 per cent carbon. Most of the reject is in this category. The remainder is fine coal reject material, also known as slimes and tailings. It is usually discharged as a slurry of fine carbonaceous shales, clays and fine coal, containing perhaps up to 55 per cent carbon. Many methods for the treatment and disposal of coal reject material have been developed; a number of these are indicated in *Figure C.3.*

Cobalt-60   A radioactive isotope of the metallic *element* cobalt, at. no. 27, at. wt. 58.933. It is usually produced by *neutron* irradiation in a *nuclear reactor*; it has a *half-life* of 5 years. Cobalt-60 is used as a gamma-ray source in the treatment of cancer and in industrial radiography.

'CO' boiler   A boiler which may be used as an integral part of a *fluid catalytic cracking unit* in an oil refinery; its purpose is to provide steam for the cracker, using for this purpose hot carbon monoxide—rich waste gases from the regenerator—and such

71

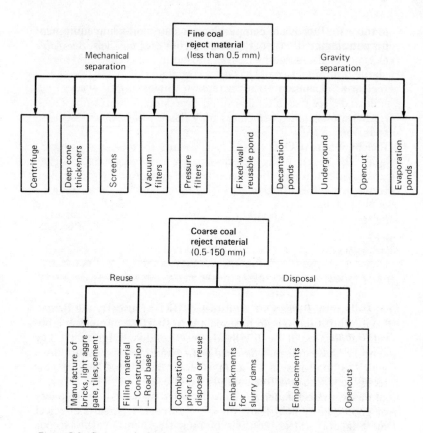

Figure C.3 Some methods of treatment and disposal for coal washery reject material

supplementary fuel as is required. The burning of the carbon coating on the catalyst with air during the regenerative stage produces a waste gas containing up to 8 per cent carbon monoxide at a temperature of about 500° C; the waste gas is discharged from the regenerator through cyclone separators to remove a high percentage of the entrained catalyst. It may then be utilised in a 'CO' boiler; supplementary fuel is always necessary, however, to give a boiler furnace temperature of about 1000° C.

Cochrane abrasion test A test for the abradability or surface hardness of coke; the test consists in rotating a prescribed quantity of coke in a drum at 18 rev/min for a set period. The product is then screened, the percentage remaining on the screen being an index of surface hardness.

COED process Or coal-oil-energy development; a project of the FMC Corporation for the production of syncrude and substitute

72

natural gas from coal. A 36 tonne per day pilot plant has been operating since 1970 at Princeton, New Jersey. Testing of a variety of coals, including lignite, sub-bituminous coals and high-volatile bituminous coals, has been completed.

Coefficient of expansion The increase in dimension of a material for each degree rise in temperature, expressed as a fraction of the original dimension; coefficients may relate to linear or to cubic expansion.

Coefficient of haze (COH) An American unit for measuring smoke stains, the result being determined by the reduction of light transmission through the filter after the collection of the smoke sample from the atmosphere. It is a measure of haze over a distance of 300 metres.

Coffinite A uranium-bearing mineral, being a hydrated uranium silicate.

COGAS process A process, involving *pyrolysis*, for the gasification of coal; gas is produced from the pyrolysis stage, and subsequently from the char gasification stage. Pilot plants have been constructed at Princeton, New Jersey, and at Leatherhead, England.

Coke The solid residue remaining after the *destructive distillation* of coal in ovens or retorts from which air has been excluded. The *volatile matter* given off during the heating process is purified if for use as *town gas*. Coke has a high *fixed carbon* content and a low volatile content (about 2–7 per cent). High-temperature coke is difficult to ignite, whereas low-temperature coke is relatively easy. The calorific value of coke is of the order of 23 kJ/g. *See carbonisation; shatter test.*

Coke breeze The residue from screening after all graded coke has been removed; it consists of pieces and particles below about a half-inch in size. A number of older power stations use coke breeze with coal on *chain grate stokers*; the mixture may contain up to 15 per cent of coke.

Coke by-product Important raw materials for the chemical industry obtained during the manufacture of coke; they include crude tar, crude benzole, ammonia liquor, toluene, xylene, naphthas and crude tar distillates. *Benzene*, derived from *benzole*, is used to make nylon, dyestuffs, drugs and photographic chemicals. Toluene is used to make explosives, saccharine, dyestuffs, food preservatives, perfumes and drugs. Xylenes are used to make paint, varnish, rubber and printing ink. Naphthas are used to make waterproof material, metal and floor polishes, varnishes and black protective coatings. Tar distillates include road tar, tar fuel oil, pitch, creosote, tar acids, anthracene, naphthalene and pyridine.

Coke oven A retort for the manufacture of metallurgical or hard coke, mainly for the iron and steel industry. A battery consists of

10–60 or more ovens; each oven is a slot-like chamber constructed of refractory brickwork, about 12 m in length, 4 m high and 40 cm in width. An oven of this size will carbonise 16–18 tonnes of coal in about as many hours. The oven ends are closed by self-sealing refractory-lined doors. Three or four charging holes, normally closed by heavy cast-iron covers, are situated in the oven roof. The refractory brickwork walls contain flues in which coke oven, blast furnace or other gas is burned to heat the oven. The temperature of the oven reaches about 1300° C. Each battery has one or two gas mains to remove the gas evolved in the ovens during carbonisation; the connection between the oven and the main is known as an ascension pipe. A charging car runs along the top of the battery discharging coal into the ovens as required; another machine incorporates a ram for pushing the coke out of the oven. One hundred units of coal yield about 70 units of coke, coke oven gas and other by-products.

Coker An oil refining unit used to convert vacuum residiuum to coke and distillates.

Coke residue Synonymous with *fixed carbon*.

Coke tree A solid mass of coke formed by strongly caking coals in under-feed stokers; it grows as it is pushed upwards by fresh fuel fed from below. 'Off' periods tend to aggravate the formation of coke trees, and if manual breaking is resorted to, smoke is created.

Coking (1) An alternative name for 'caking'. *See caking coal.* (2) The undesirable building-up of coke or carbon deposits in oil refinery equipment.

Coking coal *Bituminous coal*, suitable for the production of *coke*. *See caking coal.*

Coking or deadplate firing A method, originally devised by James Watt, of firing a boiler or furnace with soild fuel. The coal is piled on the deadplate or on the front end of the bars to a depth of about 25 cm. The heat of the furnace drives off the volatiles from the fresh coal. In passing over the incandescent firebed, they ignite and burn. When the heap of fresh coal has been partially coked, it is pushed forward over the grate. The method provides a most effective way of preventing smoke; it is emulated in the automatic coking stoker. It does not produce as much steam per hour as the sprinkling and side-firing methods but, as refuelling is done at less frequent intervals, it is useful where the fireman has other duties to perform. More skill is required than with other methods to keep the whole grate covered and prevent *blow-holes*. The main problem is at the back of the grate, which cannot be seen over the heap of fresh coal at the front. The rake must be frequently used. *See firing by hand.*

Coking stoker A *mechanical stoker* for solid fuel in which the fuel

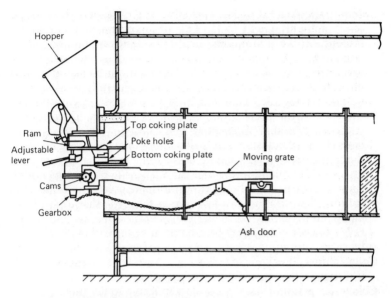

Figure C.4   Coking stoker (Source: Hodgkinson Bennis Ltd)

is fed from a hopper by a ram on to a top coking or distribution plate at the front of and above the grate; the coal is thus partially carbonised by the heat of the furnace before spilling over on to the bottom coking or deadplate. The latter holds sufficient fuel to ensure the rapid ignition of the coal as it spills over from the top plate, thus creating a condition of both overfeed and underfeed ignition. The combination of the movement of the fuel bed caused by the reciprocating fire bars and the new coal pushed in over the top plate by the ram causes the ignited fuel to fall off the bottom coking plate and travel along the grate. There are two types of coking stoker—the high ram, which is now considered obsolete, and the modern *low ram coking stoker. See Figure C.4*.

Cold end   That section of an *air preheater* where the cooled flue gas leaves and cold air enters the heater.

Cold gas efficiency   In respect of a **gas producer**, thermal efficiency calculated as follows:

$$\text{cold gas efficiency, \%} = \frac{\text{potential heat of gas}}{\text{total heat of fuel}} \times 100$$

where total heat means the sum of potential heat (calorific value) plus sensible heat due to preheating. Cold gas efficiencies range from 63 to 80 per cent. *See hot gas efficiency*.

Cold rolled steel   Steel that is passed cold through a rolling mill.

75

Coliform bacteria   Bacteria, including the bacterium *Escherichia coli*, commonly found in the human large intestine and whose presence in the environment usually indicates contamination by human wastes. Laboratory results are expressed as the number of organisms per 100 ml of sample. While the presence of *E. coli* indicates faecal pollution, the organism is generally considered to be non-pathogenic. To confirm that pathogenic organisms are associated with faecal pollution, samples should be tested for the pathogenic genera *Salmonella* and *Shigella*. If *Salmonella* is detected, further tests are needed in order to differentiate between the organisms of this group, e.g. *Salmonella typhosa* is the causative organism of typhoid fever, whereas *Salmonella typhimurium* is the causative organism only of gastrointestinal upsets. There are other coliform bacteria more or less similar to *E. coli* that exist in soil and plant materials (particularly *Aerobacter aerogenes* and *Escherichia freundii*); they may be present in water without indicating faecal pollution.

Colloidal fuel   Fuel oil containing 30–35 per cent of *pulverised fuel*; it is fired in the same manner as heavy fuel oil.

Colour–temperature scale   A scale applicable to hot metals at and above about 550° C, the temperature of which can be estimated from their colour. For iron and steel, the colour scale is broadly: dark red, 550° C; medium cherry-red, 700° C; orange, 900° C; yellow, 1000° C; white, 1300° C.

Combined heat and power generation   Systems in which both the heat and the power needs of a works are met, utilising exhaust heat from turbines and diesel engines in a manner which raises the overall thermal efficiency of the plant to as high as 80 per cent. Back pressure and pass-out power generation is used on a large scale in the oil, paper, chemical and sugar industries.

Combustion   A state of chemical activity in which the reactive elements of a fuel burn or unite with oxygen, accompanied by the evolution of heat and often light. Complete combustion means, in effect, oxidisation to the highest possible degree. However, before most of the heat available from coal can be released with the formation of the ultimate oxidation products, CO_2 and H_2O, the fuel has to be converted first into CO and H_2; this is true for all hydrocarbons. Before combustion can take place, an initiating 'ignition temperature' must be attained. The four stages of combustion of coal are illustrated in *Table C.3*.

Combustion chamber   Any enclosed, or partially enclosed, space in which the processes of *combustion* take place. The process of combustion, however, may not be completed within the confines of a single chamber, and two or more chambers may be employed. In a two-chamber arrangement the primary chamber is where

Table C.3 STAGES OF COMBUSTION OF COAL

Stage 1	Carbonisation (release of volatiles; production of H_2 and CH_4; formation of coke). Rate dependent upon: (a) heat input; (b) heat capacity; (c) thermal diffusivity; (d) particle size
Stage 2	Partial oxidation of coke and volatiles, together with pyrolysis (polymerisation) of volatiles, leading to formation of: (a) CO; (b) H_2: (c) hydrocarbons; (d) carbon
Stage 3	Combustion of: (a) H_2 and CO; (b) hydrocarbons (via CO and H_2); (c) carbon and coke (via CO) $\rightarrow H_2O + CO_2$
Stage 4	Gasification reactions: (a) $C + H_2O + CO_2$ (b) Hydrocarbons $+ H_2O + CO_2 \rightarrow H_2 + CO \rightarrow H_2O + CO_2$

(Adapted from 'Twelfth coal science lecture', D. T. A. Townend, *BCURA Gazette*, No. 48, 1963.)

ignition and burning of the fuel or waste takes place; in the secondary chamber the burning of vapours, gases and particulate matter from the primary chamber is completed.

Combustion, chemical reactions of The chemical combination of the fuel elements or constituents with oxygen; the most common basic equations for combustion are as shown in *Table C.4*.

Combustion meter A combined instrument for the measurement of steam flow and of air flow. A recorder enables the ***steam flow/air flow ratio*** to be clearly seen; adjustments are usually made so that the recording pens coincide when the desired combustion conditions are obtained.

Combustion wave The process of combustion, moving away from the point of ignition; also known as deflagration. The term 'flame'

Table C.4 CHEMICAL REACTIONS OF COMBUSTION

Fuel	Reaction (*heat of reaction is ignored*)
Carbon (to CO)	$2C + O_2 \rightarrow 2CO$
Carbon (to CO_2)	$C + O_2 \rightarrow CO_2$
Carbon monoxide	$2CO + O_2 \rightarrow 2CO_2$
Hydrogen	$2H_2 + O_2 \rightarrow 2H_2O$
Sulphur (to SO_2)	$S + O_2 \rightarrow SO_2$
Sulphur (to SO_3)	$2S + 3O_2 \rightarrow 2SO_3$
Hydrogen sulphide	$2H_2S + 3O_2 \rightarrow 2SO_2 + 2H_2O$
Acetylene	$2C_2H_2 + 5O_2 \rightarrow 4CO_2 + 2H_2O$
Ethane	$2C_2H_6 + 7O_2 \rightarrow 4CO_2 + 6H_2O$
Ethylene	$C_2H_4 + 3O_2 \rightarrow 2CO_2 + 2H_2O$
Methane	$CH_4 + 2O_2 \rightarrow CO_2 + 2H_2O$

77

may also be used, although restricted to combustion waves in which the reaction zone is luminous and propagation is taking place at relatively low velocities. 'Flames' which exceed the speed of sound are described as detonations or detonation waves.

Command economy Descriptive of an economic system dominated by central planning, as distinct from a free enterprise economy with a minimum of state interference. Command economy characteristics are not restricted to socialist economies; systems which encourage free enterprise maintain a powerful apparatus of central control and influence, although much of it may be of an indirect character through the use of monetary and fiscal policies.

Commercial economy An economy comprising industrial production, agriculture carried on for sale beyond the immediate neighbourhood, the trade through which these products are distributed, and the transport, communications, construction and services related to these processes. Such an economy is dynamic, growing and spreading, developing new institutions while adapting the old.

Commercial law A body of law establishing rights of property, providing for the making and enforcement of contracts, of conditions of incorporation of companies, and regulating relations between agents and principals, stockholders and corporations or companies, buyers and sellers, debtors and creditors.

Comminution Pulverisation; the reduction of material to a powder by attrition, impact, crushing, grinding or a chemical method.

Common banded coal Term used in the United States of America for common varieties of bituminous and sub-bituminous coals.

Common rail system A category of *fuel injection equipment* for diesel engines; the injection nozzles to each cylinder are supplied by a common fuel line, a fuel pump maintaining the pressure in the line. The fuel injection valves open and close mechanically. The *jerk system* is now more common. *See diesel engine.*

Commons, tragedy of the A tragedy of overgrazing and lack of care which resulted in erosion and falling productivity of the English commons, prior to the enclosure movement when grazing rights became restricted to the few. At one time each community had its commons, set aside for public use and used essentially for sheep and cattle grazing. Eventually the number of animals became more than the commons could support, and no one had any interest in ensuring the future productivity of this resource. Exploitation continued until productivity collapsed, and this social institution was superseded. Today, on a larger scale, the natural resources of air and water may be regarded as the 'commons' of the world; they are exploited by many who have little or no interest in the future productivity of these resources.

Hence, for example, intensive fishing and whaling threaten a dramatic loss of productivity—a repeat of the tragedy of the commons.

Commutator A device which reverses the connections from the load-circuit to the rotating coils at the instant when the electromotive force (emf) in the coils is changing its polarity; it is added to an alternator to obtain direct current (DC).

Compagnie Française des Pétroles (CFP) The French state-sponsored company involved in oil exploration.

Composting A biological process in which the organic material in refuse is converted to a usable stable material by the action of micro-organisms present in the refuse. A composting plant can range from a simple windrowing of raw refuse until it is broken down, and then screening of the material to remove rejects, to a sophisticated, fully mechanical operation. In a modern plant the refuse is fed into a digester where the refuse is broken up and commences its conversion to compost. After a period that may range up to 5 days the refuse is fed through pulverisers and screened. The reject material is normally carted away to tips, while the fine material is windrowed for several months, the windrows being turned periodically. Some plants pulverise the refuse initially and then windrow the screened material. It is quite common to add sewage sludge in the plant to serve as an additive to the compost and, of course, to dispose of the sludge. A compost plant will produce about 45 per cent compost from the refuse treated. Compost from refuse is a soil conditioner with some manurial value, but it is not a fertiliser in itself. It adds humus to the soil and helps break up the structure of the soil, improving the moisture-retaining properties.

Compounding A blending operation usually involving the addition of fatty oils to mineral lubricating oils.

Compound turbine A *steam turbine* in which steam is expanded in a number of separate cylinders. In a 'tandem compound turbine' the cylinders are in line and the rotors coupled; in a 'cross-compound turbine' the cylinders are in two lines driving two generators.

Compressed natural gas (CNG) An alternative fuel to petrol for the operation of vehicles. Natural gas from a user's piped supply is compressed and used to refuel a CNG-converted vehicle. To convert a vehicle in Australia in 1979 was costing about $A2000, and a compressor plant capable of serving a 50 vehicle fleet about $A80 000. Vehicles operate on a dual-fuel basis, with a conventional fuel reserve.

Compression-ignition engine *See diesel engine*.

Compression ratio The ratio of the volume of the fuel–air mixture at the start of compression to the volume of the fuel–air mixture at the end of compression, in the cylinders of an internal combustion engine. The usual compression ratio for cars is between 6 and 8. The ratio may be expressed:

$$\frac{V+v}{v}$$

where V is the swept volume and v the volume of space above the piston at the top of the stroke. *See engine capacity*.

Computer   A device capable of carrying out the following five main functions: (1) acceptance of information; (2) storage of information; (3) simple mathematical operations such as add, compare, etc.; (4) output of information in the form required; and (5) control of its own operation. Computers are grouped into two main types—those using analogue techniques and those using digital techniques. *See analogue computer; digital computer*.

Condensate   (1) Steam which has given up its latent heat and changed back into water; condensate forms a film on internal metal surfaces, running down such surfaces to a draw-off point at a lower level. (2) Low-vapour-pressure products that are contained in the vapour phase in natural gas in the reservoir and become liquid at standard field separation conditions; they comprise pentanes and heavier hydrocarbons.

Condenser   A chamber in which exhaust steam from, say, a *steam engine* or *steam turbine* is condensed or converted to water by the circulation or introduction of cooling water. *See jet condenser; regenerative condenser; surface condenser*.

Condensing turbine   A *steam turbine* in which all the steam used to drive it is finally expanded down to the condenser pressure, which may be a vacuum of between 725 mm and 740 mm of mercury.

Conditioning   The addition of water or steam to dusty coals or fines to improve the distribution of air through the fuel bed and reduce the quantity of dust carried forward towards the chimney.

Conductance   The reciprocal of resistance, being measured in mhos (or reciprocal ohms); a 5 ohm resistor has a conductance of 0.2 mhos.

Conduction   The transmission of heat from one part of a substance to a colder part, or to another substance or body in contact with it. *See heat transfer*.

Conduction band   *See energy bands*.

Confidence limit   A statistical expression, in which a confidence limit of x per cent means that there is an x per cent chance of the value of a variable falling between the limits defined.

Configuration control   A method of controlling a nuclear reactor by altering the geometrical arrangement of the components of the core.

Conglomerate merger A combination of firms without horizontal or vertical elements, i.e. a combination which does not bring together firms producing the same or similar products, or merge suppliers and customers in the production process. A conglomerate merger achieves diversification, combining independent risks. Risk reduction is of interest to management, while motives of prestige and power cannot be excluded; diversification may also appeal to shareholders, although shareholders can always achieve risk reduction through a diversification of shareholdings.

Coning The behaviour of a chimney plume when it expands or diffuses roughly along a cone; during such periods *efflux velocity* or momentum and thermal buoyancy tend to be dominant. *See looping*.

Conradson test A test to determine the *carbon residue* of a liquid fuel. In this test 10 g of oil is heated in a crucible by a gas burner at such a rate that oil vapours cease to be generated after 13 minutes; the temperature is then raised to full redness for a further 7 minutes, and then the crucible is allowed to cool. The residue is weighed and expressed as a percentage of the oil used. *See Ramsbottom test*.

Conservation The rational use of the *environment* to improve the quality of living for mankind. Rational use implies: (a) the use of some areas and resources for production to satisfy the requirements of humanity for goods and services; (b) the preservation or enhancement of other areas and resources for the contribution these can make to the scientific, educational, aesthetic, or recreational requirements of mankind. For the community as a whole, marginal costs should be balanced against marginal benefits over the whole scene of productive, cultural, and recreational activities, the interests of both present and future generations being taken into account. Such balancing often becomes a difficult and controversial matter: e.g. lake preservation versus electricity generation; the destruction of high-value forest land to make room for some marginal extension of grazing land; the sacrifice of park land for mining; the deployment of lowlying land for housing versus the risks of flood damage; population growth versus congestion; land reclamation versus the retention of mangroves. Conservation, therefore, is not incompatible with development, in the traditional sense, but raises difficult issues to be settled outside of the market-place.

Conservation of mass and energy A general principle, of universal applicability, based upon the mass–energy equation or *Einstein's equation*, which states that in any system the sum of the mass and the energy remains constant. This statement supersedes the statements previously held to be true in all circumstances that in any system energy cannot be created or destroyed (the law of the conservation of energy) and that in any system matter cannot be created or destroyed (the law of the conservation of mass); these two statements or laws can be applied only to systems not involving nuclear reactions or velocities approaching the velocity of light.

Constitutional formula A chemical formula which indicates the actual grouping of the atoms in the molecule. For example, normal-butane, with an *empirical formula* of C_2H_5 and a *molecular formula* of C_4H_{10} has a constitutional formula of $CH_3(CH_2)_2CH_3$. *See graphical formula*.

Constant-volume gas thermometer A device for measuring temperature. It comprises a bulb containing an inert gas and a measuring element connected by capillary tubing. The measuring device registers the change in pressure of a constant-volume gas, the pressure varying with changes in temperature. It is a useful instrument for measuring low temperatures and may be used up to 550° C. *See pyrometer; thermometer*.

Continental Shelf Act, 1964 An Act which allowed the Ministry of Power to issue exploration and production licences for the British sector of the North Sea Continental Shelf.

Continuous blow-down The draining of a constant stream of water from a boiler, often through a feed water heat exchanger. *See blow-down*.

Continuous vertical retort A closed chamber used for the manufacture of *town gas* by the *carbonisation* of coal. A retort of this type is usually about 7 m high and 3 m wide; the depth varies from 20 cm at the top to 30 cm at the bottom. It holds about 3½ tonnes of coal. The retort is heated by producer gas burnt at the sides. Coal is fed continuously into the retorts through gas-tight hoppers, coke being extracted continuously, followed by 'steaming' in an air-tight coke box. This steam quenches the coke and produces *water gas*, which increases the total thermal yield of gas per tonne of coal carbonised. While the coke is ejected continuously into the coke box, the coke box itself is emptied periodically. The steamed vertical retort produces a gas with the following typical composition, the constituents being expressed in percentages: hydrogen, 50; methane, 20; carbon monoxide, 15; nitrogen, 7; carbon dioxide, 5; hydrocarbons, 2; oxygen, 1. *See Figure C.5*.

Control rod A rod, consisting usually of steel containing a good neutron absorber such as *boron* or *cadmium*, which is used to control the reactivity of a *nuclear reactor*. Movement of the control rods enables the power level to be held constant or varied.

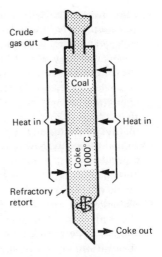

Figure C.5 Coal carbonisation: cross-section of continuous vertical retort

Convection The transmission of heat from one place to another through the medium of a heat-transforming fluid. *See **heat transfer**.*

Convector superheater *See **superheater**.*

Converter reactor A *nuclear reactor* using one kind of fissile fuel and producing a different kind of fissile material; for example the production of plutonium-239 in a reactor using uranium-235.

Convertion The process whereby neutrons are used to transmute thorium-232 into uranium-233, or uranium-238 into plutonium-239.

Coolant A heat-transport medium for removing heat from a *nuclear reactor* so that it may be used for steam raising and power production. Among the more frequently used reactor coolants are ordinary (light) water, heavy water, carbon dioxide and liquid sodium. British commercial nuclear reactors have employed carbon dioxide as the coolant; this gas is used under pressure.

Cool flame Flame produced by hydrocarbon mixtures when the temperature of the gases has risen by about 100–150° C only; combustion is incomplete, and the products of combustion include aldehydes, ethers and peroxides. Cool flames are blue in colour and slow-moving. Cool flames tend to be observed during a pre-ignition period, i.e. between the initiation of ignition and the onset of self-sustaining combustion. *See **flame**.*

Cooling pond A large tank of water in which irradiated fuel elements from a nuclear reactor are stored to allow the short-lived fission products to decay. Also an artificial lake used for the natural cooling of condenser-cooling water serving a conventional power station.

83

Cooling tower A device for cooling water by evaporation in the ambient air. A tower requires a flow of air and this may be induced by natural or mechanical means. The use of large cooling towers at power stations to dissipate recovered waste heat eliminates the problem of *thermal pollution* in masses of water; however, the make-up water for a cooling tower system is taken from the river or stream, thus eliminating its subsequent use downriver. After being used several times for cooling purposes, it is finally dissipated to the general atmosphere as steam.

Cooling tower precipitation The drizzle formerly experienced around natural draught cooling towers; this nuisance has been overcome by the general use of spray eliminators fitted to the inside of towers.

Coorongite A rubbery material first reported from Alf Flat, Coorong district of South Australia; it has a high hydrocarbon content. The material is created by the growth and subsequent desiccation of a phytoplankton, in this case *Botryococcus*. Coorongite is considered to be the peat stage of torbanite or *bog head coal*. On distillation, coorongite yields an oil similar in composition to shale oil.

Copper sweetening unit An oil refinery unit in which corrosive and unpleasant compounds in motor spirit are converted in order to give a non-corrosive product with a pleasant odour.

Core (1) The part of a *nuclear reactor* containing the fissile material. (2) Samples of geological formations penetrated; special bits are employed to cut cylindrical samples, which are examined to obtain geological information.

Corner-fired furnace A *water-tube boiler*, roughly square in plan, in which the pulverised fuel burners are arranged in vertical banks at each corner and directed at points to one side of the centre line of the furnace. This results in the formation of a large vortex with its axis on the vertical centre line; the effect is to promote a high degree of turbulence. The burners consist of an arrangement of slots one above the other, admitting through alternate slots primary air–fuel mixture and secondary air. It is usually possible to tilt the burners upwards or downwards, with a maximum inclination to the horizontal of 30°. Thus, the position of the main flame region in the furnace may be varied, affecting furnace outlet temperatures and permitting superheat control. Corner-fired furnaces are used with medium- and high-volatile bituminous coals. *See Figure C.6.*

Cornish boiler A horizontal shell boiler of simple design containing a single furnace tube. It has a brick flue setting similar to that of the Lancashire boiler. The hot combustion gases leaving the grate pass to the back end of the single furnace tube and down into a

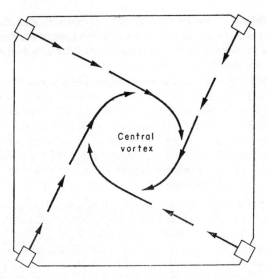

Figure C.6 Horizontal section of corner-fired water-tube boiler furnace, showing general flow pattern

bottom flue immediately below the boiler; the gases return along this flue to the boiler front and then divide into two streams passing down side flues, one being situated on each side of the boiler, to the rear. Here the gases reunite in a single main flue and pass to the chimney. *See **boiler; Lancashire boiler**.*

Corona A luminous discharge, present in the ***electrostatic precipitator***, due to air break-down when the electric stress at the surface of an electrode exceeds a certain value.

Correlation The relationship between one variable factor and another. Thus, a correlation coefficient is a measure of the degree of association between the corresponding values of two variables. Values between zero and 1 (which may be positive or negative) indicate the degree of correlation (direct or inverse). A value of zero indicates an absence of correlation, while a value of 1 indicates a perfect correlation.

Cost–benefit ratio A ratio calculated:

$$\frac{\text{gross benefits (present value)}}{\text{gross costs (present value)}}$$

The gross costs and benefits are discounted over the life of the project by a selected annual rate of interest. The difference between the two amounts is the present value of net benefits. The ratio of the two amounts is the gross cost–benefit ratio. Both considerations are relevant in choosing between projects.

Cost coefficient The cost of a unit of output, e.g. the cost of processing a kilogram of material or the cost of mining a ton of coal. A concept useful in comparisons between processes of a similar character.

Cost–benefit analysis A systemic comparison between the cost of carrying out a service or activity and the value of that service or activity, quantifying, as far as possible, all costs and benefits whether direct or indirect, financial or social. It is of particular value and relevance where public services are involved, involving both cost and benefit to the general public; thus, a public service may be justified in circumstances which a private organisation would regard as 'uneconomic'.

Costs of pollution The social costs and penalties imposed on a community or population that are not recorded in the accounts of polluters. Various monetary estimates of the losses have been made, all of them of a high order of magnitude. The British Committee on Air Pollution, under the chairmanship of Sir Hugh Beaver, reported in 1954 that the economic cost of air pollution in Britain, taking into account many direct and indirect costs but excluding the cost of medical services, was of the order of £250 million per year. Subsequently, in the Foreword to Gilpin's *Control of Air Pollution* (1963), Sir Hugh Beaver presented a revised total figure of £400 million per year. The United States Environmental Protection Agency has estimated that the average annual cost of air pollution is about $65 per person, or a total for the nation of about $13 000 million (13 billion dollars). Pollution control proposals implemented in the United States of America between 1970 and 1980 represented an increase from just under 1 per cent to a little over 2 per cent of actual and prospective Gross National Product (GNP).

Data on the estimated costs of fighting pollution in thirteen countries have been published by the Organisation for Economic Co-operation and Development (OECD). They indicate that in the framework of present environmental objectives overall costs of pollution control of all kinds may not exceed 3 per cent of GNP during the 1970s. Pollution control costs are not evenly distributed throughout the economy and some branches of industry are more affected than others.

Cottrell effect The change in the creep properties of a material caused by neutron bombardment; it is important in connection with the possible deformation of fuel elements in a nuclear reactor. *See* **creep**.

Coulomb A unit of quantity of electricity; it is the quantity of electricity transported in 1 second by a current of 1 *ampere*.

Council on Environmental Quality (US) A Council created by the US National Environmental Policy Act in 1970 for the purposes of formulating policy recommendations on environmental matters for the President. In respect of significant issues, the Council prepares a Presidential Review Memorandum which seeks to shape or influence Administration policy. An energy group set up by the Council seeks to ensure that Federal energy policy fully recognises national environmental goals. The group studies emerging technologies with a view to identifying those which hold out the most environmental, social and economic promise, and promoting their early commercial introduction. The Council has issued regulations in respect of the preparation of environmental impact statements and has studied the economic costs of environmental controls. It has concluded that environmental quality can be maintained without, in general, incurring substantial economic costs, although serious economic problems may be caused in specific instances.

Cracking furnace A furnace in which liquid or gaseous feedstock (e.g. naphtha, propane and butane) is broken down into a range of hydrocarbons under conditions of high temperature; in this example the cracked gases include ethylene, propylene and butadiene.

Creep The continuous deformation of metals under steady load. At elevated temperatures creep is exhibited by iron, copper, nickel and alloys; at room temperatures creep is exhibited by such metals as lead, tin and zinc. *See creep test.*

Creep test A method of measuring the resistance of metals to *creep*; a test piece of metal is subjected to a fixed static tensile load, the temperature being kept constant throughout the test. The elongation of the test specimen is measured accurately at regular intervals, the results being plotted on a length–time graph. Creep is divided into primary, secondary and tertiary periods, each period having its own characteristic. There are many variations of this test in terms of duration and procedure adopted.

Critical The condition of a *nuclear reactor* which is just capable of sustaining a chain reaction. *See critical mass.*

Critical Air Blast (CAB) test A laboratory test to determine the *reactivity* of solid fuels, particularly cokes; it determines the minimum rate of air supply, in m^3/min, sufficient to keep a standard bed of 14–35 BS mesh fuel alight. The lower the CAB value, the more reactive the fuel. Cokes having CAB values below 0.050 are readily lighted and maintained in approved open-fire grates.

Critical mass The minimum mass of fissile material needed for a chain reaction of neutrons to be self-sustaining. *See critical.*

Critical pressure Steam pressure below which latent heat is required to convert a liquid into vapour.

Critical pressure gauge A steam pressure gauge which gives a high reading accuracy; full-scale deflection of the pointer follows a relatively small change in pressure variation from the selected pressure.

Critical temperature Temperature above which it is impossible to liquefy a gas by pressure alone.

Cross-Channel cable A cable connecting the electricity systems of Britain and France. As the times of maximum load in each country do not coincide, electricity can be sent from one country to the other as may be economically desirable. For this purpose, *alternating current* is converted to *direct current* and re-converted to alternating current at the receiving end.

Cross-elasticity of demand *See demand.*

Cross-section, total The sum of the cross-sections of a nucleus which comprise: (a) capture cross-section—the apparent area of a nucleus for capture as 'seen' by the neutron; (b) scatter cross-section—indicating the probability of a neutron colliding with a nucleus and rebounding without altering the structure of the nucleus; (c) fission cross-section—which indicates the probability of a neutron causing fission. *See barn; neutron; nucleus.*

Cross-subsidisation The transfer of earnings from remunerative individual plants or products to unremunerative plants or products within the same enterprise, including transfers of earnings between peak and off-peak outputs. The Committee of Inquiry into the Electricity Supply Industry (the Herbert Committee), reporting to the British Government in 1956, criticised cross-subsidisation in electricity supply on the grounds that it may cause a misallocation of resources; the Committee favoured a detailed relating of prices to costs on individual services as the most fair and efficient approach.

Crude assay A procedure for determining the general distillation characteristics and other qualities of a crude oil.

Crude oil Petroleum as it occurs naturally in the earth, consisting essentially of a mixture of gaseous, liquid and solid hydrocarbons; sulphur, oxygen, nitrogen, derivatives of hydrocarbons and traces of inorganic elements are also present. *Natural gas*, a mixture of gaseous hydrocarbons, is normally present with the crude oil, accumulating in the upper parts of the oil-bearing strata. While crude oil occurs throughout the world, 90 per cent of production is mainly in North, Central and South America, the Middle East and the USSR. The composition of crude oil from all sources falls generally within the following ranges, the constituents being expressed as percentages: carbon 83–87; hydrogen, 11–15;

sulphur, 0.1–6.0; nitrogen, 0.1–1.5; oxygen, 0.3–1.2. *See petroleum*.

Crude oil, US Bureau of Mines Classification of A classification of petroleum crudes into seven categories, according to their 'base', from the distillation analysis of 800 samples of *crude oil*, from all over the world. The categories are:

Class A: paraffin base oil (wax-bearing)
Class B: paraffin intermediate base oil (wax-bearing)
Class C: intermediate paraffin base oil (wax-bearing)
Class D: intermediate base oil (wax-bearing)
Class E: intermediate naphthene base oil (wax-bearing)
Class F: naphthene intermediate base oil (wax-bearing)
Class G: naphthene base oil (wax-free)

Crude oil washing In oil tankers, the cleaning of cargo tanks with high-pressure jets of crude oil while the ship is discharging. The crude oil acts as a cleaning agent and removes oil residues; these are then pumped ashore with the cargo.

Crude unit An oil-refining unit comprising an atmospheric or low-pressure first-stage distillation unit, usually followed by a second-stage distillation unit operating under vacuum. The complete unit comprises *distillation towers, heat exchangers* and heaters.

Cryogenic distillation The separation of chemicals by distillation at temperatures down to $-120°$ C.

Cryptic damage Or hidden damge; in referring to the effects of air pollution it implies damage or impairment without visible signs, e.g. a reduced rate of growth in plants.

Culm and gob banks Inferior fuel and waste of no commercial value discharged from coal-processing plants on to land where it accumulates into hills or banks of culm (anthracite) and gob (bituminous). These banks visually disfigure the landscape and are often easily ignited, either from an external source or through spontaneous combustion.

Cupola A vertical shaft metal-melting furnace comprising a cylindrical mild steel shell lined with refractories; 'drop bottom' doors are fitted to the base. A tap-hole for the draining of molten metal is situated at the lowest point; above this on the opposite side is another tap-hole for the removal of slag formed during melting. Tuyères for the admission of an air blast are situated about 1 metre from the working bottom. Cupolas vary in height from about 3 metres up to about 25 metres. The cupola is charged with alternate layers of metal, coke and flux introduced through a charging hole high up in the shell. The output of a cupola is roughly proportional to the cross-sectional area of the melting zone; total outputs range up to 30 tonnes/h. Metal to coke ratios vary from 5:1 to 12:1. The air blast is normally cold, but a number of hot-blast cupolas have come into use.

Curie The unit of radioactivity. It is the quantity of any radioactive nuclide in which the number of disintegrations per second is 3.7×10^{10}. Sub-units are the millicurie (mc) $= 10^{-3}$ c and the microcurie (μc) $= 10^{-6}$ c.

Current The movement or 'flow' of electricity; usually measured in amperes.

Current cost accounting Or inflation accounting; a system of accounting in which assets are recorded in the accounts of a company at their 'value to the business', i.e. the loss the business would suffer if the asset were destroyed, in money terms. Accounts then reflect the effect of relative price changes on the business. In practice the system resembles replacement cost accounting, but with no adjustment for monetary items. For example, the company charges against revenue the amount it would have had to pay to buy stocks at the time they were sold, rather than the amount they actually cost when they were bought earlier. Plant and machinery is valued at replacement cost. In the balance sheet all non-monetary items are revalued annually at current costs. The balance sheet becomes then a statement about values instead of simply historic costs which have become increasingly irrelevant because of inflation. The Sandilands Committee on Inflation Accounting recommended in its 1976 report a similar approach.

Curtain wall A refractory construction or baffle which serves to deflect combustion gases in a downward direction. Also known as a 'drop arch'.

CUSEC A flow of water equivalent to 1 cubic foot of water per second. The following relationships apply:

$$
\begin{aligned}
1 \text{ cusec} &= 28.317 \text{ litres per second} \\
&= 60 \text{ cubic feet per minute} \\
&= 86\,400 \text{ cubic feet per day} \\
&= 538\,171 \text{ UK gallons per day}
\end{aligned}
$$

$$
\begin{aligned}
1 \text{ cubic metre per} \\
\text{second (cumec)} &= 35.315 \text{ cusecs}
\end{aligned}
$$

Cutback Bitumen which has been rendered liquid by the addition of a suitable diluent such as white spirit, kerosine or creosote; it is used as a means of incorporating bitumen with road-metal.

Cut-out A fuse such as the main fuse provided by an electricity board on a consumer's premises.

Cut point A specified value in respect of, say, density or size, at which a separation into two fractions takes place.

Cuts Fractions obtained in the distillation of oil.

Cycle efficiency The *adiabatic heat drop*, expressed as a percentage of the heat put into the steam. *See Rankine cycle.*

Cyclo-alkane series Two series of saturated hydrocarbons having carbon atoms linked in ring-like central structures; these are the cyclo-pentane series, having five carbon atoms in the ring, and the cyclo-hexanes, having a six-membered ring. The general formula for cyclo-alkanes (formerly called naphthenes) is C_nH_{2n}, and each series ascends by the addition of branches of carbon atoms to the outside of the ring. Members of both the cyclo-pentane and cyclo-hexane series, other than cyclo-pentane itself, can be made to release hydrogen atoms in certain refinery processes, transforming the cyclo-alkanes into aromatics. *See **aromatic or benzene series; alkane series; Figure C.7.***

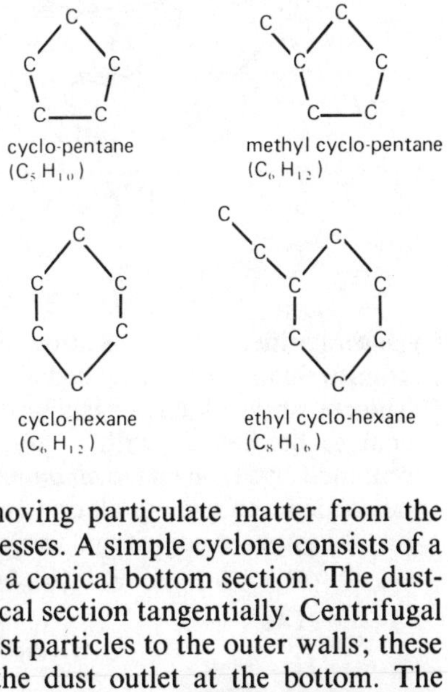

cyclo-pentane
(C_5H_{10})

methyl cyclo-pentane
(C_6H_{12})

Figure C.7 Examples from the cyclo-alkane series

cyclo-hexane
(C_6H_{12})

ethyl cyclo-hexane
(C_8H_{16})

Cyclone (1) A device for removing particulate matter from the waste gases of industrial processes. A simple cyclone consists of a cylindrical upper section and a conical bottom section. The dust-laden gases enter the cylindrical section tangentially. Centrifugal action throws the grit and dust particles to the outer walls; these particles fall by gravity to the dust outlet at the bottom. The relatively clean gases leave through a centrally situated tube within the upper section. With a pressure drop of 8 cm w.g. separation of particles down to to about 40 μm can be achieved. Cyclones may be used singly, or in groups or nests. *See Figure C.8.* (2) In meteorology, a low-pressure area with winds rotating counter-clockwise around the centre in the northern hemisphere, and clockwise in the southern hemisphere. Winds bring in moisture and rainy, windy weather prevails as rising air cools and vapour condenses. *See **anticyclone**.*

Cyclone furnace A furnace in which crushed coal is burnt in a water-cooled cyclone, the hot gases passing into a secondary furnace in which the grits and semi-molten ash are trapped before the gases continue into the boiler. *See Figure C.9.*

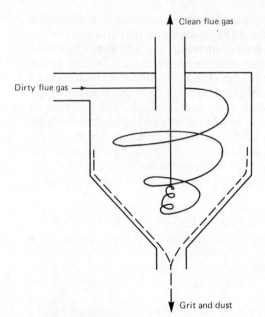

Figure C.8 Principle of a simple cyclone

Cyclone gasifier A gasification chamber in which the blast is admitted tangentially to produce a swirl.

Cyclone washer A coal-washing device which enables very small coal to be washed with a precision virtually as high as that obtained by the **dense medium process**. The cyclone washer does not depend on gravity alone to effect the separation of coal and dirt, but utilises centrifugal force; a cyclone of 50 cm diameter is capable of washing up to 50 tonnes of small coal per hour.

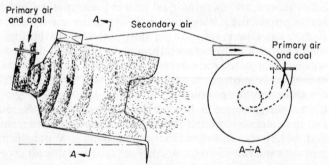

Figure C.9 Simplified diagram of the cyclone furnace

Cylinder displacement The volume of the compression space in the cylinder of an internal-combustion engine at the beginning of the compression stroke of the piston, less the compression space at the

92

end of compression; or the cross-sectional area of the cylinder multiplied by the length of the stroke.

Cylinder stock The *residuum* remaining in a *still* after the lighter parts of the crude have been vaporised.

Cylindrical screen A screen revolving at 3–4 rev/min on rollers; as coal passes through it, the smalls pass through the holes.

D

Damper A manual or automatic device to regulate the rate of flow of combustion gases through the flues of a fuel-burning appliance. *See barometric damper; butterfly damper; guillotine damper; sliding damper*.

Dark smoke Commonly defined as smoke appearing to be as dark as, or darker than, shade 2 on the *Ringelmann Chart*.

Data logger An automatic device which records events and physical conditions, usually with respect to time; it is employed where a large number of plant parameters such as pressure, temperature or flow, have to be measured, displayed and analysed. The results of the analysis are used to produce a more efficient operation of the plant or to detect abnormal conditions that may arise during the normal day-to-day running of the plant. The system usually employs a single measuring instrument which is shared between the many inputs; it may be of analogue or digital type. A data logger does not normally have storage facilities for processed information. *See computer*.

Daughter A *nuclide* formed by the disintegration of another nuclide called the *parent*.

Dayton process A process for the manufacture of gas from oil; it is a continuous and non-catalytic process. Vaporised distillate is partially combusted with air at 700–900° C; the air used is about 10 per cent of that required for stoichiometric combustion. The constituents of a typical gas produced, expressed in percentages, are: hydrocarbons, 25; carbon monoxide, 6; carbon dioxide, 6; hydrogen, 2; nitrogen, 61. The calorific value is about 20 MJ/m^3. *See Dayton–Faber process*.

Dayton–Faber process A process for the manufacture of gas from oil; it is a development of the *Dayton process*, using oxygen-enriched air in a modified form of generator. The calorific value is in the range 20–30 MJ/m^3.

Dead hearth A solid hearth; one through which no combustion air passes.

Dead plate A cast-iron plate bolted to the front of a boiler furnace tube and bevelled to carry fixed firebars; no combustion air passes through it.

'Dead plate firing' *See coking or deadplate firing.*

Dead storage Storage arrangements in which some fuel tends to be never recovered; this can occur with badly designed bunkers. Bunkers constructed with angled sides will prevent dead storage, and ensure continuous movement and passage of solid fuel. The minimum angles of hopper sides, measured from the horizontal, should be for steel and concrete surfaces, respectively: (a) coke, $45°$ and $50°$; (b) graded coal, $40°$ and $47\frac{1}{2}°$; (c) washed small coal, $55°$ and $60°$.

Dead-weight safety valve A *safety valve* the operation of which depends upon a series of weights superimposed upon it; the blow pressure can be varied by adding or removing individual weights.

Dead-weight tons The total carrying capacity of a ship when loaded to the appropriate freeboard. Dead-weight tonnage is very roughly about one and a half times the gross registered tonnage (which is in 'tons' of 100 cubic feet).

De-aerator Plant for removing the *dissolved gases*, particularly oxygen, from feed water.

Dean and Stark test A standard test for determining the water content of a fuel oil.

Decarburising The removal of carbon from iron carbide (Fe_3C), thus affecting the crystalline structure of the steel. The presence of carbon dioxide can decarburise steel. *See carburising.*

Decibel A (dBA) scale An international weighted scale of sound levels which attenuates the upper and lower frequency content and accentuates middle frequencies, thus providing a good correlation in many cases with subjective impression of loudness and sense of annoyance. Nearly all audible sounds lie between 0 and about 140 dBA; 0 dBA is the 'threshold of hearing', while sounds above about 140 dBA are not common. An increase of 10 dBA means that the noise perceived by a listener has roughly doubled in loudness. A car passing at 70 dBA sounds twice as loud as one passing at 60 dBA. The dBA measurement is widely used throughout the world for determining approximate human reaction to noise, and is the basis of legislation to control noise in many countries. Examples of typical dBA levels are as follows:

	dBA
rustle of leaves	10
quiet office	50
busy office	65
moderate traffic	70
alarm clock ringing	80
very noisy factory	90

Degree day A difference in temperature of a mean of 1 degree between ambient temperature and a given base temperature over 1 day of 24 hours. The base temperature generally adopted in Britain has been 15.6° C (60° F). The heating requirements of a building over a period are often taken to be proportional to the number of degree days, i.e. the sum of the degree day values for the days of the period.

Degrees API A scale indicating the gravity of oil, adopted by the American Petroleum Institute. The formula is:

$$\text{degrees API} = \frac{141.5}{\text{s.g. } 60° \text{ F}/60° \text{ F}} - 131.5$$

Dehydrogenation The removal of hydrogen from a chemical compound; for example, the removal of two hydrogen atoms from butane to make butylene, and the further removal of hydrogen to make butadiene.

Delayed-coking process An oil refinery process in which residual oil is heated and pumped to a reactor to coke; the coke is deposited as a solid mass of granular material and is subsequently broken up into lumps. *See fluid-coke process; petroleum coke.*

Deliquescence The absorption by a substance of moisture upon exposure to the atmosphere. Substances that are ordinarily deliquescent are concentrated sulphuric acid, glycerol, calcium chloride crystals, sodium hydroxide solid and ethyl alcohol 100 per cent. In an enclosed space these substances deplete the water vapour present to a definite degree.

Demand The amount of a commodity or service which will be bought at any given price per unit of time; effective demand as measured in the market-place by spending on consumption goods and services, and on investment.

The price elasticity of demand is the response of demand to a change in the price of a commodity. If the price is lowered, the amount demanded will normally be increased, although exceptions occur. Where a percentage increase in the amount demanded is greater than the percentage reduction in price, the demand is described as 'elastic'; if it is less, 'inelastic'. If a given percentage reduction of price leads to an equal percentage increase in the amount demanded, the elasticity of demand is unity. Many factors influence the elasticity of demand for a commodity. A commodity having no close substitutes, for examples, is likely to have an inelastic demand. It is probable that the demand for gasoline or petrol is inelastic; that is, it would require very substantial price increases to achieve a significant reduction in demand, although over time a shift to more economical vehicles might be expected.

Renault of France found in a survey that petrol costs would have to increase by 500 per cent before motorists would seriously reduce their driving.

Income elasticity of demand is the response of the demand for a commodity to changes in the real income of consumers. For most commodities increases in income lead to increases in demand, income elasticity being positive. In other cases the effect is negative, and sometimes neutral. In Britain it has been estimated that every 1 per cent increase in real personal incomes leads eventually to an increase of about $3\frac{1}{2}$ per cent in the electricity consumption of domestic consumers, i.e. the income elasticity of demand for electricity for domestic purposes is about $3\frac{1}{2}$. The cross-elasticity of demand is the responsiveness of the demand for a commodity or service to changes in the prices of other commodities. If commodities are complementary, they have negative cross-elasticities—i.e. if X and Y are complementary, a fall in the price of Y will lead to an increase in demand for both X and Y. The changes in the quantity of X and the price of Y will have opposite signs. Substitutes have positive cross-elasticities, i.e. a fall in the price of Y will increase the quantity of Y consumed, but will reduce the quantity of X consumed. Changes in the price of Y and the quantity of X will have the same sign.

The elasticity of substitution is a measure of the ease or difficulty of substituting between commodities and services by consumers, or between the factors of production by producers, in the short or long run. A change in the relative prices of fuels which appears of a long-term nature may lead consumers to substitute one fuel for another through a substitution of appliances or equipment. For example, if electricity becomes cheaper in real terms compared with domestic gas for heating purposes, then a progressive change in demand may occur, electric appliances being substituted for gas appliances; or if petrol prices significantly increase, while the prices of liquefied petroleum gas (LPG) remain stable, then a conversion of cars may occur to allow a substitution of fuels. Concurrently, a gradual switch from large cars to smaller cars may occur, together with an increasing use of more economical diesel engines in cars.

Demineralisation   An ion exchange process of water purification, in which almost all dissolved solids and gases (except oxygen and nitrogen) are removed from raw water.

Demonstrated resources   That part of a resource, the existence of which is well established with quantities, based on well data and geological projections, roughly estimated.

Dense medium process   A wet cleaning process for separating 'dirt' in the form of sandstone, shale, clay, pyrites, calcite, and so on, from *run-of-mine coal*. In this process an aqueous suspension of finely ground heavy solids, e.g. magnetite or barytes, is used to effect separation; *coal, middlings* and shale can be separated with a high degree of efficiency. Separation may be achieved in two stages; in one bath, with a specific gravity of 1.4, the clean coal is floated and removed, while in the second bath, with a specific gravity of 1.8, the middlings float while the shale sinks.

Department of Energy (UK) A department set up on 8 January 1974, to take over all responsibilities for energy previously exercised by the Department of Trade and Industry. The new department is concerned with the affairs of the National Coal Board, the Electricity Council, the Central Electricity Generating Board, the British Gas Corporation and the United Kingdom Atomic Energy Authority.

Department of Energy (US) A department of the United States federal government created in October 1977, as a result of legislation submitted to Congress by President Carter. The legislation called for an extensive reorganisation of those agencies in the executive branch of the government dealing with energy matters, and their consolidation into a new Department of Energy. The department absorbed and took over the functions of the Federal Energy Administration, the Energy Research and Development Administration (ERDA) and the Federal Power Commission. All these agencies were abolished as separate entities. The Federal Power Commission was replaced with a new Federal Energy Regulatory Commission under the Secretary of Energy. Other elements transferred to the new department included administrative bodies concerned with marketing power from federal dams and with control over offshore oil and gas operations, as well as units dealing with energy in the departments of commerce, housing and urban development, and Interstate Commerce Commission (now disbanded).

Depleted uranium *Uranium* in which the content of the fissile isotope uranium-235 is less than the 0.71 per cent normally found in natural uranium. *See uranium enrichment*.

Deposit gauge *See dust deposit gauge*.

Depreciation The diminution in the original value of an asset due to use and/or obsolescence. Depreciation can be calculated at 'historical cost' (the price which was actually paid for the asset at the time of purchase) or at 'replacement cost' (the amount that it would take to replace the asset at the present time). Historical cost is the method normally adopted, but economists tend to favour the second method. In *The Financial Obligations of the Nationalised Industries* (Cmnd 1337) it is stated (para. 19), 'depreciation should be on the historic cost basis' but 'provision should also be made

97

from revenue for such an amount as may be necessary to cover the excess in depreciation calculated on replacement cost basis over depreciation calculated on historic cost'.

In any calculation of depreciation there are three distinct elements:

(1) The length of time over which the asset is depreciated (e.g. houses are depreciated over 50 years, conventional power stations over 30 years, and motor vehicles over 5 years).

(2) The method of depreciating the asset. Several methods are employed. By the fixed instalment or straight line method, a fixed proportion of the original capital outlay is written off annually in order to reduce the asset to zero value at the end of the period; by the diminishing balance method a fixed rate per cent is written off the reducing balance of the asset account each year; by the depreciation fund method the asset is allowed to stand in the books at its original cost, and then a fixed amount, called the 'sinking fund instalment', is debited to profit and loss each year and a corresponding amount of cash is invested in securities, the amount being enough to accumulate at compound interest to the sum required to replace the asset.

(3) The difficulty of valuing the asset at present prices when the design and technical capabilities of the asset have changed considerably.

See depreciation fund method; diminishing balance method; fixed instalment method; sinking fund.

Depreciation fund method A method of calculating and allowing for the depreciation of an asset. The asset is allowed to stand in the company's books at its original cost and then a fixed amount, known as the 'sinking fund instalment', is debited to the profit and loss account each year and a corresponding amount of cash is invested in securities, the amount being enough to accumulate at compound interest to the sum required to replace the asset. *See depreciation; diminishing balance method; fixed instalment method; sinking fund.*

Derv Acronym of 'diesel engined road vehicle', a term still widely used in Britain for distillate fuel suitable for high-speed diesel engines.

Destructive distillation A distillation process in which an organic compound or mixture is heated to a temperature high enough to cause decomposition. *See charcoal.*

Desulphurisation The removal of sulphur or a sulphur compound from a fuel or flue-gas stream; the purpose is to reduce sulphur dioxide emissions to the atmosphere.

Desuperheater *See attemperator.*

Detention period   The average amount of time that each unit volume of liquid or gas is retained in a tank or chamber in a flow process. A minimum detention period is essential to complete a given stage of the process, e.g. the destruction of certain classes of bacteria or the settlement of certain size fractions of solid particles.

Detergent   A lubricating oil additive intended to prevent the formation of deposits in internal combustion engines; typical compounds are organometallic substances such as phenates and sulphonates of barium and calcium.

Detonation meter   An instrument for measuring knock intensity, detecting the rate of change of pressure in an engine cylinder. *See knocking.*

Detonation wave   *See combustion wave.*

Detoxification   The reduction of the *carbon monoxide* content of *town gas.*

Deuterium   The hydrogen isotope, mass number 2; commonly known as 'heavy hydrogen', it is obtained from heavy water. It occurs in natural hydrogen in the proportion $1:6500$. *See deuteron.*

Deuteron   The nucleus of the deuterium atom, consisting of a proton and a neutron. *See deuterium.*

Deutsch formula   A formula devised to calculate the performance efficiency of an *electrostatic precipitator:*

$$\text{efficiency} = 1 - \exp(-FC)$$

where F = specific collecting surface (projected collecting electrode surface area per unit volume of gas treated per second);

C = effective migration velocity (e.m.v.) of the particles.

Any self-consistent system of units may be used in the above equation, although e.m.v. is usually expressed in centimetres per second. The e.m.v. depends mainly on particle size and on the strength of the electric field between the electrodes.

Developed countries   A grouping of countries used by the United Nations for many purposes, including the compilation of world energy statistics; the grouping includes the United States of America, Canada, all the countries of western Europe, Yugoslavia, Israel, South Africa, Australia, New Zealand and Japan.

Developing countries   A grouping of poorer countries used by the United Nations for many purposes; the grouping includes all the countries of Central and South America, and most of the countries of Africa and Asia.

Dew point   The temperature at which moisture in air or flue gas condenses. If flue gases contain little or no sulphur trioxide, then the dew point of the gases is known as water dew point; this is usually in the region of 50° C. If, however, significant amounts of sulphur trioxide are present, the dew point may be raised as high as 150° C; this is known as the acid dew point. *See **acid soot**.*

Dewaxing The removal of waxes from lubricating oil stocks, now usually carried out by filtration at low temperature of a mixture of the oil and a solvent such as methyl ethyl ketone (MEK).

Diatomaceous-earth-base blocks An insulating material consisting of silica, uncalcined or calcined, with asbestos fibre and clay, moulded under heat and pressure. Used mainly for boiler walls behind firebrick in temperature zones, it is claimed, up to 950° C.

Dielectric heating The production of heat in non-conducting materials by their losses when subject to a high-frequency alternating electric field. The non-conducting material to be heated is placed between two electrodes across which the high-frequency voltage is applied. *See Figure D.1.*

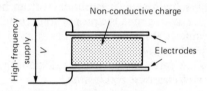

Figure D.1 Principle of dielectric heating.

Diesel engine An internal combustion engine in which air is compressed, attaining a high temperature, followed by the injection of oil which ignites; there is no carburettor or spark ignition system. Each piston in a four-stroke engine moves up and down twice to produce a power impulse:

Intake stroke—the piston descends, drawing in air only.
Compression stroke—the piston ascends, compressing the air to about one-fifteenth of its original volume, raising its temperature to about 450° C.
Power stroke—near the end of the compression stroke a very accurately measured quantity of oil fuel is injected which ignites immediately, the expansion of the gases forcing the piston down.
Exhaust stroke—the piston ascends and the burnt gases are discharged.

The air used is unthrottled when the intake valve opens, and a high compression ratio is employed to heat the air sufficiently to ignite the fuel (a higher ratio than with a petrol engine). The rate of fuel injection is varied to change the power output. With the high air/fuel ratio diesel exhaust gas contains much less carbon monoxide and hydrocarbon than the gasoline engine, but somewhat more *nitrogen oxides* and aldehydes. Engine overloading and poor maintenance can be the causes of smoke emission; basically there should be none. The thermal efficiency of a diesel engine varies between 25 and 33 per cent. *See fuel injection equipment; gasoline engine; motor vehicle exhaust gases.*

Diesel index   An estimation of ignition quality based on the aniline point and the specific gravity of a diesel fuel:

$$\text{diesel index} = \frac{\text{aniline point } °F \times \text{API gravity}}{100}$$

Diesel indices are generally about three numbers higher than the corresponding *cetane number* for fuels with cetane numbers of about 45–50, but this can vary considerably. The ignition quality of diesel fuels can be improved by using various additives which promote the oxidation mechanism of the fuels. Some of the additives used are: (a) alkyl nitrates; (b) aldehydes, ketones, ethers, esters and alcohols; (c) peroxides; (d) aromatic nitro compounds. *See degrees API.*

Diesel knock   *Knocking*, or pinking, due to an excessively rapid pressure rise during combustion in a diesel engine.

Diesel locomotive   A form of railway traction utilizing the *diesel engine*; it offers power, acceleration, efficiency and cleanliness. Diesel locomotives are of two types: (2) diesel electric locomotives, in which a diesel driven electric generator supplies power to an electric motor drive; (b) diesel hydraulic locomotives, in which the diesel engines supply power to the wheels through a torque converter in the transmission line.

Diesel oil   A general term covering oils used as fuel in diesel and other compression ignition engines. High-speed engines need high-speed diesel fuel or automotive gas oil. *See gas oil; Table D.1.*

Differential pressure meter   A meter equally suitable for measuring water from the delivery side of a feed pump and for measuring steam. It consists of a pressure generating element fitted in the main, a meter body in which the pressure difference is converted to a movement of an indicator by means of a float in a mercury-filled U-tube, together with an indicator, *recorder* and *integrator*. The differential pressure generating element consists of an orifice plate or venturi tube inserted in the steam or water main.

Diffusion   The spreading or scattering of a gas or liquid, or of heat and light. The spreading or intermixing movement of gaseous or liquid substances is due to: (a) molecular movement; (b) turbulence. In air pollution studies of the general atmosphere

Table D.1 DISTILLATE DIESEL FUEL SPECIFICATIONS

	BS 2829		ASTM D975	
	Class A1	*Class A2*	*1–D*	*2–D*
Viscosity at 37.8° C (100° F)				
cS, min.	1.6	1.6	1.4	2.0
max.	6.0	6.0	2.5	4.3
Cetane number, min.	50	45	40	40
Conradson carbon on 10% residue, % wt.				
max.	0.2	0.2	0.15	0.35
Distillation, temperature for 90% volume recovery				
min.	—	—	—	282° C
max.	357° C	357° C	288° C	338° C
Flash point, closed cup Pensky–Martens, min.	55° C	55° C	37.8° C	51.7° C
Water content % vol. max.	0.05	0.05	trace	0.05
Sediment, % wt. max.	0.01	0.01	trace	0.05
Ash, % wt. max.	0.01	0.01	0.01	0.01
Sulphur content, % wt. max.	0.5	1.0	0.5	0.5
Copper corrosion, max.	1	1	3	3
Cloud point, max.				
summer	0° C March–Nov inclusive	0° C March–Sep inclusive	—	
winter	−7° C Dec–Feb inclusive	−7° C Oct–Feb inclusive	—	—
Pour point, max.	—	—	5° C below ambient	

molecular diffusion is ignored, its effect being insignificant when compared with that of turbulence.

Diffusion flame Or jet flame; a long luminous flame arising when two streams of gas, fuel gas and air, are initially separated and the combustion reaction takes place at the interface between them. The rate of burning depends more on the rate of mixing than on the rate of chemical reaction. A common example of the diffusion flame is provided by the candle. Here the heat of the flame causes an evaporation of the wax which burns as it mixes with the surrounding air.

Diffusion flame burner A *gas burner* in which all the combustion air is provided by low-velocity diffusion in the combustion chamber.

Digest of United Kingdom Energy Statistics An annual publication by the Department of Energy, London, England, available through Her Majesty's Stationery Office.

102

Digital computer   A *computer* or fast automatic calculator in which data are represented by means of digits, i.e. the designation of one out of a finite number of alternatives by a digit or a group of digits. A digital computer consists of five basic units: (a) a central arithmetic unit where all calculations are carried out; (b) an input unit which enables all data to be read into the computer; (c) an output unit to read out the results of calculations from the computer; (d) a storage unit to store all data read into the computer, the results of calculations carried out within the computer and the instructions required by the computer to carry out the calculations; (e) a control unit which obeys the instructions written into the computer storage unit and directs the flow of information within the computer to perform the require computation. The instructions are called the program, the computer operating in a predetermined manner as specified by this program. The problems best solved by digital computers are those in which: (a) large sets of algebraic equations have to be solved; (b) large quantities of numerical data are involved; (c) many logical decisions are demanded; (d) very high orders of solution accuracy are required. Digital computers are being used in a wide variety of industries for direct control of plant and machinery, e.g. in chemical and petroleum processes, in steel mills, and in power stations both conventional and nuclear. Off-line applications have included studies of coal allocation between power stations, short-term load predictions, calculation of ground level concentrations of sulphur dioxide, transport problems, regression analyses, thermal stresses in piping systems, vibration of turbine blades, temperature distribution in a cylindrical body, stress analyses and solutions of equations, and many others.

Diminishing balance method   A method for calculating and allowing for the depreciation of an asset. In the diminishing, or declining, balance method a constant percentage of the remaining book value of an asset is written off each year. The amount to be written off each year may be calculated from the formula:

$$\frac{P - CD}{n}$$

where P is the initial asset cost; CD the cumulative depreciation charged in previous years; and n the plant life expressed in years. As the method is that of a geometric progression, it cannot provide for the complete depreciation of the plant. It is therefore necessary at a certain point to switch to the fixed instalment, or straight line, method in order to reach a point where $P - CD$ becomes zero. *See depreciation; depreciation fund method; fixed instalment method; sinking fund.*

Diminishing returns, law of A generalisation that while an increase in some inputs relative to other fixed inputs in a productive process may, in a given state of technology, cause total output to increase, there will come a point where the extra output resulting from the same addition of extra inputs is likely to diminish. This falling off in extra returns is a consequence of the fact that the additional doses of the varying resources must work with smaller proportions of the fixed resources. In the extreme, the process of additions will become counterproductive.

Dionic recorder A proprietary instrument for recording the electrical conductivity of water.

Diphenyl air heater An air heater in which the heat-conducting medium is diphenyl oxide or other suitable liquid. The liquid is heated as it passes through a *heat exchanger* situated in the flue gas stream, and cooled as it passes through another heat exchanger in the air stream.

Direct burner A *burner* in which the fuel and the oxidiser are mixed at the point of ignition. *See premix burner*.

Direct current (DC) Electrical current flowing in one direction only. *See alternating current*.

Direct-fired combustion equipment Plant in which the flame and/or the products of combustion come into direct contact with the material being processed, as in *brick kilns* and *open-hearth furnaces*. *See indirect-fired combustion equipment*.

Discharge lamp A form of electric lamp used extensively for street lighting, and in industrial premises. In the discharge lamp the current passes through a special mixture of gas and metallic vapour, forming a luminous electric discharge. The light may be greenish (mercury) or yellow (sodium), but in amount there is from two-and-a-half to five times as much available as from filament lamps of equivalent consumption.

Discounted cash flow method A method of comparing the profitability of alternative projects. This method may be subdivided into (a) yield method and (b) net present value method. Both of these techniques utilise as a measure of the 'rating' of an investment the present cash value of a sum to be received at some future date, discounted at compound interest. The present value or worth of a sum to be received at some future date is such an amount as will, with compound interest at a prescribed rate, equal the sum to be received in the future. The yield method is based upon the assumption that the best investment is that from which the proceeds would yield the highest rate of compound interest in equating the present value of the investment with future proceeds. In the net present value method an appropriate percentage rate is stipulated, the present value of the cash inflow is determined using this percentage, the original cost of the investment is subtracted therefrom, and the resulting surplus is the net present value of the investment. Both the yield and the net present value methods give reliable guidance to the profitability of alternative schemes, on the assumptions that may be used. However, errors of a serious nature may occur in the assumptions made regarding capital costs or trading returns.

Discounting The process by which future costs and revenues are converted into a present net worth. Future money is considered to be worth less than present money, a rate of interest being used for discounting.

Disintegration A nuclear transformation characterised by the emission of energy in the form of particles or photons. *See curie*.

Disintegration energy (Q) The energy evolved, or absorbed, in a nuclear disintegration. It is equal to the energy equivalent of the change of mass which occurs in the reaction. If Q is positive, the disintegration is *exothermic*; if Q is negative, it is *endothermic*. Radioactive disintegrations all have positive Q values.

Dispersoid A colloidal or finely divided substance.

Displacement pump A piston-type pump used for boiler feed purposes. The commonest type of displacement pump is the single-barrel simplex pump; the duplex pump has two water barrels, which ensure a more uniform flow of water. The steam consumption of these direct acting pumps varies between 2 and 5 per cent of the boiler output. They are reliable pumps, being mainly used where boiler pressures do not exceed about 15 bar.

Dissociation A phenomenon which occurs at high temperatures when carbon and hydrogen, after combining with their full complement of oxygen, are broken up into molecules of fuel and oxygen, and molecules of oxygen. For example, an atom of carbon combines with two atoms of oxygen to form CO_2; when the furnace temperature rises to over about 1550° C, the CO_2 may break up into carbon monoxide and oxygen, and if the temperature further increases to about 1650° C, the CO molecule may break up into carbon and oxygen. The process is endothermic. *See endothermic reaction*.

Dissolved gases Undesirable gases in feed water such as oxygen, carbon dioxide and ammonia; they may cause corrosion to feed pipes, feed-heater tubes and pumps. *See de-aerator*.

Dissolved solids Dissolved mineral salts in water supplies; in solution these salts divide almost entirely into their component parts, known as ions, which carry an electrostatic charge having one or more electrons too many or too few. The positively charged ions, called cations, include calcium, magnesium, sodium and potassium. Negatively charged ions, called anions, include bicarbonate, carbonate, chloride, hydroxide, nitrate and sulphate. Dissolved solids increase in concentration as water is evaporated into steam and tend to promote foaming in the boiler. Beyond a critical level the ions begin to recombine and the salts settle as a sludge in the bottom of the boiler, or form a hard scale on the tubes.

Distillate   A refined or semi-refined fraction of petroleum obtained in the condensation of a portion of a mixture which has been vaporised by heating. A middle distillate is that portion of petroleum boiling between 165° C and 370° C.

Distillation test: initial boiling point, final boiling point, total distillate, residue, loss   A test in which the temperature of a sample of liquid fuel is raised gradually until a temperature is reached at which vaporisation begins; at this point a certain volume of the fuel has distilled over, and both this and the point of final vaporisation are noted, i.e. the *initial* and *final boiling points*. As the volume distilled is measured by recondensation, a certain amount of *loss* is inevitable and there will also be some *residue*. The test serves as a guide to fuel volatility at atmospheric pressure. However, under working conditions, pressures completely alter the distillation and other characteristics of fuels. Nevertheless, a smooth distillation curve is considered important, as it indicates the absence of 'heavy ends' in the original fuel.

Distillation tower   Or fractionating tower; a tall cylindrical fractionating unit used in the first stage of oil-refining operations. A tower contains a number of horizontal perforated bubble-cap trays; these trays allow vapour to pass upwards and liquid to flow downwards. Crude oil is first run through coils of pipes lining the walls of a furnace and preheated to about 450° C; it then passes into the bottom of the distillation tower. At this point all but the heaviest fractions of the crude flash into vapour and pass up the tower. As the various components of the crude have different boiling points and the temperature falls steadily towards the top of the column, the rising vapours condense on the trays according to the temperature at which each becomes a liquid. The fractions of gasoline, kerosine, gas oil and diesel oil, lubricating oils and residual fuel oils are drawn off at different levels and sent to further refining processes which may include further distillation under vacuum. *See Figures D.2, D.3.*

Distillation zone   In the combustion of solid fuels, a zone in which fuel is exposed to heat, and the volatiles are distilled out of the solid material. *See flame zone; incandescent zone.*

District heating   A scheme in which both heat and hot water are provided from a central boiler plant to an entire housing estate or group of buildings; the consumer enjoys house heating at a

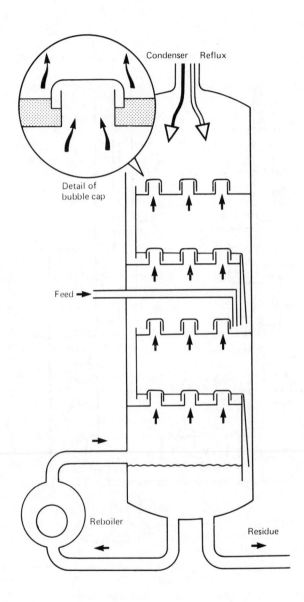

Figure D.2 Distillation (or fractionating) tower (Source: British Petroleum)

107

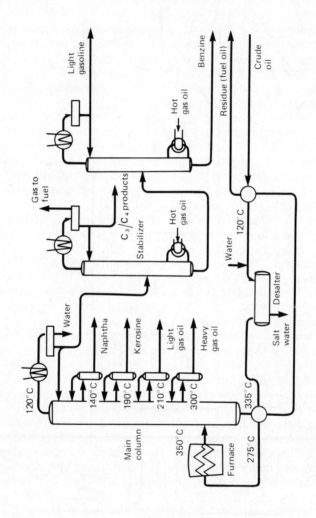

Figure D.3 Crude oil distillation (Source: British Petroleum)

comfortable level and a constant supply of hot water. There are two types of district heating scheme: (a) combined electric power and heating plant, in which steam is bled from a high-pressure turbine to a water heater; (b) central boiler plant specially designed and constructed for the district heating scheme. It is claimed that consumers enjoy lower running costs per useful therm of heat, although initial capital costs are higher. The scale of the scheme and the density of premises to be served are of great importance to the economics in a particular case. Also known as block heating (a scheme serving one or two block of flats or a shopping centre) and as group heating (a scheme serving a group of buildings or small housing estate).

Diurnal Daily, or recurring every day; the diurnal cycle of air pollution concentrations is of great interest to air pollution authorities.

Diversity factor The probability of a number of pieces of equipment being used simultaneously. For example, if 100 electrical appliances of 1 kW each rarely produce a maximum demand in excess of 20 kW, the diversity factor is said to be 1 in 5.

Doctor test A method of detecting undesirable sulphur compounds in petroleum distillates, i.e. of determining whether an oil is *sour* or *sweet*. It is a non-quantitative test.

Doctor treatment The treatment of gasoline with sodium plumbite solution and sulphur to improve its odour.

Dolomite A magnesium calcium carbonate mineral akin to limestone. It has been used because of its alkalinity for the removal of sulphur oxides from flue gases.

Dolomite brick A *refractory*, containing 38–42 per cent of magnesium oxide, 38–42 per cent of calcium oxide and 12–15 per cent of silica.

Donora smog incident An *air pollution* episode occurring in Donora, Pennsylvania, in October 1948, involving morbidity and mortality. An industrial town, Donora is situated in the valley of the River Monongahela, in hilly country, 48 kilometres south of Pittsburgh, USA. The valley is about $1\frac{1}{2}$ kilometres wide, the sides rising to some 120 metres above the river. Heavy industries in the valley have included many steel and blast furnaces, steel mills, smelting works, sulphuric acid plant, and other works. Soft coal is the primary fuel.

Visible smog until 10 a.m. has been the rule at Donora; in spring and autumn this commonly persists throughout the day, and even for several days. During the period 25–31 October 1948 very stable weather conditions produced effects closely similar to those in the Meuse Valley incident of 1930, with the fog accumulating day by day until it was terminated by rain on 31

October. About 42 per cent of the total population of 14 000 suffered some illness, 10 per cent being seriously ill.

Eighteen deaths resulted, all in persons over 50 years of age, with fourteen of these persons having had a previous history of respiratory illness. The final conclusion of the investigators was that no single substance was responsible for the episode, but that the toxic effects could have been produced by a combination or summation of the action of two or more contaminants. Sulphur dioxide and its oxidation products, together with particulate matter, were considered to be significant contaminants.

Donor solvent process Developed by Exxon, a process for coal liquefaction; a pilot plant came on stream in 1980, although a series of commercial plants could not be introduced before the mid-1990s. It is claimed that the process will yield oil at about twice the cost of imported oil at 1979 prices.

Dose The quantity of radiation delivered to a given area or volume, or to the whole body; it is usually measured in rems. *See Roentgen equivalent, man.*

Dosimeter An instrument for measuring the radiation *dose.*

Double pole A description of a switch which is inserted in the wires or 'poles' of a circuit. *See single pole.*

Dounreay fast reactor A fast nuclear reactor of experimental design located at Dounreay, Caithness, Scotland. The reactor can use **plutonium** as a fuel; the coollant is a mixture of liquid sodium and potassium. The reactor is a breeder—that is, it makes more nuclear fuel as it goes along.

Down-draught A region of severe turbulence formed on the leeward side when wind flows around and over a building. The region of down-draught begins at the top of the windward face of the building, rises to about twice the height of the building, and stretches for about six times the height of the building downwind. Chimney emissions discharged into a down-draught will be brought rapidly to the ground. Chimneys should discharge their gas high enough for them to escape the influence of down-draught. A Committee appointed by the British Electricity Commissioners recommended in 1932 that a power station chimney should be at least two-and-a-half times the height of the tallest adjacent building (usually the power station boilerhouse), plus an allowance for any difficult topographical features in the vicinity. Today some chimneys may need to be higher than this to ensure the adequate dispersal of sulphur dioxide. *See down wash; Memorandum on Chimney Heights.*

Down-fired furnace Or down-shot furnace; a *water-tube boiler* in which pulverised coal and primary air is fed vertically downwards from burners near to one side of the furnace roof. The gas outlet

110

is on the other side of the roof, and the gases follow a roughly U-shaped path. Secondary air may be introduced concentrically with the primary air jets, or through rows of ports on the side wall adjacent to the burners. This type of furnace is used predominantly with low-volatile coals and anthracite, these being more difficult to ignite than low-rank bituminous coals. Down-fired furnaces may be fired also from opposite ends of the roof, the outlet being at the centre; the result is a W-shaped flame. *See Figure D.4.*

Downstream products The products of subsequent stages of processing, starting with the initial process.

Down-wash The drawing down of chimney gases into a system of vortices or eddies which form in the lee of a chimney when a wind is blowing. Down-wash affects the visual appearance of the plume

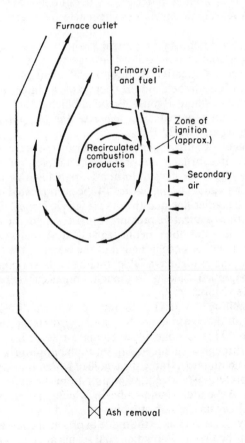

Fig. D.4 General shape of chamber and flow pattern in down-shot water-tube boiler

111

and causes blackening of the stack. In extreme circumstances it may also assist in bringing flue gases prematurely to ground level. The risks of down-wash may be minimised by: (a) using a round chimney with as small a diameter as possible; (b) avoiding the use of large overhangs or elaborate ornamentation; (c) discharging the gases from the mouth of the chimney at a sufficiently high velocity. American wind tunnel experiments have indicated that down-wash can be prevented by discharging the gases with a velocity of one-and-a-half times that of the wind passing the top of the chimney. *See efflux velocity.*

Dragon High Temerature Reactor A joint venture between the United Kingdom Atomic Energy Authority and *Euratom*, a *nuclear reactor* using enriched uranium in ceramic form and graphite as a canning material; temperatures of over 1000° C in the heat cycle are designed for. The reactor is located at Winfrith Heath.

Draught The difference in pressure which causes air and the products of combustion to flow through the flues of boilers and furnaces; to be effective it must be sufficient to overcome the resistance of the firebed and the friction of the internal surfaces of the system as a whole. It must also be sufficient to achieve the rates of combustion required. *See balanced draught; forced draught; induced draught; natural draught.*

Draught gauge An instrument for measuring the pressure, negative or positive, in a furnace tube or flue. The simplest form of draught gauge is a glass tube bent to form a U, part filled with water and fitted with a scale; one leg of the U-tube is connected to a suitable point in the boiler flue and the other leg is left open. If the flue is under suction or negative pressure, the pressure of the atmosphere will push the water down in one leg and cause the water to rise in the other; the difference in height of the water in the two legs is a measure of the draught available. Inclined tubes or pointer gauges having magnified scales give a much more accurate reading than the simple U-tube.

Draught stabiliser A device to reduce natural draught to that required at the bottom of the flue regardless of the ambient conditions. This is done by allowing air to infiltrate into the chimney through a balanced flap set to the appropriate pressure. The flap is built into a frame the whole of which is inserted into an opening preferably above the flue pipe from the boiler.

Drift mine An underground mine in which the coal is reached horizontally from the side of a hill.

'Drift' theory A theory as to the mode of origin of coal seams which states that the material of which coal seams are composed drifted there from the areas in which it grew, the site of a seam representing a lake or estuary in which it was deposited. *See 'in situ' theory*.

Driving cycle   The basis for the testing of motor vehicle emissions; vehicles under test are required to conform with standard driving cycles involving set patterns of steady speeds, acceleration and decelerations typical of urban vehicle use. Two test cycles in common use have been developed, one in the United States and the other by the Economic Commission for Europe.

Dry   A basis for reporting an analysis of coal; the values for *volatile matter, fixed carbon* and *ash* on this basis are determined by multiplying the values obtained on an *air-dried coal* by:

$$\frac{100}{(100-M)}$$

where M is the determined percentage of *inherent moisture*. *See proximate analysis; ultimate analysis*.

Dry, ash-free (d.a.f.)   A basis for reporting an analysis of coal; the values for *volatile matter* and *fixed carbon* on this basis are obtained by multiplying the values obtained on the *air-dried coal* by:

$$\frac{100}{[100-(M+A)]}$$

where M and A are the determined percentages of *inherent moisture* and *ash*, respectively. *See proximate analysis; ultimate analysis*.

Dry, mineral-matter-free (d.m.m.f.)   A basis for reporting an analysis of coal. The value for *volatile matter* on this basis is obtained by duly correcting the determined value and then multiplying the result by:

$$\frac{100}{[100-(M+M.M.)]}$$

where M is the determined percentage of *inherent moisture* and $M.M.$ the mineral matter in the coal. The value for $M.M.$ is obtained by modifying the percentage of ash by the *King–Maries– Crossley formula*. *See proximate analysis; ultimate analysis*.

Dry-back economic boiler   *See economic boiler*.

Dry-bottom furnace   A furnace in which the ash deposited is not molten. Only 10–30 per cent of the ash leaves through the ash hopper. *See wet-bottom furnace*.

Dry-cleaning process   A process for cleaning small *run-of-mine coal* without the use of water. Coal, usually less than 5 cm in size, is passed over a shaking table with a perforated deck through which a current of air flows upwards; the current of air has a velocity sufficiently high to render the coal 'fluid', the shale sinking on to the deck while the clean coal rises. The clean coal is then separated from the shale. Coal often comes to the surface too wet to be suitable for treatment by dry methods.

Dry gas meter A gas meter consisting of a rectangular box of tinned steel equipped with two compartments, each containing a bellows, and a system of valves. One valve admits gas to one of the bellows, which expands and expels a corresponding amount of gas from the outer compartment. At the end of each movement the bellows operate a sliding valve so as to reverse the inlet and exit ports; after a bellows has been filled, the reversal of the ports allows gas to enter the compartment outside the bellows, thus compressing the bellows, which discharges gas to the outlet. The valves are connected to dials indicating the volume of gas passed through. *See wet gas meter*.

Drying The removal of moisture from a solid by thermal methods in the presence of air.

Drying plant Plant designed for the drying of materials. Types include: (a) convection driers (drying chambers, rotary, tray, tunnel); (b) contact driers (stationary flat surface, film or drum); (c) spray driers; (d) pneumatic driers; (e) air-swept mills; (f) vacuum driers; (g) radiant and infra-red driers.

Dryness fraction The weight of actual steam compared with the total weight of entrained moisture and steam combined. Thus:

$$\text{dryness fraction} = \frac{\text{weight of dry steam}}{\text{total weight of steam and water}}$$

When the dryness fraction is 1.0, then the steam is called *dry saturated steam*. In practice, dryness fractions may vary from about 0.99 (1 per cent moisture) to about 0.90 (10 per cent moisture); much variation is to be expected.

Dry quenching The cooling of hot gas coke by circulating an inert gas through it, the heated gas then being passed through a *waste heat boiler*. *See carbonisation*.

Dry saturated steam Steam containing neither free moisture nor superheat; this is rarely achieved in practice. *See dryness fraction; wet saturated steam*.

Dual firing Plant with provision for the handling, storing and firing of two kinds of fuel. To equip a power station to fire, say, oil and coal increases the capital costs over a station designed to burn oil or coal exclusively. However, the higher cost may be offset by the ability to take advantage of marginal changes in the relative prices of oil and coal, and there is a larger element of security in an electricity supply system capable of flexibility in the event of an unforeseeable reduction in the supplies of one particular fuel. Kingsnorth power station (2000 MW), on the lower Medway, England, is equipped for dual firing by oil and coal.

Duff *Smalls*, usually with an upper limit of 2 cm.

Dumping plate Metal plate which controls the flow of ash over the end of a *chain grate* so that it is kept covered. Also called an ash plate.

Durain Hard and almost lustreless dull coal which presents no pronounced lamination. The two principal components are *micrinite* and *fusinite*.

Dust Particulate matter of natural or industrial origin which passes a 200 mesh BS test sieve (76 μm). Dusts which are about 5 μm in size or less are respirable and capable of reaching the alveoli of the lungs; such dusts, without necessarily being a nuisance, may constitute a health hazard. *See grit.*

Dust arrester A device for removing grit and dust from flue gases and other gas streams. In some situations two types of dust arrester may be combined in series. One example is the use of cyclones before bag filters in cleaning the exhaust gases from a copper and alloys foundry. In another example multicellular collectors may be used before electrostatic precipitators. This latter approach has been adopted by the *Central Electricity Generating Board* in respect of a number of modern power stations, although the trend is now towards using electrostatic precipitators alone. *See bag filter; cyclone; dust arrester efficiency test; dust burden; dust burden, measurement of; electrostatic precipitator; fan collector; gravel bed filter; high-efficiency cyclone; multicellular collector; settlement chamber; venturi scrubber; wet washer.*

Dust arrester efficiency test A test to determine the collecting efficiency of an arrester designed to remove grit and dust from flue gas streams. An efficiency test on an arrester plant consists essentially of measuring the amount of dust in the flue gases entering the plant and the amount in the flue gases leaving the plant over a given time. The efficiency is then:

$$\frac{\text{wt. of dust entering} - \text{wt. of dust leaving}}{\text{wt. of dust entering}} \times 100\%$$

When large plants are tested, the cross-section of the entry duct is divided into 24 imaginary rectangles. A probe is inserted through holes provided in the walls of the duct to the centre of each rectangle in turn and left there for a predetermined length of time. During this period, dust is collected by the probe and deposited in a filter; at the same time, the gas velocity is also measured. After every rectangle in the cross-section has been sampled, the total dust collected is weighed and the total mass flow of gases determined from the velocity measurements. A combination of these data and the known characteristics of the sampling probe enable the total amount of dust passing the cross-section during the test to be determined. An exactly similar procedure is followed at the exit duct from the arrestor plant; both tests take place simultaneously. The test procedure is generally in accordance with British Standard 893:1940 *Method of Testing Dust Extraction Plant and the Emission of Solids from Chimneys of Electric Power Stations.* A typical test on a modern power station plant requires 10 minutes' sampling time at each of the 24 positions or a total of 4 hours sampling. Making allowance for withdrawing and re-inserting the probe extends this time to 5 or 6 hours. Thus, one team can make only one test a day, and a further one or two days is required to work out and check the result. The practical difficulties of testing arrester plant also increase with larger sizes of unit. With 500 MW units it is normal design practice to divide the gas stream from the boiler into three or four parallel flows; each parallel flow must be tested independently and simultaneously. Allowing for supervision, about 20 men would be required to test the arrester plant of one 500 MW boiler. A 2000 MW power station would, of course, contain four such units and each may need to be tested under several different operating conditions (e.g. full load, part load, low CO_2, etc.). Simplified methods for the measurement of grit and dust emission are described in BS 3405:1961.

Dust burden The weight of dust suspended in a unit of medium, e.g. flue gas. It is customary to express this in g/m^3 measured at *normal temperature and pressure (NTP)*.

Dust burden, measurement of The determination of the *dust burden* in flue gases by appropriate sampling techniques. It is not possible to collect the whole of the material emitted from a stack, and it is necessary to obtain a representative sample of it. This implies three stages: (a) the selection of a suitable position from which one or more samples distributed over the cross-sectional area of the flue can be taken; (b) the definition of the technique by which each individual sample is obtained from the flue; (c) the study whereby the results so obtained are converted into reliable estimates of the total emission. Methods of sampling and test procedures are described in BS 893:1940 and BS 3405:1961. In respect of dust sampling equipment developed by The British Coal Utilisation Research Association, the sampling nozzle faces 'upstream' so that the dusty gases flow into it under the suction of the fan. Rate of flow of gas is metered by a flow-measuring device and controlled by a valve. To ensure a correct sample, it is essential to maintain the same velocity into the nozzle as in the gas stream. This is known as isokinetic sampling. It is achieved by first measuring the gas velocity after insertion of the nozzle by measuring the gas flow through the probe. The dust extractor consists of a small cyclone that collects all the grit and dust, and a small filter following it that collects all the fine material. The pressure drop across the cyclone is used as a measure of gas flow. When sampling is completed, the dust collected in the extractor is weighed and calculations made, either of total amount of dust emitted or of the concentration of dust in the gas, which can be calculated from the weight of the extracted dust and the volume of gas sampled. *See **Pitot tube***.

Dust deposit gauge A widely used instrument for measuring rates of deposition of grit and dust from local sources. It is of considerable value in seeking confirmatory evidence of a nuisance. A British Standard gauge is described in BS 1747 : Part 1 : 1961. It comprises a metal stand, collecting bowl of 12 inch diameter and a 10 litre collecting bottle. The gauge collects only the heavier particles and is relatively inefficient for small particles. Another type of gauge, known as a 'directional deposit gauge', has been developed by the Central Electricity Research Laboratories in the United Kingdom; it is more efficient as a dust collector, and enables a particular source of dust to be more readily identified.

Dust monitor An instrument designed to give an indication of the total dust burden in a flue gas, or the size characteristics of a dust. In the AEI Flue Dust Monitor an isokinetically collected sample of gas is drawn through the apparatus; the dust particles are electrically charged, each in proportion to its surface area. These charges are collected and in aggregate form a current proportional to the total surface area of the dust sampled in unit time. The current, after amplification, may be used either to indicate or record, and can operate a warning above a preset level. The instrument is sensitive to fine dust down to less than 0.1 μm.

Dutch oven A furnace suitable for burning solid waste chips and sawdust. It consists of a large rectangular chamber, either lined with a refractory material or made of fire-brick, which contains a horizontal grate and a fairly high bridge. The fuel is introduced through one or more holes in the roof, and forms cone-shaped piles on the grate. There are arrangements for the admission of primary, secondary and tertiary air.

Dynamic viscosity A measure of resistance to flow; it is the force required to move a plane surface of area 1 cm^2 over another parallel plane surface 1 cm away at a rate of 1 cm/s, when both surfaces are immersed in a liquid. It is the product of the specific gravity and the **kinematic viscosity** in stokes. The unit of dynamic

117

viscosity is the *poise*; a smaller unit, the centipoise, is often used, where 1 poise = 100 centipoise. *See viscosity.*

Dyne The unit of force in the *metric system* of units; a force of 1 dyne, acting on a mass of 1 gram, imparts to it an acceleration of 1 cm/s^2.

E

Ebullition chamber A chamber in an *evaporator*, in which the boiling of water to produce distilled water takes place.

Ecart probable Or probable error; one half of the density interval between 25 and 75 per cent recovery as shown in the *partition curve*.

Economic boiler A shell boiler in which the flue gases after passing through the main furnace flues return to the 'smoke box' at the boiler front through numerous small-diameter fire-tubes, situated above, alongside or below the main furnace flues. The economic boiler does not require brick flues, as in the *Lancashire boiler*, although in some cases brick flues have been added to give a third 'pass' for the flue gases; it is more common to provide this third 'pass' by means of an additional set of fire-tubes within the shell of the boiler. Economic boilers are of wet-back or dry-back design. In the wet-back the rear flue gas reversing chamber is completely surrounded by water; in the dry-back the chamber is set at the extreme rear of the boiler and lined with firebrick. Economic boilers are designed for pressures up to 20 bar (300 lb/in^2) and for evaporations up to 5 kg/s (40 000 lb steam/h). Thermal efficiencies are up to 80 per cent, being more efficient that the Lancashire boiler as well as occupying less space. *See super-economic boiler.*

Economic Commission for Europe (ECE) A body set up in 1947 by the Economic and Social Council of the United Nations to initiate action to raise the level of European economic activity, in the wake of World War II, and for maintaining and strengthening the economic relationships of European countries, both among themselves and with other countries. The members consist of the European members of the United Nations and the United States of America. The ECE has developed a test cycle for the measurement of gaseous emissions from the exhausts of motor vehicles, based upon the European patterns of driving; nevertheless, it was adopted in Australia from 1974 as the most appropriate cycle for testing motor vehicle compliance with Australian Design Rule 27, the first serious attempt to reduce hydrocarbon and carbon monoxide emissions from vehicles in that country.

118

Economic resources   Accumulations or deposits of minerals that are economically recoverable in the context of current prices and costs, and with current technology; resources the recovery of which will provide an adequate rate of return to the investor, having regard to the risks involved and earnings elsewhere.

Economic system, functions of   The essential functions to be fulfilled in the economic arrangements of any community; these functions may be defined as follows:

(a) Generally, to match supply to the effective demand for goods and services in an efficient manner.

(b) To determine what goods and services are to be produced and in what quantities.

(c) To distribute scarce resources among the industries producing goods and services.

(d) To distribute the products of industry among members of the community.

(e) To provide for maintenance and expansion of fixed capital equipment.

(f) To fully utilise the resources of society.

In a free enterprise system the fulfilment of these six economic functions is left to the profit motive and the price mechanism working within a framework of social safeguards. In a socialist society all the operations required are consciously planned by official organisations. Many countries operate a mixed economy, splitting the economy into public and private sectors, the activity of the whole being influenced by direct and indirect planning measures. All systems have tended to neglect the increasing abuse of 'free goods' such as air and water, as these social costs have not fallen within the accountancy systems normally maintained. The increasing concern expressed in recent years in respect of environmental effects has found reflection in the policies of all countries, irrespective of political complexion. Energy policies have tended to emerge in recent years at both the national and the international level, and the interrelationship between energy consumption and economics have received increasing attention. *See Carter energy policy; energy computer models; Organisation of Petroleum Exporting Countries (OPEC).*

Economic welfare   Defined by the Cambridge economist A. C. Pigou (1877–1959) as 'that part of social welfare that can be brought directly or indirectly into relation with the measuring rod of money'; in other words, those aspects of social welfare which are concerned with material as distinct from bodily, moral or spiritual well-being, although obviously these are interrelated in some ways. Pigou stressed that there is no precise line between economic and non-economic satisfaction. Economic or material satisfaction are derived from the consumption of both goods and services; and it is this which is the subject matter of economics. Pigou warned that economic welfare will not serve for a barometer or index of total welfare; this is because an economic cause may affect non-economic welfare in ways that cancel its effect on economic welfare. Non-economic welfare is liable to be modified by the manner in which income is earned and also by the manner in which it is spent: 'Of different acts of consumption that yield equal satisfactions, one may exercise a debasing and another an elevating influence'.

Economiser Or feed-water heater; a device for utilising the waste heat from boilers to preheat incoming feed-water, thus raising the overall *thermal efficiency* of the plant. An economiser comprises banks of tubes placed in the path of the flue gases as they pass to the chimney, after the *superheater*, but before any *air preheater*. The external surfaces of the tubes are liable to fouling; in the case of plain tubes scrapers are used which move up and down each tube keeping the surface clean, but in respect of gilled tubes steam or compressed air soot blowers must be used. The temperature of feed-water in an economiser may be raised to within 22° C of the water and steam temperature in the boiler.

Ecosystem The natural system through which matter and energy flow. The organisms of a community, together with the atmosphere, soil or water, form a functioning system. Matter and energy are taken up by producer organisms from the physical surroundings and passed from organism to organism through the food chain. Matter is progressively decomposed by each organism and returned to air, soil or water; energy is lost from each organism and dissipated as heat. Each ecosystem has an input of elements as gases and mineral nutrients from the atmosphere, from the weathering of rocks and from other ecosystems, and an input of energy as either light or organic matter or both. Each ecosystem has a flow of matter and energy through it, with much of the matter being recycled through the community again and again. Each ecosystem has an output of gases and heat energy to the atmosphere, and often an output of nutrients and organic matters to other ecosystems as well. *See carbon cycle; photosynthesis*.

Eddy currents Electrical currents induced in the interior of conductors carrying alternating currents, due to variations in the magnetic flux surrounding the conductor.

Eddy diffusion The process by which gases, including smoke, diffuse in the atmosphere; molecular diffusion is extremely slow by comparison and may be ignored.

Edeleanu process In the petroleum industry, a finishing process based on solvent extraction using liquid sulphur dioxide as a solvent. The process removes undesirable aromatics and other polar compounds such as sulphur, gum and colour constituents; it has been used widely in the manufacture of premium *kerosine*. The process yields an unpleasant waste *acid sludge*. The sludge may be decomposed in a special kiln and the *sulphur dioxide* converted to sulphuric acid.

Edge water The body of water underlying the oil and/or gas accumulation in anticlinal or similar geological structures.

Effective demand *See demand.*

Effective height of emission The height above ground level at which a plume is estimated to become approximately horizontal.

Effluent charge A fixed fee levied by a regulating body against a polluter for each unit of waste discharged into public waters. The fee may be uniform for all waste producers in the area, or it may be selective according to the composition of individual wastes or the absorption capacity of the local environment. The fee may be charged continuously at all times or it may be levied only when conditions deteriorate below a specified level.

Efflux velocity The speed at which gases leave a chimney or vent and escape into the general atmosphere; the velocity of discharge should be high enough to avoid any substantial risk of *down-wash*.

Einstein's equation An equation which can be used to calculate the energy released in any process, such as *fission*, in which a loss of mass occurs; it is expressed $E = mc^2$, in which E is energy, m is mass and c is the velocity of light. The velocity of light is about 3.0×10^8 metres per second.

Elasticity of demand *See demand.*

Electrical energy A capacity for doing work, measured in terms of the *joule*; 1 joule is expended when an electrical power of 1 *watt* is exerted for 1 second. A 100 W electric light bulb consumes 360 000 J of electrical energy in 1 hour.

Electrical power The rate of expenditure of *electrical energy*; it is measured in J/s, or watts. Watts = volts × amperes.

Electric arc furnace A *steelmaking furnace* in which three carbon electrodes are used to carry an electric current from a supply transformer to the steel charge in the furnace bath. Once a circuit is established, the electrodes are withdrawn slightly from the steel so that the current jumps in a lightning flash from the electrode tips to the metal; thus, an electric 'arc' is struck between metal and electrodes. *See Figure E.1.*

Electric battery An energy conversion device capable of retaining or storing chemical energy, and supplying this as electrical energy when required. There are two basic types of battery: the 'primary battery', capable of only a single electrical discharge; and the

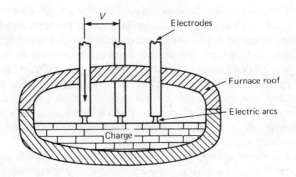

Figure E.1   Section of a 3-phase direct electric arc melting furnace

'secondary battery', designed to be discharged and recharged for many hundreds of cycles. The three main types of primary cell are the zinc–carbon or Leclanché; the mercuric oxide–zinc; and the alkaline–manganese. Secondary batteries fall into three classes: stationary batteries, transportable batteries and motive power batteries. *See fuel cell; sodium–sulphur battery.*

Electric cars   Battery-powered electrically driven vehicles often regarded as a replacement for the internal combustion engine on the grounds of both pollution abatement and conservation of petroleum. Over 100 000 battery-driven vehicles are in use in Britain, mainly for milk rounds and similar functions. Their relative silence is appreciated by residents in the early hours. These vehicles have low running costs and require little maintenance. In the role of passenger vehicles, however, a number of difficulties have yet to be overcome: there is slower acceleration, a limited top speed and a range of only about 75 kilometres before recharging, while the batteries occupy much space. Electric car development is being undertaken at the Flinders University in South Australia. In Japan, in 1979, the Daihatsu Charade EV car came into limited commercial production; it is powered by eight lead–acid batteries and a direct current motor, has a maximum speed of 70 kilometres/hour, and a range of 75 kilometres at an average of 40 kilometres/hour before recharging from a 240 V electricity supply.

Electric current   Descriptive of electrical energy when the movement of free electrons in a conductor is in the same direction. Electricity is 'static' if the electric charges do not move as a current, but collect in one place. *See ampere (A); volt (V).*

Electricity   A general name for all phenomena that arise out of the flow or accumulation of electrons in matter. The units of quantity

122

are the coulomb, the joule and the watt; of pressure, or electromotive force, the volt; and of flow, or current, the ampere.

Electricity Council A body set up under the Electricity Act of 1957 to take over the central administration of the electricity supply industry in England and Wales. The Council's statutory duty is 'to advise the Minister on questions affecting the electricity supply industry and to promote and assist the maintenance and development by the Electricity Boards of an efficient, co-ordinated and economical system of electricity supply'. It also deals with labour relations, finance and research.

Electricity supply system An array of power-generating sources and load centres, linked by an arrangement of transmission and distribution lines, forming a grid which covers a wide area, the whole operating under central management and control. Large, centrally controlled, integrated systems offer substantial advantages and economies in meeting a given demand pattern and in responding to changes in demand, compared with a number of non-integrated undertakings attempting to meet similar market conditions. The United States electricity supply industry comprises about 3600 business units which vary greatly in size, type of ownership and range of power supply functions performed; of these business units, about 2300 engage exclusively in distribution. About 100 of the largest systems account for some 90 per cent of all electricity generated in the United States. The electricity supply system of England and Wales is under the ownership and control of the *Central Electricity Generating Board*, and remains the largest integrated power supply system in the world.

Electrodes The conductors which convey an electric current into or out of a liquid or a gas.

Electrolysis The chemical decomposition of a substance when electricity is passed through it in solution or in the molten state.

Electrolytes Chemical compounds and solutions of chemical compounds, capable of being decomposed by the passage of an electric current.

Electromagnet A device comprising a coil of wire on a piece of soft iron carrying a direct current, a magnetic effect being created so long as the current flows.

Electromagnetic induction A principle discovered by Michael Faraday in 1831: that if an alternating current, or a direct current of varying strength, is passing through a conductor, then any other approximately parallel conductor in the vicinity will have an electromotive force induced in it. If the ends of the latter are joined so as to form a closed circuit, then an electric current will be induced. In practice, the respective conductors usually take the form of coils of insulated wire. Electric generators, static transformers and induction motors depend upon electromagnetic induction for their operation.

Electromagnetic pump A type of pump, without moving parts, used for pumping liquid metals; the pumping action depends on the interation between an electric current which flows through the liquid metal and a magnetic field produced over the same region by a magnet, the directions of the current and field and the required direction of flow being mutually perpendicular.

Electromotive force (e.m.f.) A difference in electrical potential which tends to cause an electric current to pass from the point of higher potential to the lower.

Electron An elementary particle of mass 9.11×10^{-28} gram carrying a charge of negative electricity of 1.602×10^{-19} *coulomb*, which, in the *atom*, moves rapidly in orbit about the *nucleus*.

Electron beam heating Works on a similar principle to the thermionic valve, with a heated filament ejecting electrons which are accelerated by a high potential and focused into a beam; the beam can be directed over the charge as required, the charge being heated by the high kinetic energy of the electrons. Very high energy intensities are possible over small areas. An important application is in welding joints; the beam gives deep weld penetration without significant heating of the metal adjacent to the weld. However, the operation must take place in a vacuum chamber.

Electronic Halarc A new domestic light bulb developed and marketed by the General Electric Corporation of America; the new bulb uses one-quarter of the electricity used by an ordinary incandescent bulb and lasts five times as long. Light is produced in a minute quartz chamber, filled with argon and mercury gases and with metal halides. A current or arc passes through this chamber, vaporizing the gases and producing an intense light. It takes about 30 seconds for the bulb to warm up; an ordinary incandescent filament is used to provide instant light, this turning down automatically when the arc lamp achieves full brightness. This energy-efficient bulb represents the new phase in domestic light bulbs. *See Figure E.2.*

Electronic valve rectifier A transformer-rectifier for converting the alternating current of normal electricity supply into high-tension direct current electricity; this unit has found wide use in the United States in electrostatic precipitators. *See electrostatic precipitator; rectifier.*

Electron-Volt A unit of energy equal to the energy acquired by an *electron* when it is accelerated through a potential difference of 1 volt. $1 \, eV = 1.602 \times 10^{-19}$ J. *See Joule.*

Electroplating A process by which a metal is deposited, usually as

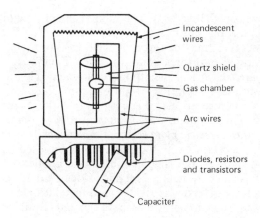

Figure E.2 Electronic Halarc Lamp. (Source: General Electric Corporation)

a relatively thin coating, upon the surface of an article which is itself made of a different metal. The main purposes for which electro deposits are used are to provide: (a) protection against corrosion of the basis metal; (b) a particular surface appearance; (c) good wear resistance. Low-voltage direct current from a *rectifier* or motor-generator is passed through a solution consisting mainly of a compound of the metal to be deposited. The article to be plated is the 'cathode', or negative electrode, in the solution; the 'anode', or positive electrode, is either an insoluble material or more usually the metal to be deposited. Metal is deposited from the solution on to the article.

Electrostatic filter   An *air filter* in which dust particles are given a positive electric charge by an ionizing screen; they are then attracted to the negatively charged filter plates. While they offer little air flow resistance, the high efficiency of these filters declines as air velocity increases. Some form of after-filtration is usually recommended.

Electrostatic precipitator   A device for the arrestation and removal of dust from a gas stream. It utilises the general principle that if a gas is passed between two electrodes, one of which is supplied with a very high negative voltage and the other earthed, the gas and any particles of dust in suspension become electrically charged; the dust is attracted to the earthed electrodes, where it collects. A mechanical rapping device dislodges the particles and they fall into a collecting hopper beneath the precipitator. The modern precipitator is undoubtedly one of the most efficient means of extracting dust particles from flue gases. Units installed at large new power stations are designed to operate at collecting

125

efficiencies of 99.3 per cent. The subdivision of precipitators into zones or sections ensures better performance and flexibility; in large units there may be three parallel banks or more, each bank being divided into three zones in series. Each zone is supplied with separate electrical and rapping equipment. This arrangement allows optimum operating conditions in each zone, and in the event of a failure in one zone ensures that the increase in dust emission will be small. *See **Automatic voltage control; Deutsch formula; dust arrester; plate precipitator; tubular precipitator**.*

Element A substance which cannot be decomposed (split up) by chemical changes into simpler substances. Elements are divided into metals and non-metals. Metallic oxides are described as basic and, if soluble in water, as alkalis. Non-metallic oxides are acidic. A compound consists of two or more elements combined together chemically in definite proportions.

Elutriation The classification or grading of particles effected by movement relative to a rising fluid. A known weight of dust is placed in a receptacle at the base of the apparatus and is then subjected to an upward current of air or water; by varying the velocity of the upward stream it is possible to obtain a series of fractions expressed in terms of the falling velocity of the dust.

Emergency core-cooling system (ECCS) A system devised to protect a nuclear reactor against the consequences of a loss of coolant, with resultant over-heating; the principle is the injection of water into the reactor when failure occurs. Some have questioned the efficacy of this technique under certain circumstances.

Emergency Gas Act (US) An emergency Act introduced by Congress early in 1977 authorizing the Federal Power Commission to direct the transfer of natural gas from one pipeline to another and to exempt emergency sales from normal price regulations. The measure was a response to the most severe winter of the century in the United States, which caused an energy crisis in many areas when natural gas and electricity resources proved inadequate. The situation was particularly severe in Ohio and parts of the south; under emergency conditions, factories were ordered to close to allow the available gas to be used in homes. Efforts to obtain emergency supplies of fuel oil and liquefied petroleum gas were hampered by snow and ice. It was estimated that the extreme cold in the 1976–77 winter raised the national bill for home heating by $5.5 billion.

Emissivity The ratio of the rate of loss of heat per unit area of a surface, at a given temperature and in certain surroundings, to the rate of loss of heat per unit area of a *black body* at the same temperature and in the same surroundings. A black body is a body with an emissivity of unity. The emission of radiant energy from a body depends upon both its temperature and its emissivity. Typical values for emissivity for various substances at 1000° C are: building brick, 0.45; chromium, polished, 0.38; fireclay brick, 0.75; lampblack, 0.96; silica refractory brick, 0.66. *See absorptivity*.

Emphysema Swelling of the air space of the lungs due to the destruction of the *alveoli*; the effect is to diminish the area exchanging oxygen and carbon dioxide between the air and the blood.

Empirical Formula The simplest possible formula to describe a chemical compound, an empirical formula indicates only the proportions in which the constituent atoms are present in the molecule. For example, both acetylene, C_2H_2, and benzene, C_6H_6, have an empirical formula of CH; normal butane, C_4H_{10}, has an empirical formula of C_2H_5. Thus, an empirical formula must be distinguished from a *molecular formula*, which indicates the actual numbers of the constituent atoms in the molecule. *See constitutional formula; graphical formula*.

Enamel firing A stage in the firing of pottery ware in which the colours or decoration are fired into the glazed ware at a temperature of about 750° C. *See biscuit firing; glost firing*.

Endothermic gas A furnace atmosphere and carrier gas used in a wide range of heat treatment processes, including gas carburizing. The use of *liquefied petroleum gases* for producing endothermic gas is well established. To produce the gas, a mixture of the hydrocarbon feedstock and about one-third of the volume of air required for its complete combustion is passed over a nickel catalyst heated to about 1050° C. A typical analysis of endothermic gas, each constituent being expressed as a percentage, is: carbon monoxide, 20–28; hydrogen, 19–41; nitrogen, 38–50. *See exothermic gas*.

Endothermic reaction A chemical reaction accompanied by the absorption of heat. An endothermic reaction occurs during the manufacture of *water gas*.

Energetics The abstract study of the energy relationships in physical and chemical changes.

Energy The capacity of matter or radiation to do work, although it may not often be possible to effect the transition from energy to work; not all energy is available energy. The unit of energy is the same as that of work—the *joule*. There are several forms or descriptions of energy. Kinetic energy is the ability of matter to do work by virtue of its motion, e.g. the energy of flowing water, which may be used to generate electricity. Potential energy indicates the ability to do work by virtue of the position or configuration of matter, the presence of stresses or the availability of a supply of electrical energy. A bent spring possesses potential energy, for it is capable of work in returning to its natural form; an electric current can turn a motor. Other classifications are: chemical energy, which arises out of the capacity of atoms to evolve heat as they combine and separate; electrical energy, which arises out of the capacity of moving electrons to evolve heat, and create electromagnetic radiation and magnetic fields; nuclear energy, which arises out of the elimination of all or part of a mass of atomic particles; radiant energy, which, after transit through space, strikes matter and appears as heat; and heat energy, an aspect of the above, arising from the kinetic energy of molecules.

Energy, free The capacity of a system to perform work, a change in free energy being measured by the variation in the work obtainable from a given process.

Energy Act 1976 A United Kingdom Act which provides for compulsory labelling and the supply of information on the fuel consumption of new cars. New cars on display in showrooms must carry a label showing official fuel consumption figures for that model; dealers must have details of fuel consumption tests for buyers to consult on request; and references in promotional literature to the petrol consumption of a new car must include the test results.

Energy audit A survey of energy usage in a system or organization and an examination of how energy may be used more economically. An audit requires the application of both engineering and economic skills. An analysis and appraisal of expenditure for the use of electricity, gas, water and oil would include a review of the impact of tariffs and taxes.

Energy Authority (NSW) A body created by the New South Wales Government, Australia, in 1975 to advise the state government on energy matters. Under the legislation, the Chairman of the Authority is always the Chairman of the Electricity Commission of New South Wales. *See Joint Coal Board.*

Energy balance *See See heat balance; Sankey diagram.*

Energy bands A change from one energy level to another taking place in quantised steps; in a crystalline solid the energies of all the electrons and atoms fall into several 'possible' energy bands between which lie 'forbidden' bands. The range of energies corresponding to states in which electrons can be made to flow, by an applied electric field, is called the conduction band. The range of energies corresponding to states which can be occupied by valency electrons, binding the crystal together, is called the valency band. *See energy levels.*

Energy barrier In chemistry, the minimum amount of free energy which must be attained by a chemical entity in order to undergo a given reaction.

Energy Bonds *See Carter energy policy.*

Energy cascading The utilisation of energy, otherwise wasted, in the performance of work matched to the characteristics of the energy, e.g. the use of waste heat from an industrial process for a variety of purposes in sequence, performing work at a lower temperature at each stage. The sequence may include air heating, steam raising and water heating.

Energy Commission (UK) Following the Secretary of State's Energy Conference in June 1976, a commission set up by the British Government to further public debate on energy policy issues. With broad representation, the Commission met for the first time in February 1978. In 1978, following discussion by the Commission, the Government published its Green Paper *Energy Policy: A Consultative Document.*

Energy computer models Computer models of the energy sector of an economy, which have been developed largely in the United States of America and Europe; the models consider both the sector itself in detail and its relationship with the national economy. Modelling activity in this field developed from about 1970, as the necessary mathematical tools and computers became available, and the impact of the 1973 oil price rises was felt. Most models are concerned with the medium-term (5–15 years) and the long term (15–50 years). Two basic types of energy analysis have evolved: one focusing on the technical side of the energy system and the other on economic analysis. Models developed in the United States can be used separately or collectively to study a wide range of economic and energy problems. A model has been developed at the US National Centre for Energy Analysis, Brookhaven National Laboratory. The Brookhaven Energy System Optimization Model (BESOM) is a linear programming model in which the total cost of satisfying a given set of national energy demands is minimized. An input–output model developed by the Brookhaven National Laboratory and the Centre for Advanced Computation at the University of Illinois gives the output of the energy sector in physical units (British thermal units).

Energy conservation The better and more efficient use of energy with proper regard to the related costs and benefits, whether these be economic, social or environmental. The *International Energy Agency* has proposed a wide range of measures which should be given serious consideration when national conservation programmes are being strengthened. These are aimed at achieving higher thermal efficiencies, higher load factors, better building standards, more effective use of wastes and the promotion of public education, all within a framework in which energy is priced at competitive levels coupled with increased taxes on certain fuels in certain circumstances. Efficiency improvements currently thought possible in the automobile are: engine (15 per cent), transmission (5 per cent), better lubrication (5 per cent) and improved design (15 per cent). It has been estimated at 1979 prices that an extra $A400–500 on the cost of a new car would produce fuel savings of about 25 per cent.

Energy costs   Or variable costs, or separable costs; costs which vary with the total amount of energy produced; at power stations they comprise fuel and fuel-handling costs, the cost of water (both light and heavy) and consumable stores, together with supervisory costs and user costs. User costs are the additional costs of running equipment compared with the cost of keeping it idle, consisting essentially of the costs of accelerated physical deterioration and additional maintenance. *See marginal costs.*

Energy Council (SA)   A council created by the South Australian Government in 1978 to develop policies and advise the state government on energy conservation; the development and co-ordination of existing energy supplies; development of exploration; and research into alternative energy sources, with particular reference to solar energy.

Energy expenditure, human   The rates of expenditure or loss of energy by human beings under a wide range of conditions, ranging from a state of rest to vigorous activity; estimates of average rates of expenditure by 'reference man' and 'reference woman' have been prepared by the UN *Food and Agriculture Organization.* Some of these estimates are indicated in *Table E.1. See basal energy requirement; energy value.*

Table E.1   HUMAN ENERGY EXPENDITURE

Activity	Average rate of expenditure at 10° C (kJ/min)	
	men	women
Working activities	10.5	7.7
Washing, dressing	12.6	10.5
Walking	22.2	15.1
Sitting	6.4	5.9
Active recreation/domestic work	21.8	14.6
Rest in bed/basal energy requirement	4.2	3.7

Source: FAO

Energy farming   The growing and harvesting of *biomass*, such as trees and field crops, for energy production, especially for liquid fuels needed in transport.

Energy flow   In ecology, the total assimilation at a particular trophic level, which equals the production of *biomass* plus respiration.

Energy in food   The energy value of foodstuffs in relation to the internal needs of the consuming organism; the 'combustion' of food uses oxygen and liberates carbon dioxide, water and energy. The amounts of energy in foods are now expressed in megajoules, formerly being expressed in kilogram-calories or kilocalories. One kilocalorie is about 4.2 kJ or 0.0042 MJ. A steak of 100 g, comprising protein 25 g, fat 22 g and water 50 g, would have an energy value of about 1.3 MJ/kg, or 1.3 kJ/g. *See energy value; Table E.2* (page 132)

Energy levels   Different intensities of energy corresponding to certain definite states found in atoms and nuclei; for each different atom or nucleus there exists a series of energy levels. The lowest stable energy level of an atom or nucleus is known as the ground state; atoms or nuclei at higher energy levels than the ground state are said to be 'excited'. Excitation energy is the difference in energy between the ground state and the excited state. *See energy bands.*

Energy Mobilization Board   *See Carter energy policy.*

Energy Policy and Conservation Act 1975   A United States measure introducing fuel economy standards for all passenger cars produced in, or imported into, the country; the legislation also contained specified standards for labelling passenger cars to show fuel consumption test results, although the labelling itself was voluntary. Late in 1976 Congress rejected proposals to suspend emission control requirements, which it was claimed would have made it easier for manufacturers to meet fuel economy standards; however, the requirement for fuel consumption labelling became compulsory. The 1985 national target for fuel economy reduces the 1975 figures by almost 50 per cent; while this is a remarkable reduction, it has been regarded as attainable because of the considerable scope for reduction of vehicle size and weight in the United States. An average 1975 car in Europe weighed 860 kilograms, and in the United States 1500 kilograms.

Energy pyramid   In ecology, a diagram showing the rate of energy flow, or production, at successive trophic levels.

Energy Research and Development Administration   *See Department of Energy* (US).

Energy Security Corporation   *See Carter energy policy.*

Energy stocks   Stocks and shares in companies operating in the energy sector of the national economy, notably in oil, shale oil, natural gas, manufactured gas, uranium, coal mining, coke, briquettes and electricity. Companies may be engaged in exploration, processing, mining, drilling, treatment, transmission, generation, storage, transport or distribution.

Energy Survey Scheme   A British government scheme, announced in 1976, offering financial assistance to energy users in the non-domestic sector in seeking advice on energy saving. The scheme seeks to assist smaller firms in meeting in part the cost of the services of a consultant, who would report on 'good housekeeping' energy savings and the potential for greater savings through longer-term measures.

Energy Technology Support Unit (ETSU) An agency of the ***Department of Energy (UK)***, responsible for the development of renewable energy sources; research, development and demonstration related to energy conservation; and the development of a strategy for energy research and development in the United Kingdom.

Energy Trends A monthly statistical bulletin published by the Department of Energy (UK).

Table E.2 TYPICAL GROSS ENERGY VALUES

Fuel	kJ/g	Btu/lb
Bagasse	9	4000
Wood		
green	10	4400
dry	17	7310
Peat		
mild	11	4800
sod	14	6200
Lignite (brown)	21	
Coke	28	12 100
Coal		
bituminous	30	13 000
anthracite	34	14 500
electricity (UK)	24	10 200
industrial (UK)	28	11 800
Oil		
heavy fuel	43	18 400
medium fuel	44	18 900
gas/diesel	45	19 200

	MJ/m³	Btu/ft³
Natural gas	39	1 050
Manufactured gas (UK)	19	500
Liquified petroleum gas	50	21 300

	Cal/g	kJ/g
Protein	4.0	17
Fat	9.0	37
Carbohydrate	3.7	16
Ethyl alcohol	7.0	29

Energy value The quantity of heat or energy developed by the complete combustion of a given weight or volume of fuel; sometimes referred to as 'heat value' or 'calorific value'. It is often expressed in kilojoules per gram, or megajoules per kilogram or cubic metre; However, the older *British thermal unit* is still widely used, expressed as so many Btus per pound or per cubic foot. *Table E.2* shows some typical energy values. These values are described as the higher or gross energy value; in practice, the steam released in the process of combustion (from inherent moisture and hydrogen in the fuel) is a loss in terms of effective heating value; hence, it is necessary to have regard to the lower or net energy value in design work.

Engine A device for converting energy of different kinds into *kinetic energy. See diesel engine; engine capacity; engine efficiency; gasoline engine; gas turbine; heat engine; ram-jet engine; steam engine; steam turbine; turbo-jet engine; turbo-prop engine*.

Engine capacity Expressed in terms of litres or cubic centimetres, the cylinder capacity of an engine multiplied by the number of cylinders (usually 4 or 6). The cylinder capacity is the 'swept' volume of a cylinder, i.e. the area of cylinder bore multiplied by the length of the piston stroke or sweep.

Engler system A system for measuring *viscosity* used in Germany, Italy and some other European countries, based on the ratio of the time taken for a given volume of oil to pass through an aperture, compared with the time taken for an equal volume of water at the same temperature to pass through a similar aperture. The results are given in terms of the Engler degree ($° E$).

Enriched uranium *See uranium enrichment*.

Ente Nazionale Idrocarburi (ENI) The Italian state company with large interests in hydrocarbons, chemicals, nuclear energy and engineering. An ENI group company is Snam, which has built a 2500 kilometre intercontinental gasline from Africa to Europe. The gasline from Algeria ensures a natural gas supply to Italy of 12 billion cubic metres annually for a period of 25 years. Italy is also linked by gasline with the Netherlands and the Soviet Union.

Enthalpy The total amount of heat that water, steam, air or other gas contains, measured from $0° C$. *See total heat*.

Enthalpy–temperature (It) diagram A diagram which enables the direct derivation of the initial *enthalpy* of any combustion gas to be made, rendering stoichiometric calculations based on the ultimate analyses of fuels unnecessary. Gas temperatures may be directly obtained from the enthalpy of the combustion gases without recourse to the composition or specific heat of the gas.

Entrained flow/fixed bed composite gasification A process involving the steam–oxygen gasification of coal at high thermal efficiency to

133

produce a substitute natural gas, or (by steam–air gasification) a low-energy-value fuel gas. The process is being developed by the **British Gas Corporation**.

Entrainment   The collecting and transporting of a substance by the flow of another fluid moving at a high velocity. For example, boiler water may become entrained in the steam leaving the boiler under certain conditions, or dust particles may become entrained in flue bases and carried out of the chimney.

Entropy   In thermodynamics, an index of the availability of heat-energy for producing power. If, in a reversible change, a substance receives or loses a quantity of heat dQ at an absolute temperature T, the substance gains or loses an amount of entropy given by:

$$d\phi = \frac{dQ}{T}$$

where, $d\phi$ = entropy change;
$\quad dQ$ = heat added or lost, Btu/lb;
$\quad T$ = absolute temperature, ° R.

This simple formula covers the case where, for example, water is converted into steam through the addition of *latent heat*, without change of temperature. If the temperature also varies, from T_1 to T_2, the change of entropy is given by:

$$d\phi = \int_{T_1}^{T_2} \frac{dQ}{T}$$

Environment   The region, surrounding or circumstances in which anything exists; everything external to the organism. The environment of the human being includes the abiotic factors of land, water, atmosphere, climate, sound, odours and radiation; the biotic factors of animals, plants, bacteria and viruses; and the social factor of aesthetics. *See ecosystem.*

Environment, Department of the   A British government department formed in 1970 which brought together the functions of the Ministry of Housing and Local Government, the Ministry of Transport and the Ministry of Public Building and Works. Its responsibilities include the reform of local government, housing policy and finance, control of environmental pollution, policies for the port industry, regional development, correlation of urban and transport planning, investment policies for the nationalised industries, responsibility for the construction industry, and conservation and amenity.

Environmental forecasting A forecasting programme capable of timely and effective warning of technologically induced perturbations of any given health-welfare parameter of the population. The need for environmental forecasting has grown with the pace of technological and community change. In making detailed forecasts concerning community hazards, models need to take account of emission levels; the transport, storage and reaction of pollutants in the *environment*; and the resulting exposure of the community and its reponse.

Environmental geology The application of geologic data and principles to the solution of problems likely to be created by human occupancy and use of the physical *environment*; geology oriented towards the planned utilisation of resources and the safeguarding of the environment. The areas of interest include metallic and non-metallic minerals; mineral fuels; groundwater; soils and soil conditions unsuitable for septic systems or sanitary land-fill use; land use; and beach preservation.

Environmental impact statement A considered report, following careful studies, disclosing the likely or certain environmental consequences of a proposed action, thus alerting the decision-maker, the public and the government to the environmental risks involved; the findings enable better informed decisions to be made, perhaps to reject or defer the proposed action or to permit it subject to compliance with specified conditions. Each environmental impact statement must include:

(a) A detailed description of the proposed action, including information and technical data adequate to permit a careful assessment of environmental impact.

(b) Discussion of the probable impact on the environment, including any impact on ecological systems and any direct or indirect consequences that may result from the action.

(c) Discussion of any adverse environmental effects that cannot be avoided.

(d) Alternatives to the proposed action that might avoid some or all of the adverse environmental effects, including analysis of costs and environmental impacts of these alternatives.

(e) An assessment of the cumulative, long-term effects of the proposed action, including its relationship to short-term use of the environment versus the environment's long-term productivity.

(f) Any reversible or irretrievable commitment of resources that might result from the action or that would curtail beneficial use of the environment.

(g) Discussion of objections raised by government agencies, private organisations and individuals.

The impact statement procedure affords the public an opportunity to participate in decisions that may affect the human environment. The preparation or discussion of statements may involve the holding of public hearings or inquiries. *Table E.3* summarises the principal environmental impacts of energy sources.

Environmental Protection Agency An Agency created by the National Enviroment Policy Act 1970 to implement the environment protection policies of the United States federal government. The agency is concerned mainly with pollution control measures and the environmental impact assessment of major projects. In respect of air pollution control, the agency sets national ambient air quality standards for oxides of sulphur, oxides of nitrogen, carbon monoxide, photochemical substances, odours and noise. The aim has been to reduce motor vehicle pollution by 90 per cent, while requiring the best available control technology for all new pollution sources. From June 1979 the Agency required that all new electric power plant be equipped with scrubbers to remove 90 per cent of the oxides of sulphur arising from the use of high-sulphur coal, and 70 per cent of oxides of sulphur arising from the use of low-sulphur coal. Wet and dry scrubbing methods are envisaged.

Environmental quality standards Levels of exposure to pollutants which should not be exceeded; standards may be statutory or presumptive. Two levels have been adopted by the US *Environmental Protection Agency*:

(a) Primary. Levels judged necessary to protect health with an adequate margin of safety.

(b) Secondary. Levels judged necessary to protect public welfare from any known or anticipated adverse effects.

These are essentially ambient environmental quality standards. Standards may also prescribe the contents of products, e.g. the amount of phosphates in detergents or of pesticide residues in foodstuffs. They may also take the form of emission standards, e.g. the upper limits of what may be emitted from the exhausts of motor vehicles or from the chimneys of industrial plants.

Environment protection policies Policies developed by governments, agencies, associations, communities, groups, companies or corporations, relating to the control of wastes, the protection of the natural environment, the improvement of the man-built environment and the protection of heritage values. Many countries have developed an effective range of environment policies governing pollution, the environmental assessment of new projects, the declaration of national parks and reserves, the

Table E.3 PRINCIPAL ENVIRONMENTAL IMPACTS OF ENERGY SOURCES

I. Fossil Fuels

	Impact on lithosphere	Impact on hydrosphere	Impact on atmosphere	Impact on biosphere
1. Coal Extraction	Land subsidence in underground mining, land disturbance in strip mining	Acid mine drainage effects on underground and surface waters	Dust in mining areas	Health hazards in underground mining
Processing		Waste water of coal processing	Coal-dust	Acid mine drainage effects on fish and wildlife
Use (in power plant)	Storage areas	Thermal discharges	Emissions from power plants (particulate SO_x, No_x, CO, HC Possible local meteorological effects (pollutants, thermal discharges)	Health effects of pollutants Impact of thermal discharges on aquatic life Effect on plants
2. Oil Extraction	Oil pollution Disposal of brines	Oil pollution in offshore activities		Effects of oil spills on marine life
Transportation	Pollution from pipelines Disturbance of land (permafrost, etc.)	Marine oil pollution		Effect on marine life Effect on wildlife (pipelines)
Processing	Oil pollution	Refinery wastes and their effects on groundwater and/or coastal waters	Emissions of SO_x, HC, NO_x, CO	Health effects of pollutants Effects of industrial wastes on aquatic life

137

Table E.3—*continued*

	Impact on lithosphere	*Impact on hydrosphere*	*Impact on atmosphere*	*Impact on biosphere*
Use (in power plants)		Thermal discharges	Emissions of SO_x, HC, NO_x, CO; Possible local meteorological effects (pollutants, thermal discharges)	Health effects of pollutants; Effects of thermal discharges on aquatic life; Effect on plants
Use (in mobile sources)	Gasoline and oil leakages; Lub-oils	Oil leakage from ships, boats, etc.	Pollutants from automobiles; Pollutants from aircraft and their effects on ozone layer	Health effects of pollutants; Effects on aquatic life; Effects of pollutants on plants in forest areas (highways)
3. Natural Gas Extraction Transportation	Same as oil; Accidents (fire, etc.) in liquefaction plants	Same as oil; Marine spills	Air pollution in case of accidents	Health effects in liquefaction and regasification plants; Effects on marine life in case of spills
Use (in power plants)		Thermal discharges	Emission of particulate SO_x, NO_x	Health effects of pollutants; Impact of thermal discharges on aquatic life
II. Renewable Sources of Energy				
1. Geothermal Energy Extraction	Land subsidence, local disturbance of land (drilling, pipelines)		Noxious gases	Effects on plants
Use		Disposal of brines; Thermal discharges	Noxious gases (e.g. hydrogen sulphide), thermal discharges; Possible local meteorological effects	Health effects of pollutants; Effects on plants; Effects on aquatic life

	Impact on lithosphere	Impact on hydrosphere	Impact on atmosphere	Impact on biosphere
2. Hydroelectric Man-made lakes	Disturbance of land, possible earth-crust disturbances. Terrestrial effects around lakes (soils, etc.)	Changes in hydrological cycle, water quality, nutrients, etc.	Microclimatic changes due to evaporation (humidity)	Aquatic life (fish, plants, etc.) Health effects
Downstream effects	Erosion of river banks Delta-shoreline erosion	Changes in hydrological cycle, water quality		Aquatic life (fish, plants, etc.) Health effects
3. Solar Energy	Land areas for solar farms	Thermal pollution from water-cooled turbines	Possible local meteorological changes	
4. Tidal Energy	Land drainage (shorelines), sediment budget Geomorphological changes			Possible effects on aquatic life
5. Wind Energy	Areas for wind mills		Possible local meteorological changes	
6. Sea-thermal Power		Possible changes in sea temperature		Possible effects on marine life
7. Renewable Fuel Sources	Land areas for photosynthetic material Disposal of wastes and storage	Disposal of wastes from digesters		Health effects of wastes (in handling) effects on agriculture, etc.

Source: OECD, Paris.

139

protection of wildlife, the preservation and restoration of heritage buildings, the conservation of forests and landscapes, the protection of wilderness and the promotion of environmental planning. Pollution control involves the control of all wastes from moving and stationary sources to air, water or land, and includes the control of noise and environmentally hazardous chemicals. The state of Victoria, Australia, has developed under its Environment Protection Act the creation of State Environment Protection Policies.

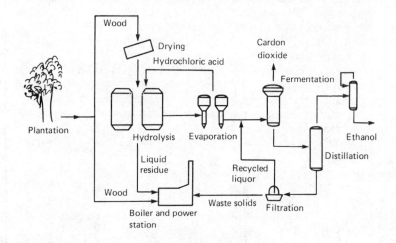

Figure E.3 Flowsheet for ethanol production

Epidemiology A branch of medical science concerned with the study of the environmental, personal and other factors that determine the incidence of disease. For example, epidemiological studies on large groups have shown that the prevalence of bronchitis is closely associated with air pollution, and that mortality and morbidity varies closely with changes in smoke and sulphur dioxide. A deterioration in health of a group of 1000 bronchitic patients was observed when smoke concentrations rose above 300 $\mu g/m^3$ and sulphur dioxide above 600 $\mu g/m^3$ (0.21 ppm). However, sulphur dioxide cannot be specifically incriminated, since the concentrations of most pollutants rise and fall together; it is perhaps best regarded as an indicator.

Equity crude Crude belonging to the oil companies operating within the territory of the *Organisation of Petroleum Exporting Countries*, which was originally the basis on which tax and royalty payments were made. *See posted prices.*

140

Equivalent free-falling diameter The diameter of a sphere which has the same density and the same free-falling velocity in any given fluid as the particle under consideration.

Erg The unit of work or energy in the *metric system* of units. It is equal to the work done when the point of operation of a force of 1 *dyne* is allowed to move 1 centimetre in the direction of the force. *See joule*.

Error curves A method, developed by Tromp, for assessing coal washing performance. He observed that the shape of the curves derived from *float and sink test* data resembled Gaussian error distribution curves. By using these curves he could demonstrate the difference between theoretical and practical results.

Ethanol Or ethyl alcohol; derived from the fermentation of starches and sugars obtained from energy crops, including wheat, barley, rye, pearl millet, grain sorghum, sugar cane, cassava and sugar beet. Cars may run on a blend of 80 per cent petrol or gasoline and 20 per cent ethanol without significant engine modification. Large-scale production would require complexes, each comprising a crop plantation with a processing plant to convert the material to ethanol. A production process is outlined in *Figure E.3*. Also known as 'solar ethanol'.

Ethylene A compound containing two carbon and four hydrogen atoms; the basic building block of the petrochemical industry, used for making ethylene derivatives, including polyethylene, vinyl chloride, ethylene oxide and vinyl acetate.

Eurocom A union of European coal merchants.

European Atomic Energy Community (Euratom) A European community formed to exploit the peaceful uses of atomic energy. It became into being at the beginning of 1958. Its members are the ten members of the European Economic Community (the 'Common Market')—Belgium, France, Germany, Italy, Luxembourg, the Netherlands, the United Kingdom, Ireland, Denmark and Greece. Euratom co-ordinates the activities in the nuclear field of its members by providing common research facilities and a nuclear supply agency, and by establishing basic standards for health protection and safety. Its main institutions are a Commission and a Council of Ministers representing member states. Agreements for co-operation in research and construction have been signed with the United States and Canada.

European Coal and Steel Community A Community for the creation of a common market for coal and steel established by a Treaty, signed in Paris on behalf of Belgium, France, the German Federal Republic, Italy, Luxembourg and the Netherlands ('the Member Countries') on 18 April 1951, and ratified in all of them with effect from 23 July 1952. The Treaty is for a period of 50 years. The creation of the Community, to which other European countries may be admitted, was the first step taken by the member countries towards European integration, and facilitated the establishment in 1958 of the European Economic Community ('the Common Market') for all products other than coal, steel and nuclear energy, and of the European Atomic Energy Community ('Euratom'). Each of these three communities constitutes a separate entity.

The powers of the Community in respect of the coal and steel enterprises under its jurisdiction were initially vested in the High Authority, subject in certain stipulated circumstances to the concurrence of a Special Council of Ministers and subject to a right of appeal to the Court of Justice of the European Communities ('the Court'). The decisions of the High Authority under the Treaty were binding on the member countries. Twenty-four countries, among them the United States of America, have established diplomatic relations with the European Coal and Steel Community. *See European Commission.*

European Commission   A single European Commission which in July 1967 replaced the Coal and Steel High Authority, the Euratom Commission and the Common Market Commission; thus, the executives of the three European communities were merged. The functions of the Commission are to prepare and submit new policies to the Council of Ministers, issue regulations for the community as a whole, police the customs union, administer the Common Market farm policy, and act as a European monopolies commission. In 1974 the Commission proposed that the Council of Ministers approve the main lines of a research and development programme entitled 'Energy for Europe'. *See European Economic Community (EEC).*

European Economic Community (EEC)   Popularly known as the European Common Market, a free trade organisation created initially by the Treaty of Rome, 1957. Its members were originally France, West Germany, Italy, Belgium, the Netherlands and Luxembourg. In 1973 the United Kingdom, Ireland and Denmark joined the Community, while Greece became the tenth member of the Community in January 1981. The purpose of the Treaty was to permit goods to travel freely, without custom duties or quota restrictions, throughout the area of the Common Market, permitting manufacturers to invest on a scale made possible by modern technology; it also sought to encourage mobility of labour. As members of the European Economic Community, nations belong also to two other communities: the *European Coal and Steel Community* and the *European Atomic Energy Community*. The European Economic community has had a role in respect of energy matters, evolving policies which are recommended to the

142

member nations. In 1978, the *European Commission* recommended that EEC investment in energy saving and alternative sources of energy be increased from $US 23 million in 1975–78 to about $US 75 million in 1979–83.

European Nuclear Energy Agency (ENEA)   An organisation set up in December 1957, as part of the *Organisation for European Economic Cooperation*, to develop collaboration between the countries of Western Europe in the use of nuclear energy for peaceful purposes. There are 18 members. The United States, Canada and Japan are associate members. The Agency is concerned with the establishment of a uniform regulating and administrative atomic régime in Europe, especially in relation to health and safety, nuclear liability and the transport of radioactive materials. It also studies the economic aspects of nuclear energy and its place in Europe's overall energy balance sheet. The Agency promotes scientific and technological co-operation between members in the field of nuclear energy. Three major joint undertakings created by the ENEA are the Eurochemic Company at Mol (Belgium), which reprocesses irradiated fuel; the Halden boiling heavy water reactor project in Norway; and the Dragon high-temperature gas-cooled reactor in Britain (at Winfrith).

European Parliament   A parliament which met for the first time in Strasbourg in March 1958, a year after the Treaty of Rome was signed; the body was essentially a forum, replacing the 'common assembly' of the European Coal and Steel Community created in 1952. In June 1979 the old appointed parliament was replaced by a new directly elected parliament. It is likely to become at least a 'voice' in European Economic Community energy issues.

Eutectic mixture   That mixture of two or more substances which has the lowest possible freezing point.

Evaporation   The conversion of a liquid into vapour, without necessarily reaching boiling point. The fastest-moving molecules escape from the surface of a liquid during evaporation, the average kinetic *energy* of the remaining molecules being reduced; in consequence, evaporation causes cooling.

Evaporative capacity   The quantity of steam produced by a boiler expressed in kg/s.

Evaporator   A device for the concentration of solutions or for the preparation of distilled water. Heating steam may pass through a coiled tube surrounded by liquid (weir type) or may surround a bank of tubes through which cold liquid flows (Calandria, Kestner and climbing film types). Evaporators fall into two categories: (a) single-effect evaporators, in which steam arising in the evaporation process is simply condensed; (b) multiple-effect evaporators, in which steam produced in the initial stage is used as a heating medium in a second evaporating unit, additional 'effects' being added until the useful range of heat in the steam is exhausted (hence, double- or triple-effect evaporators).

Everclean Window   A device developed by the Central Electricity Research Laboratories, Leatherhead, Surrey, England, for keeping the windows of smoke-indicating equipment clean. It comprises a honeycomb unit of experimentally determined proportions into which dust will not penetrate. *See smoke density indicator*.

Exajoule   A unit of heat, equivalent to 10^{18} J, or about 160 million barrels of oil.

Excess air   Combustion air supplied in excess of the theoretical or stoichiometric air required for combustion, in order to ensure complete combustion under practical conditions. It is usually expressed as a percentage of the *theoretical air*; hence,

$$\text{excess air, } \% = \frac{(W_a - W_t)\,100}{W_t}$$

where W_a = amount of air in kilograms actually supplied per kilogram of fuel;

W_t = theoretical or stoichiometric air in kilograms.

Excess air may be calculated from a knowledge of oxygen content of the waste gases, assuming no combustibles, as follows:

$$\text{excess air, } \% = \frac{100\,O_2}{21 - O_2} K$$

where K = 0.96 for bituminous coal;
0.95 for oil;
0.90 for natural gas.

Excitation energy   *See energy levels*.

Exhauster fan   Fan used in conjunction with a suction-type pulverised fuel mill to provide carrying air for the pulverised coal.

Exmouth Plateau   An extensive geological area about 500 kilometres from the north-west coast of Australia, which has been the subject of intensive investigation for oil supplies. The search represents one of the most intensive deep-water undertakings anywhere in the world, with companies operating at the frontiers of current technology. Drilling of the first well was completed in May 1979.

Exothermic gas   Gas burned with less than the stoichiometric amount of air to produce an exothermic gas; *town gas*, or *liquefied petroleum gases*, may be used for this purpose. Exothermic gas may be 'lean' or 'rich', according to the greater or smaller amounts of air used, respectively. This gas has a wide application in **heat treatment** processes, e.g. bright heat treatment of low-carbon steels and bright **annealing** of copper. *See endothermic gas.*

Exothermic reaction A chemical reaction accompanied by the evolution of heat. This occurs during combustion processes.

Exploration effort, characteristics of Characteristics which vary from country to country, and region to region, yet generally have a marked influence on the known existence of reserves and resources; there is often less diseconomy in establishing reserves and resources of coal ahead of need, than in respect of oil. Some characteristics of exploration effort in Australia are shown in *Table E.4.*

Table E.4 CHARACTERISTICS OF EXPLORATION EFFORT IN AUSTRALIA

Characteristic	Coal	Petroleum and natural gas
Exploring organisation	Government authorities	Private enterprise, employing risk capital
Size of reserves	Seams extend over a large area	Relatively small reservoirs
Depth of reserves	Generally less than 650 metres	Commonly 2000–3000 metres
Cost of drilling per equivalent energy quantity	Moderate	Very high particularly if off-shore
Diseconomy in establishing reserves and resources well ahead of time of use	Small	Very large

Explosion A violent increase in pressure which may accompany the self-acceleration of a reaction; an explosion associated with the release of heat is often called a thermal explosion. High explosives involve the maximum energy release within a minimum period of time, involving detonation rather than combustion waves. Low-order explosives are used as propellants; deflagration occurs and energy is released much more slowly.

External economy A fall in the cost of any of the materials and services which a firm requires, which are obtained from outside sources. These materials and services comprise raw materials, labour, fuel and power, transport services, and the services of specialised firms and selling agencies. Where large quantities of, say, fuel and ores are used, then any change in their prices may have a marked effect on costs of production without any change in the internal economy of a firm. External economies for society as a whole are achieved if the production of goods and services yields fortuitous benefits in the areas of education, research, health or environment.

145

External effects Or externalities; social costs and benefits caused by the activities of an industry which are not reflected in the price at which the product is sold, or do not influence the quantities purchased; costs not borne by those who occasion them, and benefits not paid for by the recipients. In the generation and supply of electricity, two externalities are: (a) those costs to the community of grit emission from old power stations; (b) those benefits to the community of replacing dirty fuels with clean electricity, thus reducing air pollution in cities at the point of use. The cost or benefit is external to the industry and does not find its way into the price charged the ultimate consumer; thus, the electrical energy market, among others, is imperfect in giving the consumer incorrect information about the cost of resources used to produce power. The economist Professor R. Turvey has suggested that externalities should be 'internalised' if they are known to have a significant effect on the demand or cost structure of a product, i.e. corrections should be made to allow for them when calculating marginal cost. Marginal cost thus becomes a social opportunity cost, or true cost.

Extraneous ash *Ash* arising from that part of the *mineral matter* associated with, but not inherent in, *coal*.

F

Fahrenheit scale A temperature scale based on three fixed points: (a) the lowest (0°), being the temperature of a mixture of ice, water and sea-salt; (b) the second (32°), being the freezing point of water; (c) the highest (212°), being the boiling point of water at normal atmospheric pressure. The Fahrenheit and Centigrade (Celsius) scales are conventional scales: $32° F = 0° C$ and $212° F = 100° C$, where 1.8 Fahrenheit degrees = 1 Centigrade (Celsius) degree. *See temperature scales.*

Fairweather calorimeter A continuous recording *gas calorimeter*; it is a modified *Boys calorimeter*, the temperature rise of the water through the calorimeter being recorded electrically in terms of *calorific value*.

Fan A pressure-producing device; fans are rated at total water gauge, i.e. actual pressure of 'static head', plus pressure equivalent to velocity or 'kinetic head'. *See axial flow fan; backward-curved fan blading; centrifugal fan; forced draught fan; forward-curved fan blading; induced draught fan; radial-tip fan blading.*

Fan collector An *induced draught fan* and *dust arrester* combined; the dust particles are thrown to the periphery of the stream by the fan and are skimmed off with a proportion of the gas into a *cyclone* for separation and collection. The clean gas returning up the centre of the cyclone body to the outlet scroll is led back to the fan inlet.

Fanning The behaviour of a chimney plume when the air is very stable, the gases quickly reaching their equilibrium level and travelling horizontally with sideways meanderings but with very little dilution in the vertical direction. Fanning produces a thin but concentrated layer of pollution which may impinge on hillsides or tall buildings. *See fumigation.*

Farad A unit of electrical capacitance; it is the capacitance of a capacitor between the plates of which there appears a difference of potential of 1 volt when it is charged by a quantity of electricity equal to 1 *coulomb*.

Faraday F A quantity of electricity associated with 1 mole of chemical charge, i.e. 96 487 coulombs.

Fast breeder reactor A type of nuclear reactor in which the number of fissile nuclei produced (or bred) exceeds the number of fissile nuclei concurrently destroyed; they have no moderator and consume plutonium-239 and uranium-235. The major incentive for fast breeder reactor development lies in expected fuel economies of a substantial order; fast breeder reactors, it is anticipated, will use about 70 per cent of the potential energy available in uranium fuel in contrast to the 1 or 2 per cent used in present reactor types. In the United States of America and the United Kingdom most of the development effort has been devoted to the fast breeder reactor. The United Kingdom has developed a sodium-cooled fast breeder reactor designed to burn the plutonium being produced as a by-product in magnox and AGR power stations. Following success with a 30 MW fast breeder reactor at Dounreay, Scotland, a 250 MW prototype has been commissioned. In 1977 President Carter announced that the United States fast breeder programme would henceforth give priority to alternative designs of the breeder other than plutonium. The effect was to reduce the priority given to the Clinch River, Tennessee, breeder reactor project, previously the centrepiece of the US breeder programme. The reason for this was the result of growing apprehension about plutonium and its ultimate disposal, or diversion into weapons, and the need to retard the progress of the world towards a 'plutonium economy'. France has been developing and constructing the first commercial fast breeder reactor of 1200 MW, the Super Phénix, near Grenoble. *See Carter energy policy; Dounreay fast reactor; plutonium; plutonium economy; super Phénix.*

FCCU feed pretreater An oil refinery unit for pre-refining the feed stock charged to a *fluid catalytic cracking unit (FCCU)*. The unit usually involves hydrogenation, which practically eliminates sulphur and metallic compounds and reduces nitrogen.

Federal Energy Administration (US) *See Department of Energy (US)*.

Federal Energy Regulatory Commission (US) *See Department of Energy (US)*.

Federal Power Commission (US) *See Department of Energy (US)*.

Feedback control system *See closed-loop control system*.

Feedback controller A device which measures the value of a controlled variable in a control system, makes a comparison with a standard representing the desired performance, and as necessary manipulates the controlled system in order to maintain the required relationships. *See closed-shop control system; open-loop control system*.

Feed preparation units Oil refining units which redistil long residuum from the primary crude distillation units, giving some 50 per cent heavy distillate feed for the catalytic cracker and 50 per cent of short residuum worked up later for bitumen or blended with other stocks to produce fuel oil. The inlet temperature in the columns approaches 400° C and the process is under vacuum. The vacuum is obtained by multi-stage steam ejectors and any gases produced are burnt in plant heaters.

Feed water Water suitable for feeding to a boiler.

Feed-water accumulator A system in which a supply of boiler feed water is 'topped up' by surplus steam not required elsewhere in a plant. By using surplus steam in this way, a reserve of preheated feed water is built up, enabling peak loads of 25–30 per cent in excess of average demand to be met.

Feed-water check valve A non-return valve to prevent water from escaping from a boiler should the pressure fall in the feed-water supply pipe. Unless a combined stop valve–non-return valve is used, a stop valve must be provided between the non-return valve and the boiler; this enables the non-return valve to be repaired or replaced while the boiler is in operation.

Feed-water heater A heat exchanger for preheating feed water before it passes to the boiler. *See economiser*.

Feed-water meter A device for measuring the supply of feed water to a boiler; it may consist of a *V-notch meter*, an *inferential meter* or a *positive displacement meter*.

Feed-water regulator A feed-water control valve automatically maintaining a constant water level in the boiler drum.

Ferric oxide Fe_2O_3. An oxide of iron emitted as a dense reddish-brown smoke during the 'after-blow' from a *Bessemer converter*, and during oxygen lancing in steel-making processes.

Ferricyanide process A wet scrubbing process for the removal of

148

hydrogen sulphide from refinery and petroleum oil gas streams. The scrubbing medium is sodium ferricyanide, $Na_3Fe(CN)_6$, the hydrogen sulphide forming ferrocyanide; the ferrocyanide is oxidised, precipitating sulphur. *See hydrogen sulphide removal.*

Ferrous metals Metals in which iron is the main constituent. *See Cast-iron; steel.*

Fertile As applied to a *nuclide*, the capability of absorbing, or capturing, a neutron and subsequently being transformed into a fissile nuclide. Uranium-238 and thorium-232 are important fertile isotopes, giving rise to plutonium-239 and uranium-233, respectively.

Fetoxicity The ability of a substance to produce toxic manifestations in a fetus.

Fibrous refractory insulation Refractory insulating materials, such as alumina and magnesia, which have been formed into fibres as a consequence of a melting and steam-blowing process. The fibres are up to 250 mm long and interlocked; the product is in the form of blankets or blocks. The blanket material may be used as *hot face insulation*.

Fick's Law A formula giving the molar rate of *mass transfer*, per unit area of a component A in a mixture of A and B:

$$N_a = -D_{AB}\frac{dC_A}{dy}$$

where, N_a = molar rate of diffusion per unit area;
D_{AB} = diffusivity of A in B;
C_A = molar concentration of A;
y = distance in direction of diffusion.

With turbulent motion, eddy diffusion taking place in addition to molecular diffusion, the formula becomes:

$$N_a = -(D_{AB}+E_D)\frac{dC_A}{dy}$$

Field gases Natural gas as produced, before treatment or extraction of any of its constituents.

Filament lamp A conventional electric light bulb, containing a fine coiled thread of tungsten which becomes incandescent when an electric current passes through it; the glass envelope of the lamp holds an inert gas such as argon or nitrogen to prevent premature failure of the filament through oxidation or evaporation. *See electronic Halarc.*

Film badge A piece of wrapped photographic film, often with parts shielded by filters against certain types of radiation, worn by workers liable to be exposed to nuclear radiation. Examination of the film after photographic development enables both the radiation dose and to some extent the type of radiation received by the worker to be determined.

Filter ratio The number of cubic metres of gas that will pass through 1 square metre of filter surface per minute at a given filter resistance; the gas-to-cloth ratio.

Filter resistance The pressure drop across a filtering surface expressed in millimetres of mercury.

Final steam temperature Steam temperature at the main steam stop valve of a boiler.

Fines Very fine coal particles, usually less than 3 mm.

Fire point The temperature of a liquid fuel that will vaporise enough oil to support continuous combustion; it is generally about 10° C higher than the *flash point*.

Firebrick A *refractory* consisting essentially of aluminosilicates and silica, comprising less than 78 per cent of silica and less than 38 per cent of alumina.

Firebridge A firebrick wall or barrier in a boiler furnace tube to prevent coal from being thrown over the back of the grate into the furnace tube and to prevent air from by-passing the grate at the back. The firebridge is 20 cm thick, and the distance from the top of the bridge to the crown of the flue varies from 24 to 36 cm, depending on the diameter of the tube.

Firebridge bearer Cast-iron support bolted to brackets fixed in the furnace plates of a boiler furnace tube; the end of the bearer is bevelled to suit the firebars.

Firing Or stoking; the process of feeding *fuel* to a furnace or combustion chamber.

Firing by hand Methods of hand-firing boilers and furnaces with solid fuel, three of which are in use: *spreader or sprinkler firing; coking or dead plate firing; alternate side- or wing-firing.*

Firing tools Tools used by a fireman in tending a furnace; they include (a) shovel, not larger than size 6; (b) poker, or pointed bar; (c) rake, for levelling the fire; (d) hoe, for removing clinker; (e) slice bar, for separating clinker from the grate; (f) pricker bar, for cleaning the spaces between the firebars from below.

Fischer–Tropsch synthesis A process in which synthesis gas at about 230–330° C and at pressures up to 50 bar is brought into contact with a suitable prepared catalyst (iron oxide, cobalt, nickel or ruthenium) to produce liquid fuels, combustible gas, waxes or organic chemicals. The Fischer–Tropsch process has not been a success economically in most countries. *See gas synthesis.*

Fishtail burner A gas burner in which gas is emitted from a slot in the blunt end of the tube. Air mixes with the outer fringes of the gas, the resulting combustion heating the gas which has not been

150

exposed to air; the effect of heating is to thermally decompose or crack the gas to carbon and hydrogen. While the hydrogen burns with no visible flame, the carbon particles become incandescent and produce a yellow luminescence. If combustion is incomplete, soot and carbon black are formed. *See bunsen burner.*

Fission product poisons Fission products which have a large capture cross-section for neutrons; the effect of absorbing neutrons unproductively causes a decrease in the reactivity of the nuclear reactor.

Fission products Elements that result from *nuclear fission*; in addition to uranium and plutonium, these may consist of more than 40 different radioactive elements, e.g. arsenic, barium, cadmium, cerium, iodine, silver and tin.

Fixed-bed gasifier A gas generator or producer in which a column of close-packed fuel is gasified.

Fixed carbon Carbon which does not pass off in *volatile matter* (tarry matter and gases) when coal is heated, remaining in the coke. In a *proximate analysis* the percentage of fixed carbon is found by adding the percentages of moisture, ash and volatile matter together and subtracting this total from 100.

Fixed instalment method A method for calculating and allowing for the depreciation of an asset. The fixed instalment, or straight-line, method distributes the cost of an asset uniformly over its depreciable life. The amount to be set aside each year may be calculated from the formula:

$$\frac{P-L}{n}$$

where P is the initial asset cost, L the expected salvage value at the end of the useful life of the asset and n the plant life expressed in years. *See depreciation; depreciation fund method; diminishing balance method; sinking fund.*

Flame A chemical interaction between gases, accompanied by the evolution of light and heat. *See cool flame; diffusion flame; neutral flame; sensitive flame.*

Flame impingement The impingement of a flame or atomised oil against the side of a combustion chamber; this may lead to the formation of carbon deposits and smoke.

Flame speed The speed of propagation of a flame under certain prescribed conditions. If the flame speed is higher than the velocity of the gas at the burner. back-firing may occur; if it is slower, the flame will be extinguished.

Flame stability The degree to which a flame maintains its correct position in relation to the burner; if a flame does not maintain its correct position, it is said to be unstable.

Flame temperature A property of *flame* which depends on the calorific value of the gases, the latent heat in steam, the heat losses due to radiation and the dissociation of gaseous molecules, and the volume and specific heat of the total gaseous products. It may be represented:

$$T_f = \frac{Q}{V}$$

where T_f = flame temperature;
Q = net calorific value of gases + sensible heat of gas and air − heat loss by radiation and dissociation;
V = volume of products × specific heat at constant pressure.

Flame traps Devices which are used to prevent flame propagation; a traditional example has been the miner's safety lamp, in which the flame is surrounded by a fine copper gauze. The flame will respond to concentrations of combustible gas, without igniting it beyond the gauze.

Flame zone In the combustion of solid fuels, a zone in which volatile matter is burning after being ignited in the *incandescent zone*. See *distillation zone*.

Flash point The lowest temperature at which a product gives off just sufficient vapour to form an inflammable mixture with air under the conditions of a standard test. It is usually a statutory requirement that flash point shall not be lower than 66° C (150° F) in respect of various types of fuel oil; in the case of inflammable products with flash points of lower than 66° C (150° F) special regulations relating to safety in storage and handling must be complied with. A flash point may be 'open' or 'closed', depending upon whether the test apparatus is used with a cover or not; the 'open' flash point is several degrees higher than the 'closed'. See *Abel flash point apparatus; open flash point; Pensky–Martens flash point apparatus*.

Flash steam Steam produced from water when the pressure is suddenly reduced.

Flasks Fifty-tonne containers for carrying irradiated nuclear fuel elements; each flask contains some $2-2\frac{1}{2}$ tonnes of used fuel elements which are transported to chemical separation plant.

Flat plate collectors Solar energy collectors suitable for temperatures up to about 85° C. They employ water as the heat transfer fluid. The simplicity of the fixed collector array has led to the development, both in the United States of America and in Europe, of evacuated tubular collectors capable of operating efficiently at temperatures as high as 150° C, and therefore suitable for the generation of low-pressure steam. In Australia the most cost-effective use of solar energy is in applications which require heat throughout the year at temperatures up to about 85° C; the simpler and cheaper flat plate collector is most suitable for these purposes. Following the devastating cyclone which destroyed Darwin in December 1974, many of the new houses constructed have solar collectors in the roofs for the provision of hot water. *See solar energy*.

Flat rate tariff A *tariff* for the supply of electricity consisting solely of a single unit charge; used chiefly by very small users of electricity for lighting and power.

Floating roof A special roof which floats upon the liquid in a storage tank, thus eliminating vapour space irrespective of the level of the liquid. If the advantages of floating roofs are to be fully realised, there must be effective sealing between the tank wall and the roof. Seals may consist of polyurethane foam or synthetic rubber tubing filled with paraffin or light fuel oil, or be of a mechanical nature, comprising perhaps a ring of steel shoes pressed firmly against the shell by springs. Floating roofs are generally considered not necessary for kerosine, gas oil, fuel oil and lubricating oils. They are necessary, however, for all fractions lighter than kerosine and for crude oil; it is from these materials that a significant contribution to oil refinery air pollution would otherwise arise.

Float and sink test A test for determining the possibility of efficiently cleaning a coal by a gravity separation process; a prepared sample of coal is suspended in a series of liquids of increasing density from 1.3 to 1.8 by increments of 0.1. The percentages, and the ash contents, of the floats are determined at each stage, the results being recorded graphically as 'washability curves'. *See error curves*.

Flowmeter A measuring instrument utilising an orifice plate, *venturi tube* or flow nozzle, to show the rate of flow in a pipeline.

Flue gas The gaseous products of *combustion* from a furnace consisting predominantly of nitrogen and carbon dioxide, with some moisture, oxygen and oxides of sulphur; carbon monoxide and oxides of nitrogen may also be present, together with particulate matter. *See Orsat apparatus*.

Fluid-bed operation A type of oil refinery operation based on the tendency of finely divided powders to settle in a gas stream of low velocity; the fluidised-powder technique involves suspending the finely divided powder in an upward-flowing stream of gas. This suspension has most of the characteristics associated with a true liquid, and it can flow through pipes and valves.

153

Fluid catalytic cracking unit (FCCU) An oil refinery unit which enables more gasoline to be obtained from the crude oil than is 'naturally' present. In this unit vaporised oil, air and catalyst of clay-like material (alumina-silica microspheroids) are circulated at high temperatures through a complex system of pipes and chambers. Cracking takes place in a reactor; the cracked gases pass to a fractionating tower for distillation and are fractionated into gas, gasoline, and light and heavy cracked gas oils. The catalyst passes from the reactor to a regenerator, where air is used to burn off carbonaceous matter deposited on the catalyst; once the carbon is burned off, the catalyst is re-activated and may be used repeatedly. Catalytic cracking has made it possible to produce more than twice as much gasoline from a barrel of crude as can be made by simple distillation. *See Figure F.1*: *thermal cracker*.

Fluid-Coke process An oil refinery process in which hot residual oil is sprayed on to externally heated seed coke in a fluid bed; the fluid coke is removed as small particles. *See delayed-coking process; petroleum coke.*

Fluidity The reciprocal of viscosity. In the *metric system* the unit is known as the 'rhe'; it equals 1/poise.

Fluidised bed A bed of solid particles through which air or gas is blown upwards so that the bed assumes the properties of a fluid.

Fluidised combustion A system of combustion in which the fuel particles are supported by a stream of air and are in rapid motion relative to one another; the bed has the appearance of a boiling liquid and shows many fluid-like properties. The combustor vessel contains a bed of granular material, such as coal ash, sand or limestone. Air is blown through a distributor in the base so as to 'fluidise' the particles. Fluidised combustion offers advantages such as: low combustion temperatures, 800–950° C; high heat transfer coefficients, with a consequent reduction in heat transfer area; high heat generation rates, resulting in smaller combustion chamber volumes; reduced fireside fouling, corrosion and erosion. *See Figure F.2; Ignifluid grate.*

Fluidised gasification of coal *See Winkler system.*

Fluorescein A liquid dye suitable, for example, for detecting condenser tube leaks.

Fluorescent lamp A tubular *discharge lamp* internally coated with a powder which fluoresces under the action of the discharge, producing a shadowless white or coloured light. Fluorescent lights achieve nearly 80 lumens per watt, compared with 14 lumens per watt from an ordinary incandescent bulb.

Fluorine F. An *element* present in rocks and soils and, hence, in fuels, fluxes and raw materials used in industry; at. no. 9, at. wt.

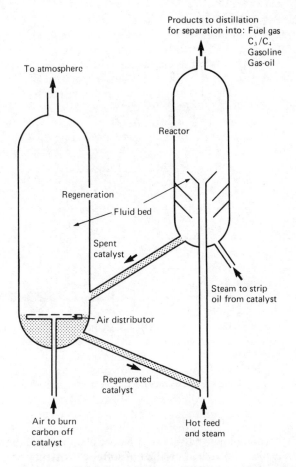

Figure F.1 Catalytic cracking using fluidisation techniques.
(Source: British Petroleum)

18.9984. Fluorides occur in British coals in concentrations ranging up to 175 ppm; the coals containing the higher proportion of fluorine are found in Kent, Staffordshire and South Wales. Brickworks and other branches of the ceramic industry emit fluorides. Fluorine is also emitted in steelmaking, in which fluorspar is used as a flux. Fluorine is an extremely active element; the use of the word 'fluorine' usually means some compound of this element.

Flux In nuclear reactor physics, the number of particles crossing unit area per second, the unit area being at right angles to the path of the particles.

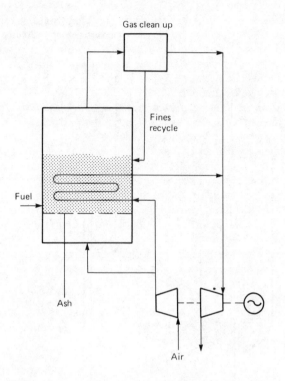

Figure F.2   Fluidised bed combustor with open cycle gas turbine

Flux oil   An oil of low volatility suitable for blending with bitumen, or with asphalt, to yield a product of softer consistency or greater fluidity.

Fly ash   Non-combustible particles suspended in flue gas. *See pulverised fuel ash.*

Food and Agriculture Organisation (FAO)   An agency of the United Nations, the essential function of which is to combat the poverty, malnutrition and hunger which afflict about half the people of the world. One of the concerns of the Organisation is with the quality of food in relation to diets and energy values. The FAO Reference Man weighs 65 kg, is aged 25 years and is moderately active; he is estimated to require an average 3200 cal (13.4 MJ) intake daily. The FAO Reference Woman weighs 55 kg, is also aged 25 years and is moderately active; she is estimated to require an average 2300 cal (9.6 MJ) intake daily. *See energy expenditure, human; energy value.*

Foot-candle The unit of intensity of illumination; it is the illumination produced by 1 candle power at a distance of 1 foot.

Forced draught The supply of combustion air to a furnace at a pressure greater than that of the atmosphere, utilising a fan. The forced draught pressure required for stokers ranges from 25 to 100 mm wg *See draught*.

Forced draught fan A *fan* through which air is drawn before entering the furnace. Air may be delivered below the furnace grate, or above and below simultaneously.

Force majeure Forces beyond the control of a company preventing it from fulfilling a contract; a company unable to deliver in these circumstances may declare *force majeure*. For example: 'Queensland Mines, the focal point of the Australian uranium share boom, yesterday declared *force majeure* on its first uranium supply contract. This means it is unable to deliver.' (*Business Age*, Melbourne, 24 March, 1977).

Forward-curved fan blading A design of blading for centrifugal fans which is considered one of the most suitable for forced draught fans; its point of maximum efficiency is at the load that it is designed to carry, and it has no reserve of pressure with which to overcome dirty boiler conditions. Initial cost is lower but running costs are slightly higher compared with *backward-curved fan blading*. *See Figure F.3; fan.*

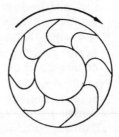

Figure F.3 Forward-curved fan blading

Fossil fuels A general term embracing coal, oil, and natural gas, which are fuels derived from organic deposits laid down in past geological periods. Such fuels comprise essentially carbon and hydrocarbons. Nuclear fuels and wood are not fossil fuels.

Four-wire distribution A system of distribution employed on three-phase ac systems; it consists of three 'phase' wires and one neutral wire.

Fractional distillation The separation of the components of a liquid mixture by vaporising the mixture and collecting the fractions which condense in different temperature ranges. *See distillation tower.*

Fractionating tower *See distillation tower.*

Free-burning coal *Coal* which does not cake in the fuel bed. *See caking coal.*

'Free-carbon' A loose description for microscopic particles of resinous material present in creosote-pitch. At normal storage temperatures of 27–32° C these remain dispersed; if overheating occurs in certain conditions, the particles tend to form a deposit. *See coal tar fuels.*

Free-falling velocity The rate of fall of a particle through a still fluid.

Free goods In economics, those attributes of the natural world which are valued by society but are not in individual ownership and do not enter into the processes of market exchange and the price system. Notable among such resources are the atmosphere, watercourses, ecological systems and the visual properties of the landscape. In reality in developed economies they are common property resources of great and increasing value, presenting society with important and difficult allocational problems, which exchange in private markets cannot resolve.

Free moisture Moisture in a solid fuel which can be removed by evaporation or by centrifugal force. Free moisture is acquired when coal is washed, and is that moisture which remains after normal draining. Free moisture within certain limits improves the fuel bed by lowering its resistance to the passage of air and stimulates combustion. *See inherent moisture.*

Free-swelling index A measure of the behaviour of coal when heated rapidly; it may be used as an indication of the coking characteristic of a coal when burned as a fuel.

Freons A range of chemicals which have proved particularly useful as refrigerants because of their non-explosive characteristics; they consist mainly of fluorine (and other halogen) derivatives of hydrocarbons.

Frequency The cycles produced per second by an alternating current electricity supply. *See alternating current.*

Friability test A test to measure the tendency of a coal to break during handling; it is determined by the use of a standard tumbler.

Frigorie A term for the 15° C kilocalorie when used in a negative sense for the extraction of heat. Abbreviation, fg.

'From and at 100° C' A basis for comparing the evaporative capacities of boilers. The total heat in steam depends upon both the pressure and the temperature of the steam, and boilers are assumed for comparative purposes to generate steam at normal atmospheric pressure i.e. from water at 100° C to steam at the same temperature. In respect of any boiler, an 'equivalent evaporation' may be calculated as follows:

$$E = \frac{W(T-t)}{L}$$

where, E = equivalent evaporation 'f. and a. 100° C';
$\quad$ W = weight of water evaporated, kg;
$\quad$ T = total heat in steam per kg, kJ;
$\quad$ t = heat in feed water per pound, from 0° C, kJ;
$\quad$ L = latent heat of steam at 100° C and normal atmospheric pressure.

Front A relatively narrow zone of transition between air masses. If a cold air mass is displacing a warm one, the transition is called a cold front, and vice versa. Fronts are associated usually with low-pressure areas and pressure troughs.

Froth flotation A wet cleaning process for separating 'dirt' in the form of sandstone, shale, clay, pyrites, calcite, and so on, from *run-of-mine coal*. In this process, fine coal and water are beaten up by an impeller with a 'frothing agent' such as creosote. Clean coal floats to the surface, while dirt sinks to the bottom of the flotation cell. The froth is separated from the fine coal in a vacuum filter.

Fuel A substance used to produce heat, light or power, usually by *combustion* in air or oxygen. Most natural or primary fuels such as coal, wood, peat, oil and natural gas are made up of compounds of carbon, hydrogen and oxygen, in association generally with mineral ash, moisture, sulphur and nitrogen. *Nuclear fuel* stands in sharp contrast by virtue of its nature and mode of heat release. *See fuels, primary and secondary.*

Fuel Abstracts and Current Titles (FACTS) A monthly publication by the UK Institute of Energy which surveys world scientific and technical literature on the production, utilisation and by-products of solid, liquid, gaseous and nuclear fuels; power; and related materials and fundamental sciences. FACTS carries a separate energy conservation section.

Fuel/air flow ratio A relationship which has been employed as a means of combustion control in gas- and liquid-fuel-fired steam boilers.

Fuel cell An electrochemical cell which operates by utilising the energy of a spontaneous chemical reaction, e.g. the combustion of a carbonaceous, hydrogen or hydrocarbon fuel by oxygen from the air. The reactants are fed to the cell at rates proportional to the amount of electrical energy which is required. The electrodes used in fuel cells are generally gas-adsorption electrodes. The only product of combustion is water. The United Technologies

Corporation has a 1 MW fuel cell linked to the American Grid; in 1979 a 4.5 MW fuel cell was under construction in New York. *See cell; electric battery.*

Fuel economics   The study and analysis of the whole pattern of forces determining the supply of, and demand for, fuel and energy of all kinds.

Fuel economy   The theory and practice of efficient fuel utilisation.

Fuel efficiency   The proportion of the potential heat of a *fuel* which is converted into a useful form of energy.

Fuel element   An assembly consisting of fissile material (for example, natural or enriched uranium) contained in a can; the latter often has fins to improve the heat transfer between the element and *coolant*.

Fuel injection equipment   In respect of the *diesel engine*, equipment to inject fuel oil into the cylinders; injection begins during the compression stroke (at 10–20° of crank angle before top dead centre) and continues during the power stroke. The equipment controls the quantity of fuel injected into each cylinder and ensures that the fuel is distributed in a suitably atomised form. Formerly achieved by using compressed air, fuel injection is today described as 'solid injection'. Injection equipment falls into two main categories, the *common rail system* and the *jerk system*.

Fuel oil   Heavy petroleum distillates or petroleum residues (or blends of these) used in furnaces for the production of heat and power. The average gross energy value for fuel oil in the United Kingdom is about 43.4 kJ/g (18 700 Btu/lb).

Fuel ratio   The ratio of *fixed carbon* to *volatile matter*; a ratio sometimes used in coal classification systems.

Fuel Research Institute   A South African research institute created by the Coal Act of 1930; it is managed by a Fuel Research Board. Research is financed in part from a levy on coal sales, and in part by the government. In addition to research, the Institute undertakes a considerable volume of repetitive testing and analysis. The organisation is divided internally into survey, chemistry and engineering departments.

Fuels, primary and secondary   Primary fuels are those forms of energy obtained directly from nature, e.g. coal, oil and natural gas. Secondary fuels are those derived from primary fuels, e.g. coke and town gas.

Fugitive leaks   Small random gas leaks to atmosphere from pipe joints, valve glands, pressure shells, and so on.

Fuidge diagram   A diagram defining the combustion characteristics of rich gases; increasing air/gas ratios are plotted against thermal inputs.

Fulham–Simon Carves process   A process for removing sulphur dioxide from flue gases using ammonia liquor as the washing medium. The process consists essentially in scrubbing the gases with ammonia liquor to produce ammonium salts which by autoclaving are converted into ammonium sulphate and sulphur. The advantage of the process is that it produces saleable products. See *Battersea gas washing process; Howden–ICI process; Reinluft process.*

Fume Airborne solid particles arising from the condensation of vapours or from chemical reactions; fume particles are generally less than 5 μm in size, respirable and visible as a cloud. They may be emitted in the following processes: (a) volatilisation, (b) sublimation, (c) distillation, (d) calcination and (e) chemical reaction.

Fumigation A rapid increase in *air pollution* at ground level caused by turbulence of the atmosphere created by a rising morning sun following a nocturnal *inversion*, in which pollutants have become concentrated aloft; very high ground level concentrations may be experienced for an hour or more, sometimes at a distance of many miles from the source of the pollution. This fumigation effect during the break-up of an inversion is due to the restoration of turbulence initially at ground level which gradually penetrates into the stable layers above which are still acting as a 'lid' inhibiting the upward dispersal of pollutants. Fumigation was originally described by E. W. Hewson in an article 'The meteorological control of atmospheric pollution by heavy industry', *Quart. J. Royal Met. Soc.*, **71**, 266–282, 1945. *See Figure F.4; fanning.*

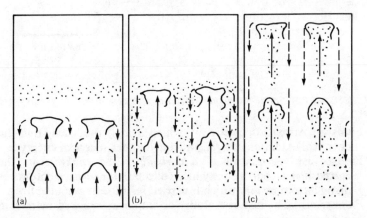

Figure F.4 Fumigation: (a) Solar heating of ground initiates mixing by up-currents; (b) Fumigation occurs as mixing involves layers of polluted air; (c) Dilution, as clean air is introduced from above (Source: Warren Spring Laboratory, Stevenage, England)

Furfural extraction A single-solvent oil refinery process in which furfural (the aldehyde C_4H_3OCHO) is used to remove aromatic, naphthenic, olefinic and unstable hydrocarbons from a lubricating oil charge stock, thereby improving the viscosity index and other characteristics. *See Figure F.5.*

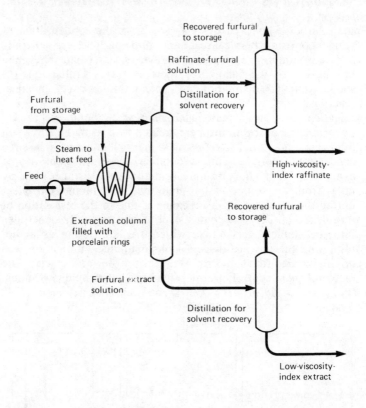

Figure F.5 Solvent extraction with furfural. (Source: British Petroleum)

Furnace An enclosed space in which heat is produced from the chemical oxidation of a *fuel* or from another source of energy.

Furnace ash A mixture of *ash, riddlings* and *clinker* from the *combustion* of coal or coke which collects in the furnace bottom, as distinct from *fly ash*, which is carried forward through the flues.

Furnace oil The heaviest grades of natural or cracked petroleum oils.

Furnace rating The maximum heat input of a furnace expressed usually as $kJ/s\ m^3$ furnace volume.

Fusain A dull, very friable, charcoal-like coal *lithotype* of silky texture, occasionally fibrous; an important constituent of coal. Known also as 'mineral charcoal'.

Fuse A safety device consisting of a few inches of relatively fine wire, mounted in a suitable holder or contained in a cartridge and connected as part of an electrical circuit. If the current exceeds a predetermined value, the fuse wire melts (i.e. the fuse 'blows') and thus obviates damage to the circuit protected by the fuse.

Fusible link A piece of low-melting-point metal forming part of a wire connected to a shut-off valve in an oil supply line. A safety device; if a fire occurs and the link melts, the valve automatically shuts.

Fusible plug A plug designed to give warning of overheating due to insufficient water in a boiler. The plug consists of an outer body of bronze or gunmetal, with a central conical passage of up to $\frac{3}{8}$ inch diameter. The passage is closed with a core secured by an annular lining of fusible alloy so that the plug may drop clear if the lining melts. The fusible metal should melt at a temperature of not less than 70° C in excess of the saturated steam temperature at the design pressure of the boiler. If the water level falls, causing undue heating, the metal melts, the core falls out and water and steam escape into the furnace. BS 759 describes fusible plugs and specifies the positions in which they should be placed.

Fusing temperature In relation to ash, the temperature at which it softens. Ashes that fuse in the range 1050–1200° C are designated 'low-fusing'; those fusing in the range 1200–1425° C as 'medium-fusing'; and those above 1425° C as 'high-fusing'. In general, ash with a low softening temperature is likely to form *clinker*.

Fusinite A powdery, black mineral charcoal; a subordinate component of *clarain*, in which it may occur as irregular bands or as disseminations in *micrinite*, but a principal component of *durain*.

Futures contract A commitment to deliver (in the case of a sold contract) or to take delivery of (in the case of a bought contract) a specified weight of a standard grade of a commodity at a fixed price and for a specified delivery date.

Futures market A market in which goods are sold for delivery at some future date—say in 3 months' time.

'Fyrite' CO$_2$ apparatus A portable instrument for measuring the *carbon dioxide* content of flue gases. A known sample of gas is introduced into the apparatus by means of a rubber bulb. When the instrument is turned upside down, the sample is bubbled through a solution of potassium hydroxide, a final reading for CO$_2$ being obtained in only 1 minute. Accuracy is within $\frac{1}{2}$ per cent of CO$_2$, the scale being graduated 0–20 per cent.

G

Galloway boiler An early type of boiler similar in outward appearances to the *Lancashire*; however, after the firebridge the two furnace flues combined into a single flue which contained water tubes set across the flue at various angles throughout the remaining length of the boiler. It was claimed that these tubes, up to 30 in number, promoted a better circulation of water while increasing heating surface and steaming power. Also known as a 'breeches-flued' boiler.

Gamma rays Electromagnetic radiation emitted by a *nucleus*, similar to X-rays but usually of shorter wavelength. Gamma rays are exceedingly penetrating.

Gas Act 1965 An Act giving the *Gas Council* a monopoly to distribute gas in Britain; thus, all North Sea natural gas produced must be offered to the Council.

Gas burner *See aerated burner; bunsen burner; diffusion flame burner; direct burner; fishtale burner; neat gas burner; nozzle-mix burner; post-aerated burner; pre-aerated burner; premix burner; proportionate control mixing gas burner; tunnel mixing burner; venturi mixing gas burner.*

Gas Calorimeter An instrument for the determination of the *calorific value of fuel gases. See bomb calorimeter; Boys calorimeter; Fairweather calorimeter; Junkers calorimeter; sigma calorimeter.*

Gas cap Natural gas which accumulates above an oil reservoir; the gas cap may be tapped to yield gas only.

Gas centrifuge process A technique for the enrichment of uranium; a plant developed by the Urenco consortium (a grouping of the United Kingdom, West Germany and The Netherlands) came into service in September 1977. The basic principle of the process is the use of a cylindrical rotor which is rotated at very high speed; UF_6 gas is introduced to the rotor, the result being the separation of isotopic species differing in density by less than 1 per cent; isotopic separation is obtained between layers at different radii. The gas centrifuge process had been under consideration during and after World War II. Work on and interest in the method waned in favour of the *gas diffusion process*. However, interest was renewed in the later 1960s and led to the signing of the Treaty of Almelo in 1970 between the United Kingdom, West Germany and The Netherlands. *See uranium enrichment.*

Gas coal *Bituminous coal* possessing a relatively high *volatile matter* and suitable for the manufacture of *coal gas* and *coke.*

Gas coke Coke produced in the manufacture of *town gas* by the *carbonisation* of coal.

Gas diffusion process A technique for the enrichment of uranium;

large plants using this technique have been in operation in the United States of America since the late 1940s. Using uranium hexafluoride gas, UF_6, isotope separation relies upon the difference in mean molecular velocities of the principal isotopes; because of the differences, the lighter isotopes diffuse more rapidly through a porous membrane. The holes are of the order of a hundredth of a micron, while the pressure is around atmospheric. Gas which has passed through the membrane must be recompressed before passing to the next stage; however, a commercial plant needs about 1000 stages to complete the process. The United States has constructed three diffusion plants with a capacity of about 17 000 tonnes of separative work a year; they are being extended to a capacity of 27 000 tonnes to meet civil needs. *See gas centrifuge process; uranium enrichment*.

Gas engine An internal combustion engine in which a gaseous fuel is mixed with air to form a combustible mixture in the cylinder and fired by spark ignition. *See heat engine*.

Gas governor A device for controlling gas pressures or volumes in mains or appliances. The purpose of a constant-pressure governor is to maintain a constant pressure at the outlet which is independent of pressure fluctuations at the inlet. Simple pressure governors are of a diaphragm or bell type. A constant-volume governor maintains the same gas rate irrespective of small pressure fluctuations.

Gasification The conversion of the combustible material in a solid or liquid fuel into a combustible gas.

Gasifier Any unit for equipment in which the process of *gasification* is carried out. *See Lurgi gasifier*.

Gas Industry Standard A standard of temperature, pressure and humidity used by the United Kingdom gas industry of 60° F, 30 inHg, saturated with water vapour. *See standard temperature and pressure (STP)*.

Gas laws Laws indicating the relationships between pressure, temperature and volume, while two of those quantities are changing. *See Boyle's law; Charles's law*.

Gas meter *See dry gas meter; wet gas meter*.

Gasohol A blend of petroleum and organic alcohol, in a ratio of nine parts to one part, for use in cars. The carbohydrates in organic materials such as grain, wood, whey and sugar beet can be fermented to produce alcohol for use as a fuel; a concentrated protein meal suitable for cattle is also produced as a by-product. Brazil has a large programme to produce gasohol from sugar cane.

Gas oil A petroleum distillate having a viscosity and distillation range intermediate between those of *kerosine* and light lubricating oil. Its name was derived from its extensive use in the gas industry

165

to produce **carburetted water gas**. Its use has been extended to gas turbines and furnaces generally; suitable gas oils are also used as fuels for high-speed diesel engines. *See **diesel oil**.*

Gasoline or petrol; a complex mixture of light hydrocarbon blending stocks, the characteristics of which are varied to suit all motorcars and aircraft in all seasons and climates. During refining, motor gasoline distils within the temperature range of 38–205° C; aviation gasolines are blends of high-octane number stocks and distil within the temperature range of 40–163° C. Four grades supply the needs of virtually all conventional aircraft, the grades being based on octane number requirements. Fuels for jet turbine engines boil in the 40–315° C range and have no octane number requirement.

Gasoline additives Small amounts of compounds of substances added to gasoline to reduce the tendency to knock, i.e. to improve the **octane number**. The compounds react with the activated oxygenated intermediate compounds whose decomposition products result in knocking. Generally, the most effective additive has been found to be tetraethyl lead (TEL); about 2 ml/gal is added to gasoline in the United Kingdom, and up to 4 ml/gal in America and Europe. TEL is used with a scavenger to prevent the formation of deposits; ethylene dichloride and dibromide are used, and these form volatile lead halides. Another additive, tetramethyl lead (TML), is now used in some super-grade gasolines. A Joint report by the US Public Health Service and the Petroleum Industry in 1965 concluded that lead in gasoline resulting in atmospheric lead contamination was not a hazard to health in the three cities studied—Cincinnati, Philadelphia and Los Angeles. Other additives include organic phosphates to prevent spark plug fouling; phenols and amines as anti-oxidants to reduce gum formation; and unsaturated organic acids and orthophosphates as anti-rust agents.

Gasoline engine An internal combustion engine in which a gasoline–air mixture, provided by the **carburettor**, is drawn by vacuum through an intake manifold to the combustion chamber of each cylinder; it is then compressed and ignited by an electric spark. Each piston in a four-stroke engine moves up and down twice to produce a power impulse:

(a) Intake stroke—the piston descends drawing into the cylinder a gasoline–air mixture produced by the carburettor.

(b) Compression stroke; the piston ascends compressing, the gas to about one-sixth or one-eighth of its original volume in the cylinder.

(c) Power stroke—following ignition by an electric spark the expansion of the gases forces the piston down.

(d) Exhaust stroke—the piston ascends and the exhaust gases are expelled to atmosphere via exhaust manifold, muffler and tail pipe.

The thermal efficiency of a petrol engine varies between 18 and 25 per cent. *See diesel engine; motor vehicle exhaust gases.*

Gas passes In respect of a steam generator or boiler, the routes by which flue gas passes from the combustion chamber to the stack; the number of gas passes normally includes the combustion chamber as the first pass and the reverse flow from the smoke box as the second, with a third pass being through the tube bank.

Gas producer Plant for the production of *producer gas*; it usually comprises a vertical cylindrical water-cooled shell fitted with a charging hopper at the top and a fire grate at the bottom containing tuyères for the admission of an air–steam blast. The plant may be followed by cleaning equipment for the removal of dust and tar. Fuels suitable for gas producers include gas coke and oven coke, anthracite and dry steam coals, and non-caking or weakly caking bituminous coals. The reactions in the fuel bed of a producer are shown in *Figure G.1*. Steam is added to the air blast to assist in

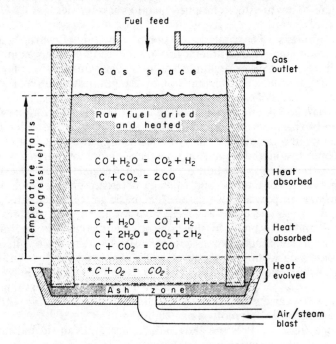

Figure G.1 Thermal effects in a gas producer
* Reaction shown in italics is exothermic. All other reactions are endothermic

167

reducing clinker formation and to promote the formation of carbon monoxide and hydrogen. In the secondary distillation and reduction zone a reaction known as the *water gas shift reaction* takes place whereby carbon monoxide is replaced by its own volume of hydrogen.

Gas purification   The removal of unwanted or injurious components of a gas, particularly *hydrogen sulphide*.

Gas recycle hydrogenator (GRH) process   A *Gas Council* pressure process for manufacturing rich gas from *naphtha*; the gas may be supplied direct to the consumer after blending.

Gas reduction process   A process in which a gas (e.g. carbon monoxide, hydrogen or methane) is employed to reduce ores to metals.

Gas separation units   In oil refining, units which receive cracked gases from the catalytic cracker and by a sequence of absorption and distillation under pressure, yield streams of propane-propylene, butane-butylene and light gas. These streams are treated in a Girbotol or similar plant to remove from them hydrogen sulphide and other undesirable sulphur compounds. These gases may be liquefied by compression and are known as *liquefied petroleum gases*. They are stored in spherical tanks at refineries. *See Girbotol process*.

Gas synthesis   The catalytic reaction of a mixture of hydrogen and carbon monoxide (synthesis gas) to produce liquid fuels, combustible gas, waxes or organic chemicals. The synthesis gas may be made by reforming methane or by the gasification of coal with steam. *See Fischer–Tropsch synthesis*.

Gas turbine   A *heat engine* working on the principle of compression and expansion of a gas, normally air. The essential difference between a gas turbine and other forms of heat engine is that the compression and expansion take place across rotating parts, not by reciprocating motion as in a *diesel engine*. The process is generally continuous, and this characteristic gives a very even torque for power transmission. The basic gas turbine consists of a compressor, a combustion chamber for heating the compressed air and a turbine. In the 'open cycle' turbine the air, after compression and expansion, exhausts to atmosphere. In the 'closed cycle' turbine the working fluid, generally air, is continuously recirculated, the heating necessary for expansion being effected by an air heater. The gas turbine is capable of running on a wide variety of fuels—gases, liquids and solids. *See steam turbine; turbo-jet; turbo-prop*.

Gauge pressure   Pressure above atmospheric pressure as indicated on a *pressure gauge*.

Geiger–Müller counter   A device for counting the number of charged particles or photons by the ionisation which they produce in a gas between two electrodes. The operating voltage is sufficiently high for the primary ionisation to cause breakdown of the gas, resulting in the production of a relatively large output pulse. *See photon; radiation detector.*

Geological time scale A scale of time which serves as a reference for correlating events in the history of the Earth; the time scale is divided into three main 'eras', based upon the general character of the life which they contained, each era being subdivided into periods. *See Table G.1.*

Geomorphology The study of the form and development of the earth, especially its surface and physical features, and the relationship between these features and the geological structures underneath.

Table G.1 GEOLOGICAL TIME SCALE

Era	Period	Epoch	Age in years
Cenozoic (recent life)	Quaternary	Recent	10 000
		Pleistocene	
			2 000 000
	Tertiary	Pliocene	7 000 000
		Miocene	
			25 000 000
		Oligocene	40 000 000
		Eocene	
			60 000 000
		Paleocene	
			70 000 000
Mesozoic (intermediate life)	Cretaceous		135 000 000
	Jurassic		
			180 000 000
	Triassic		
			225 000 000
Palaeozoic (ancient life)	Permian		270 000 000
	Carboniferous		
			350 000 000
	Devonian		400 000 000
	Silurian		
			440 000 000
	Ordovician		500 000 000
	Cambrian		
			600 000 000
Precambrian			3 500 000 000

Geosphere The solid, non-living portion of the earth; a concept which excludes the atmosphere, the hydrosphere and the biosphere.

Geothermal energy Energy obtained, in the form of heat, from the depths of the earth. This heat may sometimes be harnessed, through the medium of hot water and steam, for heating and other purposes; this is done in the Rotorua are of the North Island of New Zealand. Sources have also been tapped in Italy, the United States of America, Japan and the Soviet Union. However, the total world potential of 'naturally occurring' geothermal energy remains very small. *See Figure G.2.*

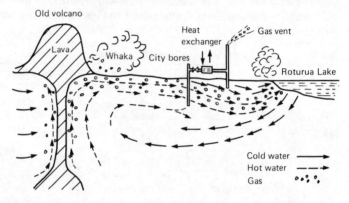

Figure G.2   City Bores, Rotorua, New Zealand. (Source: Department of Health, New Zealand)

GeV Giga-electron-volt = 10^9eV.

Gieseler Plastograph Test A laboratory test which measures the variation of viscosity of a sample of coal when in the plastic phase.

Gigajoule A unit comprising 1000 MJ, or 10^9J.

Gigawatt A unit comprising 1000 MW, or 10^9W.

Girbotol process A wet scrubbing process for the removal of *hydrogen sulphide* from refinery and petroleum oil gas streams. The scrubbing medium is an aqueous solution of ethanolamines, usually diethanolamine, the reaction being:

$$(CH_3CH_2OH)_2NH + H_2S \rightarrow [(CH_3CH_2OH)_2NH_2]HS$$

Absorption is carried out in packed towers, and regeneration in a bubble-cap tower using stripping stream. *See hydrogen sulphide removal.*

Gland steam Steam used to prevent air from entering a *turbine cylinder* between the turbine shaft and the casing.

Glass wool An insulating material, but not suitable for surfaces above about 500° C; it is available in mattress form or in rigid semicircular sections.

'Gloco' A *gas coke* meeting British Standard specification 3142 as a fuel suitable for coke-burning domestic open-fires.

Glost firing A stage in the firing of pottery ware in which the glaze is fired on to the biscuit ware at a temperature of about 1050° C. *See biscuit firing; enamel firing.*

Go-devil A device for cleaning out the bore of a pipe; it consists of a piston-type scraper which is usually pumped through the line.

Gonad The sexual or reproductive organs of an animal or plant; important within the concept of exposure to radiation hazards.

Grab sample A sample of gas or liquid taken over a very short period of time, a time insignificant compared with the total duration of the operation.

Grain A unit of weight equal to 64.798 milligrams, being the same whether avoirdupois, apothecaries' or troy weight. The grain has often been used in air pollution work, e.g. emissions have been measured in grains per cubic foot of gas at a prescribed temperature and pressure. Emission concentrations are now expressed in most countries in grams per cubic metre. 1 grain/ft^3 = 2.3 g/m^3.

Gram The metric unit of mass; the kilogram (1000 grams) is the basic SI unit of mass.

Graphical formula A chemical formula which indicates the position of every atom and linkage in the molecule. *See constitutional formula; empirical formula; molecular formula.*

Graphite A dense, rigid, allotropic form of carbon used as a *moderator* in a *nuclear reactor*.

Graphite sleeve In respect of a *nuclear reactor*, a hollow graphite cylinder used to contain and support the fuel rods in the channels of the reactor.

Grate A device to support the fuel bed, allow sufficient *primary air* to pass through it with as even a distribution as possible, and assist in the separation of ash from the burning fuel. Grates may be horizontal or inclined, stationary or movable, and operated manually or automatically. With stationary grates the ash and clinker has to be removed with hand tools; a rocking bar grate discharges the ash through the bars into the ashpit by mechanical action; with a self-cleaning grate the fuel moves to the rear, the ash and clinker being discharged over the back end.

Gravel bed filter A *dust arrester* consisting of filter beds of one or more layers of abrasion-resisting material such as gravel; the filtering material effectively removes dust from the gas stream passing through it. The captured dust is removed from the filter

171

bed by a vibrating system which shakes the spring-supported filter bed containers. A gravel bed filter may be used in continuous operation at temperatures up to 350° C; the pressure drop is low.

Gravity feed system   A system for supplying oil to an *oil burner*; oil is pumped to an overhead tank from which it flows by gravity to each burner. As the oil is not under high pressure, the system is suitable only for blast atomisers. *See blast atomiser.*

Gray–King Assay   A test for determining the caking and swelling properties of coal. As described in BS 1016, 20 grams of coal is heated in a silica tube in a standard furnace at a defined rate of temperature increase to 600° C. The appearance of the resulting residue is then compared with a series of standard cokes, to which the letters A to G are allocated; the series ranges from non-caking up to highly caking coal. Coals of type A are non-caking; those which give a hard strong coke but have not swollen are described as type G. For coals more strongly swelling than those of type G subscripts are added—G_1, G_2, and upwards. The subscripts are determined by the addition of inert material to the coal.

Greenhouse effect   The property of selective absorption used in the construction of greenhouses which finds a parallel in the general atmosphere. Water vapour and *carbon dioxide*, although only a minute fraction of the mass of the atmosphere, exercise considerable influence over the heat balance of the atmosphere and ground. While relatively transparent to incoming short-wave solar radiation, they are relatively opaque to long-wave back-radiation from the earth; hence, they exercise a warming or greenhouse effect. A secular increase in carbon dioxide in the atmosphere, arising from progressive industrialisation and the combustion of fossil fuels, could raise the mean temperature of the atmosphere, effecting profound climatic changes. On the other hand, an increase in aerosols in the atmosphere, also due to industrial emissions, could cool the earth through a reflection of solar radiation. The arguments concerning the actual trends and effects in the atmosphere are at present inconclusive, but this aspect of the effects of consuming fossil fuels merits the closest study.

Grey cast-iron   A *cast-iron* in which the carbon is in a free state as graphite flakes; it is easily machined.

Grindability index   The relative ease of pulverising a coal. *See ball-mill method; Hardgrove machine method.*

Grit   Particulate matter of natural or industrial origin retained on a 200 mesh BS test sieve (76 μm); such particles are visible when deposited and likely to cause eye irritation. The particles are non-respirable and do not penetrate to the depths of the lungs; they can, however, be a source of nuisance. *See dust.*

Grit carry forward   The amount of grit and dust passing through a system or process suspended in a medium, e.g. flue gas. For example, in *Gas Purification Processes*, by Nonhebel (Newnes, 1964) Dransfield and Lowe given the following figures for the carry forward of *fly ash* in the gas stream from various types of furnace, expressed as a percentage of the ash in the original fuel: cyclone furnaces, 15–20; slag tap furnaces, 45–55; dry-bottom pulverised fuel furnaces, 80–85.

Grizzly Screening equipment used in the surface preparation of coal. It consists of a number of sloped parallel bars. The separating size is determined by the width of the openings between the bars, the length of the bars and the slope. This type of screen is used at mine-preparation plants for the production of modified *run-of-mine coal* or for the removal of undersize coal ahead of a crusher. Also known as a gravity bar screen.

Gross energy (or calorific) value *See energy value.*

Gross Domestic Product at constant prices The Gross Domestic Product (at either factor cost or market prices) taken over a series of years and adjusted to discount changes in the value of money, thus giving a measure of the real change in national income over the period. The trend in the Gross Domestic Product at constant prices is not, however, a comprehensive measure of changes in the national well-being, or in the progress towards an improvement in the *'quality of life'*. For example, things tangible and intangible that cannot be purchased in the market place are excluded from the statistical measurement of Gross Domestic Product; public expenditure to relieve traffic congestion while adding to the real leisure time of commuters adds nothing, on that score, to the GDP. Expenditure on cars causes the GDP to apparently grow faster than if the same amounts were devoted to education, health and culture. The problem is to maximise the real welfare of the community, not simply a statistical measure of growth.

Gross Domestic Product at factor cost The value of goods and services produced within the nation, representing only the sum of the incomes of the factors of production. It is equal to the *Gross Domestic Product at market prices* minus indirect taxes plus subsidies. Valuation at factor cost displays the composition of the Gross Domestic Product in terms of the factors of production employed, the contributions of the factors being measured by the incomes they receive as wages, salaries, profits, interest and rent.

Gross Domestic Product at market prices The value of goods and services produced within the nation, charged at ruling prices. Prices include all taxes on expenditure, subsidies being regarded as negative taxes.

Gross National Product (GNP) The total value of the goods and services produced within the nation charged at ruling prices, together with the net income from abroad. *See Gross Domestic Product at constant prices; Gross Domestic Product at factor cost; Gross Domestic Product at market prices.*

Ground furnace An enclosed but roofless furnace for safely burning-off temporary surpluses of gas at moderate rates.

Ground-level concentration The concentration, expressed as an amount per unit volume, of a pollutant in the atmosphere, usually at about breathing level.

Ground state *See energy levels.*

Group heating *See district heating.*

Guillotine damper An adjustable plate normally installed vertically in the flue between the furnace and the stack, and counterbalanced for easier operation; it may be operated manually or automatically. *See damper.*

Guillotine door A door which is fitted to the hopper outlet of a *chain-grate stoker* which controls the thickness of the fuel bed on the grate. Also called an 'outlet gate'.

Gulf HDS process A process developed for the hydro-desulphurisation of residual stocks in oil refineries.

Gum In the petroleum industry, a rosin-like insoluble deposit formed during the deterioration of petroleum and its products, particularly gasoline.

Gunmetal An alloy of copper and tin; lead and nickel are often added. It is used where resistance to corrosion or wear is required, e.g. as in gears, bearings and steam pipe fittings.

Gun-type burner A *pressure jet burner* suitable for central heating boilers; it is fully automatic, operating under the command of a thermostat. Ignition is by an electric spark.

Gustiness A form of turbulence set up near the ground by obstacles presented to the direct flow of air by the surface and its irregularities.

H

Hafnium Hf. An *element*; at. no. 72, at wt. 178.49, with a high neutron-capture cross-section. It is found in zirconium ores and must be removed before *zirconium*, is used as a canning material for the fuel elements in a *nuclear reactor*.

HAGA process A process for *uranium enrichment*, in the developmental stage in the late 1970s; it is a patented application combining elements of gas centrifuge and separation nozzle technology. It is claimed that the process will require less electricity, while offering lower unit costs and shorter construction times. *See separation nozzle process.*

174

Haldane apparatus   Fuel gas analysis apparatus in which the gas is confined in semi-capillary tubes over mercury and the volumes before and after each absorption are read through a travelling lens to ± 0.0001 ml.

Half-life   The time taken for one-half of the atoms of a radio-active *isotope* to disintegrate. Each isotope has a unique half-life which lies between less than 10^{-6} s and more than 10^6 years, according to the isotope. Iodine-131 has a half-life of 8 days; caesium-137 a half-life of about 30 years; radium a half-life of 1580 years. The half-life is related to a disintegration constant (λ), as follows: $t_{\frac{1}{2}} = 0.693/\lambda$. The disintegration constant is the fraction of the number of atoms of a particular radioactive isotope which disintegrate in unit time.

Hall oil gasification process   A cyclic and non-catalytic process for the manufacture of a rich gas from oil. The plant comprises two generators containing *chequer-brickwork*, connected to each other. During the blow stage, air is admitted to one generator, in which it burns off the carbon deposited during the previous 'make' before passing through the system restoring temperatures; the waste gas passes through a *waste-heat boiler* and then to the chimney. Steam is then admitted to the system in the same direction, initially for purging purposes and then to mix with an incoming oil spray; the cracking process is completed as this steam–oil vapour mixture passes down through the chequer-brickwork of the second generator. After the oil is shut off, the system is again purged with steam. The whole process is then repeated in the reverse direction. The constituents of a typical gas, expressed as percentages, are: hydrocarbons, 32; methane, 28; hydrogen, 20; carbon dioxide, 5; nitrogen, 15. The calorific value is over 39 MJ/m^3.

Halo method   A geochemical method used in exploration for oil. Soil surveys are conducted by measuring the hydrocarbon content and mineralisation in subsoil samples; eight samples per square mile are preferred. Significantly high values are plotted; if they form a pattern such as an aureole or halo, the area is considered positive, the petroleum deposits being roughly outlined by the pattern.

Hand picking   The cleaning of *run-of-mine coal* by hand; large screened coal exceeding 10 cm in size is placed on a slowly moving 'picking belt', while workers pick out obvious pieces of shale or dirt.

Hard asphalt (or asphaltenes) test   A chemical test to determine the quantity of asphaltic material in a liquid fuel. The test is made by determining the percentage of material which cannot be dissolved by a certain reagent, usually a petroleum ether of specified characteristics. The quantity of asphaltenes indicated varies with

175

the type of reagent. In practice, high-quality distillate fuels have a negligible asphaltene content. Fuels containing asphalt are liable to form carbon deposits in the cylinders of high-speed engines, causing deterioration of exhaust valves, piston ring sticking and liner wear. Slow-running engines are less prone to this problem, owing to the longer period available for combustion.

Hard coal   Defined by the United Nations as coal having a gross energy value of over 24 kJ/g (about 57 kcal/kg); brown coal has an energy value below that level. The United Nations definition is wide, viewing coal in only two classes; the term 'hard coal' encompasses all ordinary bituminous coal as well as anthracite. *See Coal.*

Hard coke   Metallurgical coke produced in a *coke oven* to meet the requirements of the iron and steel industry.

Hardgrove machine method   A method for determining the *grindability index*, of a coal; the Hardgrove machine is a grinding mill of ring-roll design. The index is based on a standard coal that is assumed to have an index of 100. For British bituminous coals, the index falls between 45 and 80; anthracites have an index between 30 and 40.

Hardness   A characteristic of water representing the total concentration of calcium and magnesium ions. Hardness is expressed fundamentally in terms of the chemical equivalents of metal ions capable of precipitating soap; it is also expressed in terms of the equivalent amount of calcium carbonate. Hard water requires a good deal more soap than soft water to make a good lather. Where hardness is due mainly to the presence of bicarbonates of calcium and magnesium, it is described as 'temporary'; if due to the sulphates and chlorides of calcium and magnesium, it is described as 'permanent'. However, both kinds of hardness may be reduced substantially by the use of appropriate techniques, e.g. the lime–soda process, followed, if necessary, by the ion-exchange process. Hard water is considered unsuitable for some industrial processes, as it leaves scaly deposits in pipes and steam generators; pretreatment is essential. On the other hand, for irrigation purposes a hard water is preferred, for it reacts more favourably with soils and readily reaches the root zone.

Harrisburg nuclear power plant incident   An incident in March 1979 involving the release of radioactive material to the atmosphere from the Three Mile Island nuclear power station, owned and operated by the Metropolitan Edison Utility. The plant, situated some 16 kilometres from Harrisburg, the Pennsylvania state capital, comprised reactors of a pressurised water reactor (PWR) design. Some 636 000 people lived within a radius of 32 kilometres of the plant. The radioactive material released to the atmosphere

176

as a result of a series of mechanical failures in one of the 800 MW reactors was detected 16–33 kilometres away from the site, although no excessive amounts of radioactivity were detected. For a few hours a state of emergency was declared and plans for an evacuation were drawn up. The main anxiety was the formation of a hydrogen bubble above the core which restricted circulation of cooling water and limited the opportunities to cool the overheated core. A core melt-down was feared, and possibly a massive gas explosion. In the event, the gas bubble subsided and the damaged reactor slowly cooled. President Carter visited the plant and ordered a full inquiry into the incident.

Hartridge smokemeter An instrument developed to give an instantaneous direct reading of smoke density on a 0–100 scale, specially designed for use with diesel and other internal combustion engines. The principle of the meter is based on a comparison of the density of a column of smoke with a column of clean air. *Figure H.1* shows the smoke tube and the clean air tube, which are optically and dimensionally identical. At one end of the tubes is a 12 V, 48 W light source, and at the other end a photoelectric cell;

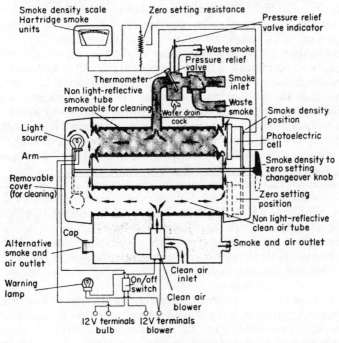

Figure H.1   The working principle of the Hartridge-B.P. smokemeter (Source: Leslie Hartridge Ltd.)

177

these are so mounted that they can be moved together from the smoke tube to the clean air tube by a control lever. The photocell is connected to a microammeter with a smoke density scale 0–100, representing the percentage of light absorbed. A sampling pipe connects the instrument to the vehicle exhaust. The smoke density scale is set to zero for the clean air tube; the control lever is then moved to the smoke tube to give an immediate reading.

H-Coal process A process for the production of fuel oil and naphtha from bituminous coal and brown coal; it uses an ebulliated bed reactor system for the coal–oil slurry, catalyst and hydrogen. While fairly high efficiencies are claimed for this process in Australia, it has yet to be demonstrated on a commercial scale.

Health Defined in the preamble of the *World Health Organisation* as '. . . a state of complete physical, mental and social well-being and not merely the absence of disease or infirmity'. Notwithstanding a wide acceptance of this definition, objective methods of measurement have not evolved; for this reason most assessments of health effects rely upon mortality and morbidity statistics.

Health physics The branch of physics concerned with the effects of ionising radiation on living matter, and the protection of personnel from the harmful effects of such radiation.

Heart cut A narrow range cut usually taken near the middle portion of the oil being distilled or treated. *See Cuts.*

Heat balance A means of determining the thermal efficiency of a process and indicating the origin and magnitude of heat losses. It takes account of the heat supplied to the system in the heating value of the fuel, and in the preheating of fuel, combustion air, feed-water, furnace or charge; and of the heat output by way of absorption in water, steam or charge, radiation losses, normal stack losses, losses due to inefficient combustion, loss due to combustible matter in the ash, and other losses not readily accounted for. *See Sankey diagram.*

Heat drop The difference between the total heat in the inlet steam and the total heat in the outlet steam of an engine or turbine. In a ideal engine, which would incur no losses, the energy that should be available is called the *adiabatic heat drop*.

Heat engine. A machine for converting heat energy into mechanical energy. *See diesel engine; gas engine; gasoline engine; gas turbine; steam engine; steam turbine.*

Heat energy *See energy.*

Heat exchanger A device for arranging the transfer of heat from a hot substance to a cooler substance, while keeping the two substances separate. They fall into two groups: (a) those in which the heat transfer process is continuous within the same heat exchanger, e.g. recuperators, condensers, air preheaters and economisers; (b) those in which the transfer is intermittent within the same heat exchanger, e.g. regenerators and blast-furnace stoves, the heat storing solid being alternately heated by the hot fluid and cooled by the fluid to be heated. *See air preheater; blast-furnace stove; cascade heat exchanger; condenser; economiser; Howden–Ljungstrom air preheater; pebble stove; recuperative air preheater; regenerator; superheater.*

Heat meter (BCURA) A meter for the measurement of heat consumption by the occupants of individual flats in a block, or of houses in a district heating scheme, under development by the British Coal Utilisation Research Association. The basic principle of the meter is that a small fixed proportion (say 1/200) of the 'return' water is diverted through a side arm, and reheated, by means of a small electric heater and control system, to the temperature of the water at entry to the dwelling. The proportionate flow is achieved by means of two orifices in parallel, a small one in the side arm and a larger one, across which the side arm is connected, in the main return pipe. An electricity meter measures the energy input to the small heater, and this reading can be directly converted to heat consumption.

Heat pump A device for transferring heat from a low-temperature medium to a higher-temperature medium. For example, heat may be extracted from flowing water by a refrigerant which is then compressed, transferring its heat to another flow of water; the refrigerant passes through an expander to begin the process afresh.

Heat rate In respect of a power plant, the heat consumption per kWh of current produced:

$$\text{heat rate, MJ/kWh} = \frac{\text{heat supplied in fuel, MJ}}{\text{Energy generated, kWh}}$$

In a large conventional power station with a thermal efficiency of 36 per cent, the heat rate is approximately 10 MJ. One kWh = 3.6 MJ.

Heat release rate The amount of heat liberated in a *combustion chamber*, usually expressed as $kJ/h\,m^3$.

Heat-retaining arches Arches of refractories often built over grates at the front, intended to radiate absorbed heat on to incoming green fuel to aid *ignition* and *combustion*.

Heat transfer *See conduction; convection; inverse square law; radiation; Stefan–Boltzmann law.*

Heat transfer medium A substance such as water, or oil, used to transfer heat between processing areas.

Heat treatment The thermal treatment of metals and alloys after production to improve their physical properties, e.g. the treatment

179

of steel by normalising, hardening and tempering, or the *annealing* of metal after cold working. Heat treatment may involve the chemical treatment of surface layers, e.g. *carburising* and *nitriding*. Furnace atmospheres often need careful control to prevent surface attack by way of oxidation, scaling, *decarburising* or sulphur penetration.

Heavy ends   The highest-boiling portion present in the distillation of petroleum.

Heavy metal   Or toxic metal; a term loosely applied to a whole range of elements, not all of them metals, which may contaminate the *environment*.

These elements include arsenic, beryllium, cadmium, chromium, cobalt, germanium, iron, lead, manganese, molybdenum, nickel, selenium and zinc.

Heavy water D_2O. Water in which the molecule contains two atoms of *deuterium*. It is present in ordinary water to the extent of about 1 part in 42 000 000. Substantially pure heavy water is used as a *moderator* in some types of *nuclear reactor*.

Heavy water-moderated nuclear reactors   Nuclear reactors using heavy water as the moderator, although not necessarily as coolant, with natural or enriched uranium fuel. *See Canadian pressurised heavy water reactor (Candu); steam generating heavy water reactor (SGHWR). See also light water-moderated reactors; nuclear reactor.*

Heliostats   Reflectors that, controlled by a central computer, are permanently oriented towards the sun, projecting its rays on to a boiler site at the top of a tower. A 1 MW helioelectric power station has been constructed at Contrasto, Sicily. *See Solar One.*

Helium   He. An *element*, at. no. 2, at wt. 4.0026; it is the lightest inert gas. It has a very low neutron-capture cross-section and good heat transfer properties, and has been considered as a *coolant* for nuclear reactors.

Henry   A unit of electrical inductance; it is the inductance of a closed circuit in which an electromotive force of 1 *volt* is produced when the electric current in the circuit varies uniformly at a rate of 1 *ampere* per second.

Hertz   Unit of frequency; the number of repetitions of a regular occurrence in 1 second.

Heterogeneous reactor   A *nuclear reactor* in which the fissile material and the moderator are arranged as discrete bodies, usually in a regular pattern. The fissile material is contained in the fuel elements. *See fuel element.*

High-efficiency cyclone   A development of the simple *cyclone*; the inlet takes the form of a true volute, as distinct from the tangential entry in the simple cyclone, thus reducing the disturbance to the gases already rotating within the body. The rotational movement within the cyclone is also prolonged by a lengthened body and cone. This type of cyclone may involve a pressure drop of up to 100 mmHg, but particles down to about 20 μm can be collected with a high degree of efficiency. *See dust arrester.*

High-energy fuels Special fuels with performance characteristics superior to those of hydrocarbon fuels; designed for the propulsion of space rockets and missiles. Such fuels include hydrogen, beryllium, boron, diborane, pentaborane, decaborane and alkylborane.

High-heat value (HHV) Synonymous with *gross value*, a term most often used in the United States of America. *See energy value.*

High-level inversion On occasions an *inversion* of temperature is formed well above the earth's surface (i.e. upwards of 300 metres or more). This acts as a lid, preventing the ascent of chimney plumes. It is formed by the slow descent of air which becomes warmer by adiabatic compression, since the descent is inevitably into a level of higher pressure. This is known as a 'subsidence inversion'. It is associated with anticyclones in which air descends at a rate of perhaps 1000 metres per day. Cloud may form under such an inversion and will be carried inland by the wind. If the wind speed is sufficiently large, the cloud and the inversion will be maintained and the sun will be unable to 'burn' the cloud away. Effluent will be trapped under such an inversion, but the existence of the cloud presupposes air movement. If the wind drops, the cloud disperses and the inversion also. However, thermal convection may bring some of the accumulated pollution to ground level, by way of *fumigation.*

High-level waste The most highly radioactive waste from nuclear fuel reprocessing, containing most of the fission products from spent fuel and typically containing millions of curies per cubic metre when first separated. It contains also small amounts of unseparated uranium and plutonium, and the greater proportion of the other actinides produced in the reactor.

High-speed mill A mill for the production of *pulverised fuel*; coal is ground by attrition and impact applied by hammers or pegs rotating inside a casing at speeds ranging from 1000 to 4000 rev/min. This type of mill is most suitable where the output required is less than 1 tonne/h.

High-steam and high-low-water alarm A safety mounting, often incorporated in a lever type *safety valve*, designed to give audible warning if the water level in a boiler becomes either dangerously high or low. The two main classes of alarm are: (a) internal, for shell boilers; (b) external, for water-tube boilers. An internal alarm consists of two floats, one at each end of a long pivoted arm suspended from the crown of the boiler shell. The low-water float lies on the surface of the water at normal level, the high-water float being suspended clear of the surface. If the water level falls, the low-level float drops; the pivoted arm moves and operates a steam whistle. If the water level rises to an excessive degree, the top float becomes buoyant, moving the pivoted arm in the same direction as before, the alarm again sounding. External alarms may be of the float or the thermostatic type. In the float type a separate chamber is mounted at normal working level and connected to the steam and water space. A float in the chamber responds to changes in the level of the water in the boiler, operating a steam whistle at predetermined high and low positions. In the thermostatic type two rods are utilised, one of these being normally immersed in the steam and the other in the water. Should the water level rise, covering the upper rod, the rod contracts operating an electric alarm; if the water level falls, immersing the lower rod in steam, this rod expands and similarly operates the alarm.

High-temperature deposits Deposits on boiler steam tubes and superheater tubes consisting of re-fused ash, alkali-matrix deposits and phosphate deposits. These deposits are associated with high fuel bed temperatures. Stoker-fired boilers, with high fuel bed temperatures, are much more liable to high-temperature deposits than are pulverised fuel boilers, which operate with lower combustion temperatures and a high *grit carry forward*. *See low-temperature deposits*.

High-temperature gas-cooled reactor (HTGR) A type of *nuclear reactor* using gas, instead of water, as the *coolant*; it has been considered as a possible successor to the *advanced gas-cooled reactor*. An experimental unit of this type has been in operation since 1966, at Winfrith Heath, England, using uranium carbide as the fuel. It was developed by the UK Atomic Energy Authority as part of a joint European project under the auspices of the *European Nuclear Energy Agency*. In March 1977 a 330 MW commercial unit began generating energy at a site 72 kilometres north of Denver, USA. The plant uses helium as the reactor coolant and a uranium–thorium mixture as fuel.

High-temperature plastics An insulating cement made of mineral-wool fibres processed into nodules and then dry-mixed with clay. It is mixed with water for application and sets as a tough fibrous insulation. It is claimed that the temperature limit for this material is 1000° C.

High voltage Means a voltage normally exceeding 650 volts.

Hilt's law In respect of coal deposits, a statement of fairly general application that where there has been relatively little earth movement, coal seams having remained more or less undisturbed, the oldest and lowest seams contain less volatile matter than the relatively younger seams above; regular progression being the general rule. Valid as a generalisation, important exceptions have been noted.

Histogram A diagram showing the frequency of occurrence of values of a variable in various ranges. Columns or rectangles proportional in height to the frequencies of occurrence are set along a time scale.

Historic costs The actual monetary cost of raw materials, machinery, products, etc., at the time they were bought or made. *depreciation*, based upon historic costs, will often be insufficient to cover replacement costs, particularly during inflationary periods. *See current cost accounting*.

Hoffmann kiln A continuous ring tunnel **brick kiln** of annular longitudinal arch design; the kiln has the appearance of an endless tunnel, access to which is gained through a dozen or more doorways or 'wickets'. The latter are assumed to divide the kiln into a number of 'chambers' in which the bricks to be fired are set. Fuel is fed through holes in the roof or arch of the kiln; there may be 16–20 fireholes to each chamber. The kiln may be coal- or oil-fired. A kiln may measure from 70 m to over 120 m in length. Firing proceeds round the kiln so that the bricks in each chamber are subjected to three stages: (a) preheating; (b) firing; (c) cooling.

Holford rules Basic rules proposed by the late Lord Holford, Professor of Town Planning, University College London, as a guide to minimising the environmental effects of transmission lines; these rules are set out in *Table H.1*.

'Homefire' A six-sided fluidised char binderless briquette, some 8 cm in diameter and 5 mm thick; a reactive solid fuel burning with a bright flame, and suitable for open domestic fires. It is produced by the *National Coal Board* at Coventry, England. *See authorised fuels; smoke control area*.

Homogeneous reactor A *nuclear reactor* in which the fissile material and the moderator are uniformly mixed in the solid state or dispersed in a solution or slurry.

Horizontal boiler A shell boiler, which may be equipped with simple internal furnace tubes as in the examples of the Cornish and Lancashire boilers, or be equipped with multi-fire-tube systems as in the examples of the *Economic, Marine* and *Package boilers*. A horizontal boiler may or may not require a brick flue setting.

Horizontal firing Systems of furnace or boiler firing in which the burners are located in the front or side walls, compared with the *down-fired furnace*.

Horizontal retort A closed chamber used for the manufacture of

TABLE H.1 Holford Amenity Rules for Transmission Lines

(1) Avoid altogether, if possible, the major areas of highest amenity value, by so planning the general route of the line in the first place, even if the total mileage is somewhat increased in consequence

(2) Avoid smaller areas of high amenity value or scientific interest by deviation, provided that this can be done without using too many angle towers

(3) Other things being equal, choose the most direct line, with no sharp changes of direction and thus with fewer angle towers

(4) Choose tree and hill backgrounds in preference to sky backgrounds wherever possible, and when the line has to cross a ridge, secure this opaque background as long as possible and cross obliquely when a dip in the ridge provides an opportunity

(5) Prefer moderately open valleys with woods, where the apparent height of the towers will be reduced and views of the line will be broken by trees

(6) In country which is flat and sparsely planted, keep the high-voltage lines as far as possible independent of smaller lines, converging routes, distribution poles, and other masts, wires and cables, so as to avoid a concatenation or 'wirescape'

(7) Approach urban areas through industrial zones, where they exist, and where pleasant residential and recreational land intervenes between the approach line and the substation, go carefully into the comparative cost of undergrounding, for lines other than those of the highest voltage

Source: Central Electricity Generating Board, London, England.

town gas by the carbonisation of coal; the retort is oval or D-shaped, constructed in silica brick or fireclay, with metal doors at either end. Often grouped in double columns of five retorts, each group being known as a 'bed'; a series of beds forms a retort house bench. Each retort holds about 1 tonne of small coal. The retort is heated by *producer gas*, generated from coke in a step grate producer serving each bed. A heating cycle takes about 12 hours, a temperature of about 1000 C being attained. *See carbonisation.*

Horsepower   The power which raises 75 kilograms against the force of gravity through a distance of 1 metre per second; it is the equivalent of 735.499 watts (J/s). The older British unit was the power needed to raise 550 pounds through a height of 1 foot in 1 second (550 foot pounds-force per second); it was equivalent to 745.700 watts, and about equal to the original estimate of the power of one horse in pulling a load. The term horsepower is still occasionally applied to boilers, although an inappropriate term in that context. Brake horsepower (bhp) is the horsepower generated by an engine which does useful work; this may be measured at the crankshaft or the flywheel. The term 'brake' is used because the force can be measured by the braking power necessary to stop the engine.

Horton sphere   A spherical, pressure-type tank used to store volatile liquids; its purpose is to prevent excessive evaporation loss which occurs when such products are placed in conventional storage tanks.

Hot face insulation The insulation of the internal surfaces of a furnace or hot gas duct; it became possible through the development of bricks which combine high refractoriness with good insulation properties.

Hot gas efficiency. In respect of a *gas producer*, thermal efficiency calculated as follows:

$$\text{hot gas efficiency, } \% = \frac{\text{total heat of gas}}{\text{total heat of fuel}} \times 100$$

where total heat means the sum of potential heat (calorific value) plus sensible heat due to preheating. Hot gas efficiencies are of the order of 90 per cent. *See cold gas efficiency.*

Hot-rolled steel Steel that is passed, while red-hot, through a rolling mill.

Hot soak That portion of the *gasoline*, remaining in the *carburettor* of a *gasoline engine* which, after operation, is evaporated by the heat remaining in the stationary engine

Hot water accumulator A technique for economically meeting the hot water demands of such industries as brewing, dyeing, abattoirs and others; surplus steam from the steam mains is discharged to an insulated hot water storage tank which serves as a heat accumulator. The system can be used to suppress peak demands for steam, by using steam only at off-peak times. *See Ruths accumulator.*

Hot well A sump tank to which hot *condensate*, may be returned and from which the boiler feed pump draws supplies.

Howden–ICI process A process for removing sulphur dioxide from flue gases. In operation at Fulham power station in London before World War II, it consisted of a closed circulation system using water with considerable additions of chalk. The process removed over 97 per cent of the sulphur dioxide in the flue gases; converting it to calcium sulphite which oxidised to calcium sulphate. The sulphur extracted was not recoverable in useful form, and several hundred tons of sludge were produced every day. The process has not been resumed. *See Battersea gas washing process; Fulham-Simon Carves process; Reinluft process.*

Howden–Ljungstrom air preheater A regenerative *air preheater* consisting of a revolving rotor containing heating surfaces. The flue gases are directed axially through one side of the rotor, while air for combustion passes through the other side, the rotor revolving slowly; the heat transferred to the rotor heating surfaces by the flue gases is absorbed by the incoming air.

Humic coal *Coal* derived from plant debris; includes most **common banded coal**.

Humidity *See **absolute humidity; relative humidity**.*

Hydrazine N_2H_4. An oxygen-removing agent suitable for treating boiler feed-water; it does not add to the **dissolved solids**. *See **sodium sulphite**.*

Hydrocarbons Organic compounds consisting of carbon and hydrogen only. They are subdivided into aliphatic and cyclic hydrocarbons according to the arrangement of the carbon atoms in the molecule. The aliphatic hydrocarbons are in turn subdivided into: (a) alkanes, (b) olefins or alkenes, (c) diolefins, etc., according to the number of double bonds in the molecule. The cyclic hydrocarbons are subdivided into: (a) aromatics, (b) naphthenes or cyclo-paraffins. In all types of hydrocarbon, hydrogen atoms may be replaced by other atoms, making the formation of a virtually endless number of compounds possible. *See **alkane series; aromatic or benzene series; cyclo-alkane series; olefin series**.*

Hydrocracking unit A process for cracking heavy hydrocarbons to light products in the presence of a high partial pressure of hydrogen and a special catalyst. This process can convert gas oil completely into gasoline and lighter fractions, or convert gas oils into highgrade middle distillates. All hydrocracked products are virtually sulphur-free.

Hydroelectric power The use of water power to drive water turbines, which in turn drive electricity generators; the degree of utilisation of water power varies from nil in some countries to over 50 per cent in Switzerland. *See **pumped storage**.*

Hydrofining A fixed-bed catalytic process to desulphurise and hydrogenate a wide range of charge stocks from gases to waxes in oil refinery processes. The catalyst consists of cobalt and molybdenum oxides on an extruded alumina support, and may be regenerated *in situ* with air and steam, or flue gas.

Hydroforming An oil refinery unit which converts or 'reforms' low-octane heavy naphtha from the crude distillation process into motor spirit of higher quality. This operation is carried out in an atmosphere of hydrogen at high pressure utilising a fluidised catalyst; the catalyst is regenerated after use.

Hydrogasification The **hydrogenation** of coal or oil to produce mainly gaseous hydrocarbons.

Hydrogen H. An **element**, at. no. 1, at. wt. 1.00797; it combines with oxygen to give the reaction:

$$2H_2 + O_2 \rightarrow 2H_2O$$

Hydrogenation A chemical reaction with **hydrogen** to produce a product containing an increased proportion of hydrogen. *See coal hydrogenation*.

Hydrogen economy   A concept for the future, in which hydrogen is generated on a vast scale, possibly from the electrolysis of water. It has been proposed that hydrogen be used instead of gasoline for internal combustion engines, as a source of energy in fuel cells, in the home for domestic purposes, and as a feedstock for hydrogen-based industrial processes. The concept envisages a very cheap production process for hydrogen, not yet realised.

Hydrogen or H_2 treater   A hydrogenating plant, often used in oil refineries to desulphurise or otherwise purify hydrocarbons.

Hydrogen seals   Oil seals fitted at the ends of the shaft of a hydrogen-cooled alternator to prevent the escape of hydrogen.

Hydrogen sulphide H_2S. A colourless gas with a density of 1.19 in relation to air and a characteristic foul odour of rotten eggs. It arises in the decomposition of organic material. Other sources include oil refineries, sulphur recovery plants, some metallurgical processes, and various chemical industries using sulphur-containing compounds. Hydrogen sulphide is New Zealand's principal air pollution problem, both from indigenous sources and from organic wastes associated with the primary industries (timber pulping, meat packaging, skin curing, etc.). It is emitted from the ground in the Rotorua area of the North Island, and is present in the mineral waters of that area. Hydrogen sulphide is irritating to the eyes and respiratory tract; it leads ultimately to death through paralysis of the respiratory centre of the brain. Though the odour of hydrogen sulphide is readily recognisable in low concentrations, the detection of dangerous concentrations by smell is unreliable and unsafe, as olfactory fatigue occurs quickly at high concentrations. The gas is corrosive to many metals, and even when present in the atmosphere in concentrations below the level of physiological significance, it discolours lead paints. Hydrogen sulphide tends to be a localised problem; in ordinary combustion processes the gas is readily burned to sulphur dioxide. Significant concentrations of hydrogen sulphide, expressed in parts per million by volume, are: odour threshold, 0.13–1.0; powerful sickening smell like rotten eggs, but harmless, 2–3; maximum allowable concentration in an atmosphere in which a man works for 8 hours at a time, 20; headache, giddiness, sickness, loss of energy, disturbance of vision, above 50; maximum allowable concentration for 1 hour, 170–300; loss of consciousness and death in a short time may occur in some people, 500; a few breaths will kill at once by action of gas on central nervous system, 1000.

Hydrogen sulphide removal   In respect of *town gas*, the removal of *hydrogen sulphide* by reaction with moist ferric hydroxide in the form known as 'bog iron ore'. The ore is placed in trays in shallow purification boxes or in towers. The reaction is:

$$2Fe(OH)_3 + 3H_2S \rightarrow 2FeS + S + 6H_2O$$

The hydrogen sulphide concentration in the gas should be reduced to 1ppm by volume. *See Claus kiln; ferricyanide process; Girbotol process; Manchester process; Seeboard process; Shell phosphate process; Thylox process.*

Hydrograph A graph indicating the level of water, e.g. in a watercourse or well; or the rate of flow of water through time.

Hydrological cycle The continual exchange of water between the earth and the atmosphere. Since the height of the ocean surfaces remains essentially unchanged from year to year, evaporation from the oceans must equal the rainfall over the oceans plus the run-off from rivers and streams and effluent discharges. *See Figure H.2.*

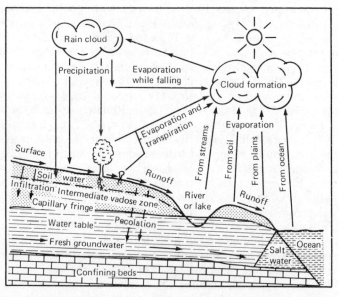

Figure H.2 The hydrological cycle (Source: Report of the Australian Senate Select Committee on *Water Pollution*, Canberra: Australian Government Printer, 1970)

Hydrolysis The decomposition of compounds by interaction with water.

Hydrometer An instrument for the measurement of *specific gravity* or density. A hydrometer may be graduated in degrees, each degree indicating half an ounce of solid matter dissolved per gallon; or graduated to show parts of solid matter dissolved in 10 000 parts of water or more. The instrument is therefore of direct use in measuring the density of boiler water, which changes with the total amount of dissolved solids in it.

Hydroxylation The formation of hydroxylated compounds by the reaction of hydrocarbons and oxygen, as in a *premix burner* or a bunsen burner; these compounds become aldehydes which during combustion become CO_2 and H_2O.

Hygas process A process for the gasification of coal, using steam and oxygen streams, developed by the Massachusetts Institute of Technology. Features of the process are illustrated in *Figure H.3*.

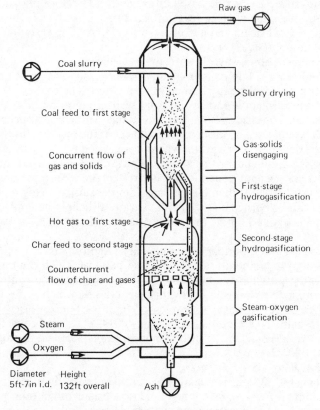

Figure H.3 Hygas Gasifier, with Steam-Oxygen Gasification (Source: MIT)

Hygroscopic Readily absorbing and retaining moisture.
Hyperkinesis Excessive motility (activity or movement) of a person or of muscles.

189

Hypolimnion The bottom stratum of water in a lake, below the thermocline, which, like the epilimnion shows a temperature gradient of less than 1 degree C per metre of depth.

I

ICI steam naphtha reforming process An oil gasification process using the light distillate naphtha; the process has two basic stages: (a) sulphur removal; (b) primary reforming with steam at pressure. A hydro-desulphurisation process first reduces the more complex sulphur compounds to hydrogen sulphide. The sulphur is then removed by zinc-oxide-based absorbents. The reforming part of the process takes place in a tubular furnace in the presence of superheated steam and nickel catalyst. The reaction is continuous and no periodic regeneration is required. All steam requirements are derived from waste heat recovery systems.

Ignifluid grate A grate, similar to a *chain grate*, over which coal fines may be efficiently burned in suspension. The coal enters the furnace through a port in the front wall; a pulsating flow of air from the underside of the grate holds the fuel bed in suspension. Ash is deposited on the combustion faces of the furnace, fuses and flows down on to the grate, where it is carried to the back of the furnace and deposited as clinker. The *grit carry forward* is high; efficient mechanical centicell grit arresters are required to collect the grit, which is then blown back into the furnace. Any type of coal can be burned on the Ignifluid grate, although the upper size limit is about 13 mm; the ash content can be as high as 35 per cent.

Ignition The beginning of combustion; ignition is effected by raising the temperature of a fuel to a point at which the rate of burning provides the heat essential for the process to continue. Burning does not occur below this temperature, which varies with different fuels. Some examples of ignition temperatures are: anthracite, 500° C; bituminous coal, 400–425° C; hard coke, 500–650° C; gas oil, 336° C; carbon monoxide, 664–658° C; hydrogen, 580–590° C.

Ignition delay See *ignition lag*.

Ignition lag Or ignition delay; the period between the injection of fuel into a *diesel engine* cylinder and its subsequent ignition. Fuels which have a short ignition lag are described as 'high-ignition quality' fuels. Engine factors such as compression ratio, fuel and air inlet temperatures, fuel injection time, air turbulence and engine speed also influence ignition lag.

Ignition line A characteristic feature of the fuel bed on a *chain-*

grate stoker, a distinct line across the surface of the fuel bed representing the division between the distillation stage and the combustion stage of the fuel. *See ignition plane.*

Ignition plane   A characteristic feature of the fuel bed on a **chain-grate stoker**, a plane along the grate towards the rear starting at the **ignition line**, caused by the ignition penetrating into the fuel bed. After the ignition plane has reached the grate, the coke burns on the after-part of the grate with a non-luminous flame.

Immature coal   A general term for low-rank *coal. See mature coal.*

Immission   A term commonly used in Europe referring to the level of pollution to which a receptor is exposed; it must be clearly distinguished from the word 'emission', which indicates simply what is discharged and not the level to which a receptor is exposed.

Imperfect competition   A market situation in which neither absolute monopoly nor perfect competition prevails; a situation closest to real life in most circumstances. It is characterised by the ability of sellers to influence demand by product differentiation branding and advertising; restriction of entry of competitors into various lines of production either because of the size of the initial investment required or because of restrictive practices; the existence of uncertainty with imperfect knowledge of profits earned in similar or other lines of production; and the absence of price competition in varying degrees.

Import parity pricing   Generally, the fixing of the prices of domestic goods and services at the same level as world prices. For example, it is the policy of the Australian Government to base the pricing of domestic crude oil on world prices as revealed in the price of oil imports. Historically, the prices for petroleum products in Australia had been relatively low; domestic crude was available at less than half the world price. With the need to conserve supplies and stimulate oil exploration, the prices of domestic oil and petrol were progressively raised during the late 1970s to reflect the price of Arabian crude. In 1977 the United States Congress rejected the concept of import parity pricing for new domestic oil discoveries, as proposed by President Carter.

Impulse-type turbine   A turbine machine in which steam is expressed in fixed blades or nozzles, the change of direction giving an impulse to the moving blades.

Incandescent zone   In the combustion of solid fuels, a zone in which the non-volatile fixed carbon is above ignition temperature. *See distillation zone; flame zone.*

'Incidents at Nuclear Installations'   A quarterly statement by the Health and Safety Executive of the United Kingdom. The most frequently occurring type of incident reported involves small spillages and leakages of radiation activity, and cases or potential cases of radiation exposure of workers exceeding the permissible levels recommended by the International Commission on Radiological Protection.

Incinerator Equipment in which solid, semi-solid, liquid or gaseous combustible wastes are ignited and burned. An incinerator may consist of one, two or more refractory-lined chambers, interconnected by flues. The purpose of an incinerator is to reduce the bulk of the waste products; after incineration the ash residue may be buried, or carried out to sea and dumped.

Income elasticity of demand *See demand*.

Indirect-fired combustion equipment Plant in which the flame and all products of combustion are separated from direct contact with the material being processed by means of metallic or refractory walls, as in heat exchanges, melting pots and muffle furnaces. *See direct-fired combustion equipment*.

Induced draught The extraction of flue gases from a furnace by means of a fan situated between the back end of a plant and its chimney; *induced draught fans* are used when the natural draught produced by a chimney is not sufficient to draw in all the air required for combustion, or insufficient to remove the products of combustion as quickly as they are formed. *See draught*.

Induced-draught fan A *fan* placed near the base of a chimney to draw air through a furnace and remove the products of combustion. As all the hot products pass through the fan, about one-and-a-half times the volume of gases has to be handled compared with an equivalent *forced-draught fan*. In addition to requiring more power, induced-draught fans have to withstand the comparatively high temperature of waste gases, some of which are corrosive.

Induction heating The production of heat utilising the principle that an eddy current will be induced in an electric conductor which is subject to a changing magnetic field. The magnetic field

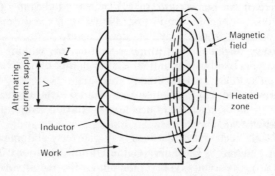

Figure I.1 Principle of induction heating

is produced by a coil carrying an ***alternating current*** the work to be heated being placed within the coil. The efficiency of conversion of electrical energy into heat by this method rarely exceeds 50 per cent, but the method offers many advantages. Frequencies used for induction heating range from fifty up to a million cycles per second. *See Figure I.1.*

Inert gas system   A system which permits oil tankers to replace the inflammable gases in ullage spaces above cargo when loaded, and those that remain in empty cargo tanks, with gases that are low in oxygen and therefore non-combustible. Frequently ship's boiler gas is cleaned and piped to cargo tanks. The installation of inert gas systems has greatly reduced the chances of cargo tank explosions, which can injure crew and endanger vessels. *See Figure I.2.*

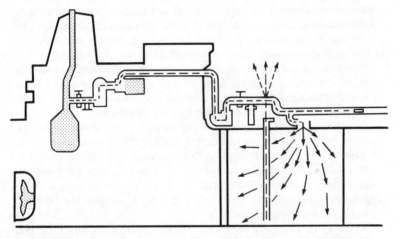

Figure I.2   Inert gas system (the diagram shows how an inert gas system works: the flue gas is conducted from the boiler through cleaning systems and into the tanks, while the air is expelled through outlet pipes). (Source: IMCO)

Inert gases   Gases such as nitrogen and carbon dioxide which do not support combustion and which are non-reactive.

Inertia   The reluctance of a body to change its state of rest or of uniform velocity in a straight line. Inertia is measured by mass when linear velocities and accelerations are considered; and by moment of inertia for angular motions, i.e. rotations about an axis.

Inferential meter   A self-contained meter for measuring feed-water flow on the delivery side of a feed pump; the meter may be of the disc or rotary type. *See feed-water meter.*

193

Inferred resources Resources associated with a known accumulation of minerals, but whose existence and quantity are considered uncertain because of the geological situation.

Inflammability, limits of In respect of gases, the lowest and highest ratios of fuel–air mixture at which *ignition* can take place. Thus, a coal gas–air mixture will not ignite if the gas is less than 5.3 per cent by volume, or higher than 31 per cent by volume, of the total mixture. These are known as the 'lower' and 'higher' limits of inflammability, respectively.

Inflation A condition in which the volume of purchasing power in money terms is constantly running ahead of the output of goods and services, with the result that as prices rise the value of money falls. Inflation may be stimulated by the growth of demand for goods and services, or by an increase in costs, e.g. as a consequence of the higher cost of imports such as petroleum. An increase in the volume of purchasing power may be both a cause of and an effect of inflation; a cause in so far as it leads to higher prices through increased demand and costs of production, and an effect in that rising prices lead to more pressing demands for higher money incomes. Any marked inflationary spiral may be quite harmful to an economy, particularly in respect of a country where the price of exports must be kept competitive. While many economists tend to regard a gentle upward trend in prices and incomes as stimulating to an economy, it operates to the disadvantage of those on fixed incomes and more generally to lenders.

Infrastructure Services and facilities considered essential for the creation of a modern economy, e.g. power, transport, housing, education and health services.

Ingot A solidified steel form made by pouring molten steel from a steel furnace into a mould.

Inherent ash Incombustible mineral matter inseparable from the coal substance prior to combustion. In the course of the mechanical cleaning of coals none of the inherent ash can be removed. *See adventitious ash; ash*.

Inherent moisture Moisture in the pores of coal which cannot be removed by mechanical means; inherent moisture increases with decrease in rank, and coals of the lowest rank may contain from 15 to 16 per cent. *See free moisture*.

Inhibitors Chemical substances added to oil to check or retard the occurrence of undesirable properties.

Injector A device for feeding water into a boiler which is under pressure; the system utilises a steam jet and a suitable arrangement of nozzles. Its operation is based upon the conversion of some of the kinetic energy of the steam into pressure energy; about 1 kilogram of steam is required to inject 10 kilograms of water into the boiler. The injector is suitable for small boilers only, and it cannot feed water with an initial temperature above 30° C. *See boiler feed pumps and injectors.*

Inside mix burner *Oil burner* in which the steam or air and oil impinge and mix within the burner and issue as a foam or fog. *See outside mix burner.*

In situ coal gasification A technique of gasifying coal for the production of low-energy-value gas which has been developed at a commercial scale in the Soviet Union; it has been used for the winning of sub-bituminous coal in the seam at depths of up to 300 metres. A vertical borehole combustion-linkage system is used in the Moscow region and at Angren. Air enters one borehole, passes through a channel in the coal and leaves through another borehole at the other end. A pear-shaped combustion zone develops at the air inlet point; the hot combustion gases are reduced to carbon monoxide and hydrogen as they pass along the channel to the exit borehole. The gasifier is extended by linking a new borehole in the virgin coal to the old gas production borehole.

'In situ' theory A theory as to the mode of origin of coal seams which states that a coal seam occupies more or less the same site on which grew the original plants from which it was derived. *See 'drift' theory.*

Installed capacity In respect of electricity generation, the total generating capacity for which a plant is designed, measured in kilowatts (thousand watts) or megawatts (million watts). Installed capacity is generally about 7–8 per cent greater than output capacity (measured in kilowatts or megwatts 'sent out'), because of the power consumed at the power station for lighting and auxiliary plant.

Installed load tariff A *tariff* for the supply of electricity, consisting in its simplest form of a unit charge plus a standing charge related to the number of kilowatts of installed electrical load on a premises.

Institute of Energy Formerly the Institute of Fuel, an institute founded in 1927 and incorporated by Royal Charter in 1946. It is a constituent member of the Council of Engineering Institutions (CEI). The Institute of Energy has 6000 members, most of whom work in the United Kingdom, while others are to be found in South Africa, Australia, New Zealand, India and elsewhere. In 1977 the Australian branch was dissolved, becoming the *Australian Institute of Energy.* The Institute is the premier professional body in the field of energy, granting professional recognition to university graduates with energy and engineering academic training and appropriate practical experience. Corporate members of the Institute of Energy are registered with the CEI as chartered engineers. The Institute also caters for the needs of energy managers, students and others with an involvement in energy matters, who are eligible for non-corporate membership of the institute. On 6 February 1979, at a meeting of Her Majesty's Privy Council, formal approval was given for the change of name from the Institute of Fuel to the Institute of Energy.

Insulating firebrick A fireclay product of high porosity, which withstands high temperatures and yet possesses good heat-insulating value. Of low heat-storage capacity it enables a furnace to be brought quickly up to its working temperature; however, because of the porosity of the brick, it can only be used under 'clean heat' conditions.

Insulation The use of materials of low thermal conductivity in order to retain heat within a furnace or flue. *See* **conduction**.

Integrated circuits Semi-conductors that can amplify, switch, store and control electrical signals; made of silicon and germanium, they are halfway between metals and insulators. On an integrated circuit thousands of such semi-conductor circuits are imprinted on a tiny silicon chip measuring perhaps only four millimetres square. The basis of *microprocessors*.

Integrator An instrument which adds up, or integrates, the momentary rate of flow to give a total quantity of whatever is being measured over any period. Instruments are available, for example, to count or integrate the total quantity of steam produced or water evaporated.

Inter-governmental Maritime Consultative Organisation (IMCO) An agency of the United Nations, concerned solely with maritime affairs; 86 countries are members of IMCO. The first IMCO Assembly met in London in 1959; the Assembly consists of representatives from all the member countries. In 1973 the Assembly established a Marine Environment Protection Committee which deals with questions of marine pollution caused by ships; the prevention of pollution now forms one of the major constituents of IMCO's work programme. In 1954 an *International Convention for the Prevention of Pollution of the Sea by Oil* was agreed upon; in 1973 an *International Convention for the Prevention of Pollution from Ships* was agreed upon, which upon entering into force will supersede the earlier convention.

Intermittent kiln In the heavy clay industry, a non-continuous kiln, the two main types being: (a) the rectangular; (b) the round or 'beehive' kiln. Each kiln is normally connected to one or two external stacks; some round kilns have a stack in the centre which projects through the crown. They are fired by coal or oil. *See* **brick kiln**.

Intermittent vertical retort A closed chamber used for the manu-

196

facture of *town gas* by the *carbonisation* of coal. A retort of this type is usually about 7 m high and 3 m wide; the depth varies from 20 cm at the top to 30 cm at the bottom. It holds about $3\frac{1}{2}$ tonnes of coal. The carbonisation period lasts between 10 and 12 hours, including 2 hours' steaming at the end. Heat is provided by producer gas.

Internal water softening The softening of water inside a boiler. Sodium carbonate added to boiler water at low pressures will dissolve to give carbonate ions which will combine with calcium and magnesium ions as a loose, soft sludge. For higher pressures, sodium phosphate may be used; it is stable and does not add carbon dioxide to the steam. The phosphate can be supplied in various forms, such as trisodium phosphate, disodium hydrogen phosphate and Calgon.

International Atomic Energy Agency A United Nations agency whose main objective is 'to accelerate and enlarge the contribution of atomic energy to peace, health and prosperity throughout the world'. The statute for the International Atomic Energy Agency was approved on 26 October 1956 at an international conference attended by representatives of 82 countries held at the United Nations headquarters, New York. The statute came into force on 29 July 1957. The first session of the Agency was held in Vienna, Austria, in October, 1957. The Agency has at present about 76 members. The Agency administers international safeguards in respect of nuclear material, employing a nuclear inspectorate, in an effort to prevent the diversion of materials to the manufacture of nuclear weapons.

International Coal Classification (ECE) A classification system for coals prepared by the Classification Working Party of the Coal Committee of the Economic Commission for Europe. It is a two-dimensional scheme, the horizontal parameter being based upon *volatile matter* content up to 33 per cent; and on *calorific value* for coals with more than 33 per cent of volatile matter on the *dry ash-free* basis. This parameter divides the coals into nine classes; the classes are then divided into groups and sub-groups according to the coking or caking properties of the coals. *See coal classification systems.*

International Convention for the Prevention of Pollution from Ships A convention concluded in London in November 1973, at the end of the International Marine Pollution Conference, convened by the Intergovernmental Maritime Consultative Organisation (IMCO). The conference—the largest ever held on the subject—was attended by 500 representatives of the leading maritime nations.

The new Convention will enter into force twelve months after it has been ratified by 15 countries constituting at least 50 per cent of the world's merchant shipping. When it does enter into force, the 1973 Convention will supersede the International Convention for the Prevention of the Pollution of the Sea by Oil 1954, which, however, will remain in force meantime. The 1973 Convention forms part of the body of maritime law.

The new instrument contains provisions aimed at eliminating pollution of the sea by both oil and the noxious substances which may be discharged operationally, and at minimising the amount of oil which would be accidentally released in such mishaps as collisions or strandings.

The Convention applies to any ship of any type whatsoever, including hydrofoil boats, air-cushion vehicles, submersibles, floating craft and fixed or floating platforms, operating in the marine environment. It covers all aspects of intentional and accidental pollution from ships by oil or noxious substances carried in bulk or in packages, sewage and garbage. But it does not deal with dumping, which is covered by the Convention for the Prevention of Marine Pollution by Dumping from Ships and Aircraft signed in Oslo in 1972, or with the release of harmful substances directly arising from the exploration, exploitation and associated offshore processing of sea-bed mineral resources.

International Convention for the Prevention of Pollution of the Sea by Oil (1954) A Convention which prohibits the deliberate discharge of oil or oily mixtures from all sea-going vessels in specific areas called 'prohibited zones'. In general, these extend at least 50 miles from all land areas, although zones of 100 miles have been established in areas which include the Mediterranean and Adriatic Seas, the Gulf and Red Sea, the coasts of Australia and Madagascar, and some others. The participating countries undertake to promote the provision of facilities for the reception of oil residues; the Convention also prescribes that every ship which uses oil fuel and every tanker shall be provided with a book in which all the oil transfers and ballasting operations shall be recorded, including entries on accidental or other exceptional discharges and escapes. Although the restrictions imposed by the 1954 Convention have been very effective, the enormous growth in oil movements since then has necessitated more stringent regulations. *See Inter-governmental Maritime Consultative Organisation; International Convention for the Prevention of Pollution from Ships.*

International Energy Agency (IEA) An autonomous body established within the framework of the **Organisation for Economic Co-operation and Development** to implement the International Energy Programme adopted by the participating countries in November

1974. The Agency carries out a comprehensive programme of energy co-operation among 20 countries, and seeks to achieve co-operation with oil-producing nations and other oil-consuming nations. The 20 countries are Australia, Austria, Belgium, Canada, Denmark, Germany, Greece, Ireland, Italy, Japan, Luxembourg, the Netherlands, New Zealand, Norway, Spain, Sweden, Switzerland, Turkey, the United Kingdom and the United States of America. The basic aim of the IEA group of countries is to bring about a better structure of energy supply and demand over the near and longer term. Collectively, these countries consume over half of the world's energy supply. The International Energy Programme was the result of an Agreement reached in the wake of the 1973–74 energy crisis. In 1977 a further agreement was reached to limit the total oil imports of IEA countries to 26 million barrels per day by 1985. The *Twelve Principles for Energy Policy* were endorsed. An *oil market information system* has been developed.

International Energy Bank   Based in London, England, a specialised bank for financing energy projects world-wide. Its shareholders include the Bank of Scotland, Barclays Bank International, Banque Worms, Canadian Imperial Bank of Commerce, Republic National Bank of Dallas and the Société Financière Européenne of Luxembourg.

International Energy Conservation Month   A concept promoted by the *International Energy Agency* to provide an international focus for national efforts to stimulate greater public awareness of the continuing and long-term need for energy conservation. It gives member countries an opportunity to plan events which will culminate during the month, or to use the month as a springboard for programmes. In 1979 the month of October was chosen.

International Institute for Energy and Human Ecology   Also known as the Beijer Institute, a research body of the Royal Swedish Academy of Sciences. Established in March 1977, the Institute was supported initially by a grant from the Kjell and Märta Beijer Foundation. The aim has been to promote research concerned with long-term issues of international importance. Studies undertaken have included improved energy utilisation in developing countries, impact assessment methodologies for energy options, environmental standards and the institutional aspects of the introduction of new energy systems. The Institute collaborates with other national bodies such as the Swedish Energy Commission, and with overseas bodies such as the Kenya Academy of Sciences.

International joule   An energy unit which became obsolete in 1948; the old joule was about equal to 1.000 19 SI joules. *See joule.*

International majors Oil industry terminology for companies operating internationally and accounting for a large share of all production outside the communist countries and North America. Of the seven international majors, five have their headquarters and an overwhelming majority of shareholders in the United States of America. The seven are the Exxon Corporation (formerly Standard Oil of New Jersey), which trades through most of the world under the Esso sign; Standard Oil of New York, trading as Mobiloil; Standard Oil of California, operating under the chevron sign; Gulf Oil, with headquarters in Pittsburgh; Texaco, operating from Texas; the Royal Dutch–Shell group; and British Petroleum. These seven companies also own and control a major share of refining capacity and internationally operating tankers. Increasingly, developing countries and others have viewed the presence of these major companies as a form of economic colonialism incompatible with national sovereignty. Known also as the 'seven sisters'.

International Nuclear Fuel Cycle Evaluation A major international study launched in 1978 as part of President Carter's nuclear initiatives; the purpose was to examine and critically review all aspects of the nuclear fuel cycle. Some 40 nations participated in the study, the aim of which was to find ways to lessen the risks of proliferation.

International Petroleum Industry Environmental Conservation Association (IPIECA) An organisation set up by the oil industry to respond to the United Nations Environment Programme (UNEP) in respect of the problems of the industry.

International practical temperature scale A temperature scale adopted in 1948 by the General Conference on Weights and Measures; temperatures on this scale are designated as °C (Celsius). The international temperature scale is based on a number of fixed and reproducible equilibrium temperatures (fixed points) to which numerical values are assigned. The fundamental points are:

	°C
melting point of ice	0
boiling point of water	100

The primary points are:

boiling point of oxygen	−182.970
boiling point of sulphur	444.600
freezing point of silver	960.8
freezing point of gold	1063.0

There are also a number of secondary fixed points. In 1968 units became expressed in both kelvins and degrees Celsius, the temperature interval being the same. *See **temperature scales**.*

International Referral System (IRS) A system introduced as a consequence of the ***United Nations Conference on the Human Environment*** which aims to put in touch those who seek information on environmental issues with those best able to provide such information. The system became fully operational in 1977.

International System of Units (SI) A metric system adopted in 1954 by the Tenth General Conference on Weights and Measures from the basic group of units from which could be derived in a coherent and rational way practically all the units required by the physical sciences and engineering technology. At the Eleventh General Conference in 1960 it was formally resolved that this system be designated as 'The International System of Units', with the symbol SI. The system has also been accepted by the International Organisation for Standardisation (ISO). The system provides for supplementary and derived units, e.g. for velocity, the metre per second; for force, the newton ($N = kg\ m/s^2$). *See Table I.1; **British system; metric system; MKS system**.*

Table I.1 SI UNITS

Basic units for	Name	Symbol
Length	metre	m
Time	second	s
Mass	kilogram	kg
Electric current	ampere	A
Temperature	kelvin	K
Luminous intensity	candela	cd
Amount of substance	mole	mol

Derived units for		
Energy	joule	J
Force	newton	N
Power	watt	W
Electric charge	coulomb	C
Electric potential difference	volt	V
Electric resistance	ohm	Ω
Electric conductance	siemens	S
Electric capacitance	farad	F
Magnetic flux	weber	Wb
Inductance	henry	H
Magnetic flux tensity	tesla	T
Luminous flux	lumen	lm
Illumination	lux	lx
Frequency	hertz	Hz

International Union of Air Pollution Prevention Associations A union of non-governmental organisations established in 1965 to promote collaboration and the exchange of information relating to air pollution and environmental problems generally, and to organise international congresses. The six founder members were Argentina, France, Japan, the United Kingdom, the United States of America and West Germany. The first congress was held in London, England, in 1966; since then congresses have been held in Washington, DC, Dusseldorf and Tokyo.

Interruptible contract A contract for the supply of *natural gas* to boiler plant which permits an interruption of supply, if necessary, to meet the domestic home heating demand which has priority; a common feature of United States fuel contracts in industrial areas. Firms likely to be affected require alternative fuels such as oil or coal as 'second fuels' during the winter months.

Inter-tube superheater *See superheater*.

Inverse square law In relation to radiant heat, a law governing the variation of heat received per unit area (λ) from a given source at varying distance (d); the heat received at any point varies inversely with the distance squared:

$$\lambda = \frac{1}{d^2}$$

This means that equal areas of material held at distances of 2, 7 and 10 m from a source of radiation will receive energy at $\frac{1}{4}$, $\frac{1}{49}$ and $\frac{1}{100}$ of the rate of the heat received at a distance of 1 metre from the source. The law also applies to light waves. In air pollution control the ground level concentration of a pollutant varies inversely with the square of the *effective height of emission*; hence, a doubling of the effective height reduces ground level concentrations by 75 per cent.

Inversion A temperature inversion in the atmosphere, in which the temperature, instead of falling, rises with height above the ground. With the colder and heavier air below, there is no tendency to form upward currents, and turbulence is suppressed. Inversions are often formed in the late afternoon when the radiation emitted by the ground exceeds that received from the sinking sun. Inversions are also caused by katabatic winds, i.e. cold winds flowing down the hillside into a valley, and by anticyclones. *See Figure I.3; lapse rate; subsidence inversion*.

Ion An *atom* or molecule which has lost or gained one or more electrons, thus possessing a net positive or negative charge. The loss of electrons produces a positive ion (cation); the gain of electrons produces a negative ion (anion). *See electron*.

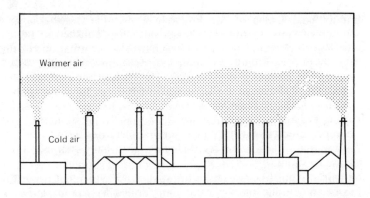

Figure I.3 Temperature inversion. A pall of pollution trapped within a stable layer. (Source: Warren Spring laboratory, Stevenage, England)

Ion exchange process A method of softening water which depends on the property of certain synthetic resins which enables them to give up ions in exchange for ions from the water; thus, calcium and magnesium cations can be replaced by sodium cations. When the resin has given up most of its sodium ions, the efficiency of the process begins to fall and the exchange material has to be regenerated. To regenerate the resin, a strong solution of brine is used from which sodium ions are given up to the resin. Calcium and magnesium ions take their place and remain in solution with the chloride anions. *See Figure I.4.* Also known as the base exchange process. *See **lime soda process**.*

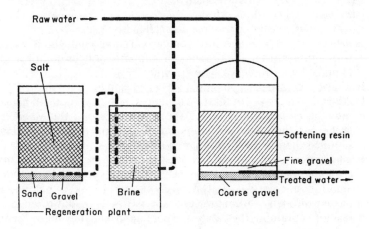

Figure I.4 Ion or base exchange process. (Source: How Group Ltd.)

203

Ionisation   The process by which a neutral atom or molecule loses or gains electrons, thus becoming electrically charged. *See ion.*

Ionisation chamber   A device for measuring the quantity of ionising radiation consisting of a chamber containing two electrodes and a gas. The radiation is measured by collecting the ions which it produces in the gas at the electrodes under the action of an applied electric field. *See radiation detector.*

Ionising radiation   Radiation which, by reason of its nature and energy, interacts with matter to remove electrons from the atoms of the matter absorbing it, producing electrically charged atoms known as ions.

Ion pair   A positive *ion* and a negative ion having charges of the same magnitude but opposite sign formed by the *ionisation* of neutral atom or molecule.

IP burning test   A test for the burning quality of kerosine. The deposit-forming tendency of a kerosine is determined by the char formed on a wick during a 24 hour test.

Iranian oil crisis   A major interruption in oil exports from Iran as a result of a political crisis beginning in October 1978 with a series of strikes in the oil industry. By December oil production ceased completely. Iran had previously been producing nearly 6 million barrels a day, of which 5 million barrels were exported. This represented about 10 per cent of the free world supply. By July 1979 Iran had resumed production with the Ayatollah setting an immediate target of 60 per cent of what the Shah had been exporting.

Irradiation   The exposure of matter to radiation.

Isenthalpic   Of constant *enthalpy.*

Isentropic   Of constant *entropy.*

Isobaric   Of constant pressure; also called isopiestic.

Isobutane C_4H_{10}. A specific hydrocarbon compound which reacts with light olefins in the presence of sulphuric or hydrofluoric acid to make high-octane alkylate gasoline.

Isochoric   Of constant volume; also called isometric.

Isokinetic   In respect of taking a sample of flue gas or other exhaust product, a situation in which the flow of gas into the *sampling probe* has the same flow rate and direction as the gas being sampled; hence, 'isokinetic sampling'.

Isomerisation   An oil refinery process for producing isobutane from *n*-butane, or isopentane from pentane. In respect of butane an aluminium chloride catalyst is employed. Liquid *n*-butane is passed through a drying tower, vaporised and passed through a reactor containing the catalyst, where approximately 40 per cent of the *n*-butane becomes isobutane.

Isoparaffins   A range of hydrocarbons characterised by branched carbon chain molecular structures which modify some properties such as *cetane number* and *pour point*, compared with corresponding normal paraffins.

Isopentane C_5H_{12}. A component of high-grade gasoline, it is obtained in practically its pure form by highly efficient fractionation of light straight-run gasolines.

Isothermal Of constant temperature.

Isotopes Atoms possessing nuclei with the same number of protons in each but with different numbers of neutrons. Isotopes are distinguished from one another by means of their mass number, which is the sum of the numbers of protons and neutrons in the nucleus.

Ixtoc 1 incident A major blow-out from an oil well off the coast of Yucatan, Mexico, during the period June–September 1979, in which oil escaped into the ocean at the rate of up to 30 000 barrels a day. The huge oil slick eventually stretched 500 kilometres northwestward from the well across the Gulf of Mexico. This massive incident has implications for offshore oil exploration around the world. *See Royal Commission on the Great Barrier Reef.*

J

Jerk system A category of *fuel injection equipment* for diesel engines; separate fuel pumps are used to deliver fuel to each cylinder during the injection period only. *See diesel engine.*

Jet condenser A steam condenser in which cooling is achieved by mixing the steam with a spray of water.

Jet propulsion A general name given to forward motion produced by the discharge of a stream of gas or liquid at the rear of an engine. The four main types of jet engines used in aircraft are the turbojet, the turboprop, the pulse-jet and the ramjet.

Joint Coal Board Established by the Australian Government and the Parliament of New South Wales in 1947, a board given wide powers and responsibilities concerning the production and distribution of coal and the welfare of the workers in the industry in that State. In 1950 only about 40 per cent of coal mined in New South Wales was cut or loaded mechanically, compared with a mechanical cutting rate of 90 per cent and a loading rate of 60 per cent in the United States, with similarly high rates in the United Kingdom. By 1966 practically all coal mined underground in New South Wales was mechanically cut and loaded. Working conditions were also greatly improved.

Joint European Torus (JET) A major nuclear fusion project located at Culham, England, following a decision of the *European Economic Community* in 1978 in respect of this joint product.

Jojoba bean Pronounced ho-ho-ba, a bush which grows in arid and semi-arid areas and has a high tolerance for salinity; it can live for more than 150 years, producing an olive-shaped bean. The bean yields half its weight in oil; it is one of the finest lubricants known and chemically almost identical with the oil from sperm whales. American Indians in California and Arizona have used the jojoba oil for cooking for centuries. Israel has established a jojoba industry in the Negev Desert and Dead Sea regions. Plantations are planned in Australia.

Jones oil gasification process A cyclic process for the manufacture of gas in which oil is thermally cracked with steam to produce a gas having a calorific value in the range 15–20 MJ/m³. The process yields appreciable quantities of carbon black, about half of which is used in the process itself. The plant consists of a two-vessel generator and two-filter gas producers. One blow and run cycle normally take 7–10 minutes and then the flow is reversed. During the blow, primary air is blown through one of the gas producers, the resultant producer gas being burned with preheated secondary air above and through the chequer work in the generator; the waste gas is discharged to atmosphere. Steam is then introduced through the gas producer, the resulting blue water gas passing to the generator; oil is sprayed into the generator and the oil vapours crack in the atmosphere of blue water gas and steam; half the suspended carbon black is filtered out in the other filter gas producer, which contains a bed of mechanically agitated refractory spheres; this carbon black is consumed when the blow-and-run cycle is repeated from the opposite direction. The remaining carbon is removed in a wash box.

Joule Unit of energy, including work and quantity of heat; the work done when the point of application of a force of 1 newton is displaced through a distance of 1 metre in the direction of the force:

$$1 \text{ joule} = 1 \text{ watt second}$$
$$= 10^7 \text{ erg}$$
$$= 0.238\ 846 \text{ calorie}$$
$$= 0.526\ 565 \times 10^{-3} \text{ Chu}$$
$$= 0.947\ 817 \times 10^{-3} \text{ Btu}$$
$$= 0.737\ 562 \text{ foot pound force}$$
$$= 1 \text{ newton metre}$$

*See **dyne; erg**.*

JP4 Term used in the United States of America for aviation turbine gasoline (AVTAG).

Junkers calorimeter A *gas calorimeter* in use in Europe and the United States; the *calorific value* of a known volume of gas is

206

calculated from the increase in temperature of a measured volume of water heated by the gas under test.

K

Kaldo furnace A *steelmaking furnace* of Swedish origin; it consists of a cylindrical furnace lined with refractory bricks, and as it rotates, a jet of oxygen is blown against the surface of the molten metal in the vessel.

Katabatic wind A wind caused by cold air flowing downhill. When a sloping land surface cools by night-time radiation, the cold air in contact with the ground flows downhill and along valley bottoms. Opposite of an anabatic wind. *See Figure K.1.*

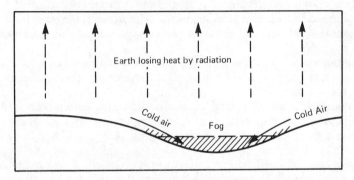

Figure K.1. Katabatic wind. Cold air accumulates in a valley to form a persistent temperature inversion (Source: Warren Spring Laboratory, Stevenage, England)

Katharometer principle A principle used in measuring instruments, depending on the different thermal conductivity of various gases; hydrogen is commonly adopted as a reference gas, its thermal conductivity being much greater than that of most gases.

Kelly In respect of oil drilling, a square hollow shaft of length slightly greater than that of the sections of drill pipe, which engages in the rotary table imparting rotation to the drill pipe.

Kelvin scale An absolute temperature scale; the kelvin (K) is the SI unit of thermodynamic temperature defined as the fraction 1/273.16 of the thermodynamic temperature of the *triple point of water*, the triple point of water containing exactly 273.16 K. The temperature of melting ice is 273.15 K (zero on the Celsius scale), while the boiling point of water at normal atmospheric pressure is 373.15 K (100 on the Celsius scale). The units of the kelvin and Celsius (centigrade) temperature scales are identical. Both the kelvin and the Rankine scales are absolute scales; zero on each is 'absolute zero'; 1 kelvin degree = 1.8 Rankine degrees. The name 'degree kelvin' (symbol K) was discontinued by international agreement in 1967. Named after Lord Kelvin (1824 – 1907). *See* *temperature scales.*

Kenotometer   An instrument in common use in power stations for measuring condenser back pressure by means of a column of mercury.

Kerosine   A petroleum fraction of light distillate fuel, with a boiling range 150 – 280° C and a specific gravity of 0.78 – 0.82. It is used for heating, in aviation gas turbines, and in some industrial gas turbines. It is similar to paraffin as used in domestic burners; it is very largely a mixture of paraffins, isoparaffins and naphthenes containing up to 18 per cent weight of aromatic compounds.

Kick's law   A statement that assumes that the energy required for the subdivision of a definite amount of material is the same for the same fractional reduction in the average size of the individual particles.

Killing a well   The filling of an oil well bore with fluid of a suitable specific gravity, usually *mud*, which overcomes the tendency of the well to flow.

Kilocalorie (Cal)   The large calorie, or dieticians' calorie; this is the heat energy required to raise the temperature of 1 kilogram of water through 1 degree Celsius, from 14.5° C to 15.5° C. One Cal = 4.1855 kJ.

Kilogram   A basic SI unit, equal to the mass of the international prototype of the kilogram, which is in the custody of the Bureau International des Poids et Mesures (BIPM) at Sèvres, near Paris, France. This is a unit of mass and not of weight or force; it is approximately the same as the mass of 1 cubic decimetre (litre) of water at maximum density (4° C). For everyday purposes, mass and weight are synonymous, so that the kilogram is now a basic unit in many countries of the world for commercial purposes.

Kilovolt-ampere, (kVA)   A unit applying to *alternating current* only; it is the product of the pressure in voltage and current in amperes which, when multiplied by the *power factor* gives the power in kilowatts. It is sometimes referred to as the 'apparent' power.

Kilowatt.   One thousand watts; the larger unit of electrical power. In respect of a *direct current* supply, or a single-phase *alternating current*, supply at a *power factor* of unity, the number of kilowatts is obtained by multiplying the pressure in volts by the current in amperes and dividing by 1000. If the power factor is less than unity, the answer obtained must be multiplied by the power factor. In three-phase systems the voltage across the phases, current per phase, and the power factor must be multiplied together and further multiplied by the square root of 3 (i.e. 1.73) to give an answer in kilowatts.

Kilowatt-hour (kWh)   The unit of electrical energy; 1 kilowatt-hour = 3.6 megajoules (MJ) = 3600 kilojoules (kJ).

Kinematic Viscosity   The ratio of the *dynamic viscosity* to the density of a liquid, the units being stokes or *centistokes*. Kinematic viscosities are quoted at 50° C. *See U-tube viscometer; viscosity.*

Kinetic energy   *See energy.*

King–Maries–Crossley formula   Or 'KMC formula'; a correction formula for determining the *mineral matter* content of coal from laboratory data (*J. Soc. Chem. Ind.*, **55**, 277T, 1936). This formula was revised in 1956 by the *National Coal Board* and in its revised form states:

$$M.M. = 1.13A + 0.5\,S_{pyr} + 0.8\,CO_2 - 2.8\,S_{ash} + 2.8\,S_{sulph} + 0.5\,Cl$$

where, as percentages of the air-dried coal,

$M.M.$ = mineral matter in the coal;
A = ash obtained on incineration;
S_{pyr} = pyritic sulphur in the coal;
CO_2 = carbonate CO_2 in the coal;
S_{ash} = total sulphur in the ash;
S_{sulph} = sulphur present as sulphates in the coal;
Cl = total chlorine in the coal.

A simplified formula subsequently derived by the British Coal Utilisation Research Association is:

$$M.M. = 1.10A + 0.53\,S + 0.74\,CO_2 - 0.32$$

where $M.M.$ = mineral matter;
A = ash;
S = total sulphur;
CO_2 = total carbon dioxide.

Kingsnorth high-voltage direct current link   A direct current interconnection completed in 1975 between load centres on the west side of the London area, England, and the main generation points to the east; it was the first scheme in the world to integrate relatively short-distance direct current transmission with a closely interconnected alternating current system in an urban area. Up to 640 MW can be transmitted from Kingsnorth power station in Kent by underground cables some 58 kilometres to Beddington and a further 24 kilometres to Willesden.

Kirchoff's radiation law   A statement that the radiating capacity of a body is proportional to the absorbing capacity of the body, at any given temperature and any given wavelength. A perfect emitter and absorber is a **black body**, with an emissivity and absorbtivity of unity.

Knocking Detonation in engine cylinders, accompanied by a knocking sound, associated with decrease in power output and overheating; combustion in a cylinder should proceed in a smooth manner at a moderately fast rate, not instantaneously. Also called 'pinking'.

Knock limited power The maximum power an engine will produce under a prescribed set of conditions without 'knocking'.

Koppers–Hasche process A cyclic process for reforming methane and natural gases to make them more suitable for use as town gas; the process consists of partial oxidation of the gases with air using an alumina catalyst:

$$2CH_4 + O_2 \rightarrow 2CO + 4H_2$$

The process is carried out in two catalyst-packed horizontal chambers, reversal of flow taking place every minute. The amount of air used determines the **calorific value** of the gas produced.

kPa Abreviation for kilopascal, or one thousand pascals; a unit of pressure in the SI system. 100 kPa, or $10^5 Pa$, equals 1 bar. One atmosphere of pressure (standard atmosphere, atm) equals 101.325 kPa.

Krypton Kr. An inert gas; at. no. 36, at. wt. 83.80. It is an important product of **fission** in a **nuclear reactor**.

L

Labile substance A substance which is unstable or reactive.

Lagging Insulating material applied to the outer parts and surfaces of equipment to prevent heat loss, and distortions arising from unequal temperatures.

Lambert's law of radiation A statement that the radiation from a hot surface in a direction at an angle with the surface varies as the cosine of the angle ϕ between the direction of the radiation and the normal to the surface. The radiation in a given direction is therefore:

$$q = q_n \cos \phi$$

where q_n is the radiation normal to the surface. *See Figure L.1.*

La Mont boiler A **water-tube boiler** employing forced circulation. The boiler heating surface comprises a number of tube elements working in parallel, the inlet end of these tube elements being expanded into the distribution header of the boiler. Controlled circulation by pump affords protection against the overheating of

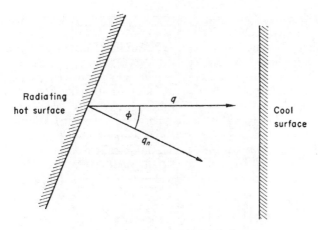

Figure L.1 Lambert's law of radiation

boiler pressure parts; it permits the use of small-bore tubes, which increases the efficiency per square foot of tube surface. The La Mont design has been adopted for all classes of plant, ranging from low-pressure water boilers up to large power station units incorporating all types of firing equipment, and utilising all classes of fuel. *See Figure L.2.*

Lancashire boiler A horizontal shell boiler, the shell containing two plain or corrugated furnace tubes running through the boiler from end to end; it has a brick flue setting. The hot combustion gases leave the combustion zone and pass through the furnace tubes to the rear of the boiler, where they enter a downtake in which a superheater is sometimes located. From the downtake the gases enter a bottom flue and return again to the furnace front; there they divide into two streams passing down side flues, one being situated on each side of the boiler, to the rear. Here the gases reunite in a single main flue and pass to the chimney. The working pressure of a Lancashire boiler rarely exceeds 17 bar. Fitted with a feed-water economiser and mechanically fired, the thermal efficiency may be raised to about 75 per cent. Evaporative capacities range from about 0.63 kg/s to 2.5 kg/s. The Lancashire boiler has been superseded by economic boilers and package boilers of higher efficiency. *See boiler; Cornish boiler.*

Land breeze The movement of air from land to sea. On clear nights the land cools faster than the water, cooling more quickly the lowest layers of air. The heavier land air spills seaward, displacing the warmer sea air.

Land-fill The most common form of disposal of refuse; it appears

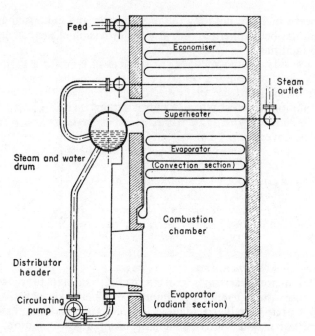

Figure L.2   La Mont forced circulation water-tube boiler

that between 80 to 90 per cent of the world's refuse will be disposed of by this method for several years to come. The type of sites being used for such disposal are: mineral excavations; low-lying land; valleys; areas involving the reclamation of land from water; and flat land to build up a feature. Generally, in a sanitary land-fill scheme, refuse is tipped in trenches or cells prepared to such a width that the daily input of refuse can be effectively covered, presenting a clean face each day. The refuse can be tipped either at the bottom of the face and dozed into the face, or tipped on top of the previous fill and dozed over the face. It is essential that the refuse be adequately covered and compacted to allow traffic over the fill.

The land-fill method is known in the United Kingdom as 'controlled tipping' and in the United States by the fuller title of 'sanitary land-fill'; while the content of both expressions is identical, the former emphasises the system by which the waste is deposited, while the latter emphasises the hygiene aspects.

Landscape or amenity conservation The safeguarding for public enjoyment of scenery or landscape, and of opportunities for outdoor recreation, tourism, field sports and similar activities; the

212

concept includes not only the preservation and enhancement of what has been inherited, but also the provision of new amenities and facilities.

Land use   The deployment of the services of land for a variety of uses, e.g. agriculture, industry and commerce, housing and recreation. Land use planning has as its object the spatial co-ordination of such activities on a national, regional and local level. Through zoning, performance specifications and building codes, governments can not only specify the uses to which land can be put, but also place restrictions on those uses. Controls can prevent the establishment of industries with a high air pollution potential in poorly ventilated basins and valleys; and the establishment of noisy or odorous processes near residential districts. The land use implications of proposed transport developments may be properly taken into account at all levels of planning.

Lapse rates   The rate of decrease of temperature with increasing height in the *atmosphere*. From the surface to a height of about 11 kilometres, known as the troposphere, the lapse rate is, on average about 6–8° C per kilometre. There are considerable departures from this average at all times and at all levels, and it is these departures which characterise the type of weather experienced. If the temperature shows no change with height, the condition is described as isothermal. In the stratosphere, the layer of the atmosphere above the troposphere, the lapse rate is isothermal, or very nearly so.

If a parcel of air is displaced upwards without any heat being supplied or removed, the air will expand and cool at a rate which is known as the dry adiabatic lapse rate (DALR), the value of which is 10° C per kilometre. This is also known as a 'neutral gradient'. If a parcel of saturated air is similarly displaced, the cooling will be less than with dry air, owing to the latent heat made available from the condensation of moisture. The rate at which it cools, known as the saturated adiabatic lapse rate (SALR), is about 6° C per kilometre. The saturated adiabatic lapse rate is not constant at all levels. An atmosphere in which the temperature decreases with height more quickly than the DALR is said to be unstable or superadiabatic. If the lapse rate is negative, i.e. temperature increases with height, the atmosphere is said to be stable; hence the term 'temperature inversion', or simply 'inversion'. Sometimes the word 'stable' is attached to any atmosphere in which the lapse rate is less than the DALR. The actual lapse rate in the atmosphere is known as the environment lapse rate (ELR). These temperature gradients and their characteristic climatic conditions are important factors in determining the rate of diffusion of air pollutants in the atmosphere. *See Figure L.3; **air pollution; inversion; static stability; superadiabatic lapse rate***.

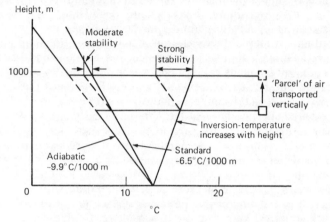

Figure L.3 Vertical stability in different atmospheres

Large coal Coal with no specified maximum size, but with a specified minimum size of, say, 8 cm or more.

Latent heat Hidden heat; the number of heat units absorbed or released by unit weight of a substance during a change of physical state, e.g. from water into steam or from water into ice. The heat transfer involved is not shown by a thermometer; hence the description 'hidden' heat. *See **latent heat of fusion; latent heat of vaporisation***.

Latent heat of fusion Heat required to melt unit mass of a substance while its temperature remains constant.

Latent heat of vaporisation Heat required to convert unit mass of a liquid into the gaseous state while its temperature remains constant.

Lattice In a *heterogeneous reactor*, a geometrical arrangement of the fuel elements and the *moderator*.

Lead Pb. A heavy metallic *element*; at. no. 82, at. wt. 207.19. It is widely used as a radiation shielding material because of its high density. Lead and its compounds can be retained in the human body with clinical and subclinical effects. Lead in the atmosphere arises mainly from the anti-knock additives used in petrol. However, the major part of human intake of lead comes not from air but from food and drink. *See **gasoline additives***.

Lead peroxide candle An instrument for measuring the relative concentrations of *sulphur dioxide* in the general atmosphere. It consists of exposing a lead peroxide surface; the rate of sulphation is proportional to the sulphur dioxide. The instrument is usually exposed for a period of 1 month. The casual errors are relatively large and the instrument may be affected by humidity; the relationship of the readings expressed in mg/SO$_3$ 100 cm^2 day, to actual concentrations of SO$_2$ is only approximate. *See National Survey of Air Pollution; volumetric SO$_2$ apparatus.*

Lead susceptibility The degree of response by way of an increase in *octane number* shown by a gasoline to which tetraethyl lead (TEL) has been added; the response is not uniform but depends on the composition of the gasoline, e.g. the presence of sulphur compounds reduces lead susceptibility.

Lean gas Gas of relatively low calorific value. *See rich gas.*

Lean mixture An air–fuel mixture in an internal combustion engine which contains a low proportion of fuel. A mixture with a high proportion of fuel is described as rich.

Lean oil *Absorption oil* from which dissolved natural gasoline fractions have been removed.

Leaving loss The loss of energy in the velocity of turbine exhaust steam.

Lever or steelyard safety valve A type of *safety valve* which possesses one weight only equal to the maximum load when the weight is at the end of the lever. The weight should be attached in such a way that it cannot be moved inadvertently.

Lidar (Light Detection and Ranging) A technique used in *air pollution* research to detect and track chimney plumes at considerable distances, long after they have become invisible to the naked eye. Lidar was developed for meteorological use in 1963 by M. G. H. Ligda and co-workers at the Stanford Research Institute. The development of the pulsed ruby laser has permitted the direct extension of radar techniques to optical wave lengths. The term 'lidar' (light detection and ranging) has come to be applied to such optical radar. Lidar can readily detect the small particles which make up the aerosol content of the atmosphere and smoke plumes.

Lidar equipment comprises: a laser transmitter which emits very brief, high-intensity pulses of coherent light; a receiver, which detects the energy at that wavelength back-scattered from the atmospheric aerosol as a function of range.

Life system concept That part of an *ecosystem* which determines the existence, abundance and evolution of a particular population. A life system comprises a subject population and its effective *environment*; the latter includes all those biotic and abiotic agencies influencing the population. Population and environment are regarded as interdependent elements which function together as a system.

Light ends The lower-boiling components of a mixture of hydrocarbons.

Light energy A type of wave motion which has different lengths between the crests of each wave; light which has different wavelengths appears different in colour. The shortest wavelength of light that can be detected by the human eye is violet in colour; intermediate wavelengths are greenish to yellowish; the longest wavelength of light visible to the human eye is red. Neither ultra-violet light, which has a wavelength shorter than violet, nor infra-red light, which has a wavelength longer than red light, is visible to the human eye.

Light meter A direct-reading instrument for measuring illumination, incorporating a photoelectric cell and an indicating meter calibrated in lumens per square foot, or in foot-candles.

Light-sensitive pyrometer A device for measuring temperatures, utilising a light-sensitive element such as a *photoelectric cell* to measure the intensity of light emitted by a hot object. The magnitude of current through the cell is determined by the intensity of radiation; the current is amplified electrically and recorded. The instrument is suitable for temperatures of 700° C upwards. *See radiation pyrometer*.

Light water-moderated nuclear reactors Nuclear reactors using enriched uranium and light (ordinary) water as both moderator and coolant. Two variations are in commercial use: the *pressurised water reactor* and the *boiling water reactor*. *See heavy water-moderated nuclear reactors; nuclear reactor*.

Lignite A solid fuel intermediate between *peat* and other coals; it shows a definitely woody structure. The carbon content is lower than for *brown coal*, while the moisture content may be as high as 50 per cent. Lignite is extensively used in many countries for heating and steam raising.

Lime-base grease A grease with water-resistant properties used for lubrication under wet or moist operating conditions.

Lime kiln In its most usual and modern form, a vertical steel cylinder lined with firebrick. Heat is supplied from the combustion of coal, oil or producer gas. Limestone is crushed and loaded continuously into the top of the kiln, the lime being withdrawn from a hopper at the base; each lump of limestone is in the kiln several hours, gradually falling towards the hot zone, which has a temperature of about 1100° C. During the process, carbon dioxide is driven off from the stone, the result being quicklime; the chemical reaction which takes place is:

$$CaCO_3 \rightarrow CaO + CO_2$$

The problem of completing the combustion process in an

atmosphere consisting largely of nitrogen and carbon dioxide gives rise to *dark smoke* emissions. Limestone is also calcined or burned in rotary kilns, similar to those used in the cement industry.

Lime silica block   An insulating material consisting of reacted hydrous calcium silicate. It is used on plant and pipes usually below 650° C.

Lime soda process   A method of softening water in which most of the calcium and magnesium is removed by controlled additions of lime and soda ash; the unwanted calcium and magnesium salts fall out as a sludge, the softened water being filtered from the top. Lime soda treatment needs a hard water of more than 100 ppm, but with not too high a proportion of magnesium. The process will not remove all the hardness, however, and must be followed by the *ion exchange process*.

Limnetic zone   The region of open water beyond the littoral zone of a lake, down to the maximal depth at which there is sufficient sunlight for *photosynthesis*. This is the depth at which photosynthesis balances respiration, known as the compensation depth. Rooted plants are absent in this zone, but there is a great abundance of phytoplankton.

Line reversal method   A method of measuring the temperature of clear flames. To obtain visible spectrum lines, it is necessary to add a metal compound such as a sodium salt; this is done in the sodium line reversal method. The emission spectrum of a flame containing salt shows the two yellow D lines of sodium in the emission. If a bright source emitting a continuous spectrum is viewed through the flame, the D lines appear as dark lines against the continuum. If the brightness of the background is adjusted, then there is a condition at which the sodium lines just disappear. Under prescribed optical conditions the brightness temperature of the background and the flame temperature are equal. The brightness temperature of the background source is determined by a separate measurement using a disappearing-filament optical pyrometer. It has been possible by this method to make measurements as high as about 9000° R (5000 K). *See pyrometer; thermometer*.

Linz–Donawitz converter   A *steelmaking furnace* of Austrian origin utilising oxygen; an L–D vessel resembles a normal *Bessemer converter*, but without provision for air injection in the base. After charging with scrap steel and molten iron, a long copper-tipped lance is lowered automatically until its nozzle is 1 metre above the surface of the metal. For about 18 minutes a jet of oxygen with a velocity of 600 m/s is projected down into the molten metal. Steel is produced without the nitrogen brittleness of normal basic Bessemer steel.

Liquefied natural gas (LNG) *Natural gas* stored as a liquid, 1 cubic metre of liquid natural gas being approximately equal to 600 cubic metres of gas at 0° C and atmospheric pressure. In the United Kingdom LNG is stored in liquid form at key terminal points in the national transmission system, partly to meet peak demands and partly to provide maximum security of supply in all parts of the system. North Sea natural gas is fed into the system at five input terminals on the east coast: LNG from Algeria is imported through a terminal at Canvey Island. Peak shave installations are planned for various areas of the country by the British Gas Corporation.

Liquefied petroleum gases (LPG) Paraffin hydrocarbon gases comprising propane, butanes and pentanes, obtained from natural gas wells and in petroleum refining. For efficient transportation, storage and use, propane and butane are liquefied under pressure and distributed in cylinders—hence the description 'bottled gas'. Apart from domestic use, these gases are used for the manufacture of chemicals and with other hydrocarbons used as components of aviation fuels. *See **butane; pentanes; propane**.*

Liquid-in-glass thermometer A device for measuring temperature; the temperature is indicated by the differential expansion of a liquid with respect to its glass container. The liquid may be mercury, alcohol, toluene or xylene, choice depending upon the temperature range over which the instrument will be used. For ordinary purposes mercury is suitable; if the temperatures to be measured are likely to fall below − 38.8° C, then ethyl alcohol or some other organic liquid is employed. The overall range for instruments of this type is − 200° C to + 500° C. *See **pyrometer; thermometer**.*

Liquid solvent extraction (LSE) A process for producing synthetic crude oil from coal, the oil being suitable for further refining into transport fuel and chemical feedstocks; in the process, coal is treated with a hot liquid solvent (derived from coal) to give a thick, tarry solution. This is filtered to remove ash and carbon residue; the solution is then hydrogenated and separated into synthetic crude oil product and recycle solvent. Products also include high-value hydrocarbons such as benzene, toluene and xylene. In 1979 pilot plants were in operation in both the United States of America and the United Kingdom.

Liquid-in-steel thermometer A device for measuring temperature; the measuring element consists of a Bourdon tube gauge acting as a pressure-measuring device, the expansion of the liquid altering the shape of a hollow, flexible metal coil which is connected to an indicator. Mercury is the usual liquid. The instrument is suitable for the range 300–700° C, although by using an inert gas lower temperatures can be reached satisfactorily. *See* **pyrometer; thermometer**.

Lithotype A macroscopically recognisable banded coal type, greater than 3 mm thickness; it includes **clarain**, **durain**, **fusain** and **vitrain**. *See* **microlithotype**.

Litre One cubic decimetre; this is the SI definition effective from 1964, superseding the 1901 definition, which made the litre equal to 1.000 028 cubic decimetres.

Live steam accumulator A vessel containing water in which surplus steam from the main steam generator is stored under a gradually rising pressure, and regenerated as steam under a falling pressure. A steam accumulator is one method of meeting fluctuating demands for steam; under ideal conditions it means that the main steam generator can be fired at a rate corresponding to the average steam consumption throughout the day. It is claimed that peaks may be met as high as 60–80 per cent above average demand. *See* **Ruths accumulator**.

Load characteristics The load pattern for plant, as revealed by variations in demand over time; such variations may be diurnal, seasonal or secular.

Load factor, plant Actual output expressed as a percentage of what would have been produced had all the plant been operating continuously throughout a specified period. In respect of electricity generating plant, it is the average hourly quantity of electricity sent out during the year, expressed as a percentage of the average output capacity during the year. *See* **availability**.

Load factor, system In relation to an electricity supply system, the ratio, expressed as a percentage, of the average load throughout the year to the highest load on the system. In order, for purposes of comparison, to allow for year-to-year weather variations, the **Central Electricity Generating Board** adjusts the actual system load factor to a basis estimated on 'average cold spell' conditions; on this basis the annual load factor is about 52 per cent. In the United States load factors of 60–65 per cent are achieved. In comparing system load factors in different countries it is necessary to take into account differences in consumer requirements determined by, among other things, social, climatic and economic conditions. System load factor is a measure of the fluctuating pattern of consumer demand, not of the efficiency with which supply industries use their plant.

Load/rate tariff A **tariff** for the supply of electricity in which the consumer pays a standing charge related to the demand he subscribes for, and when his demand is less than the subscribed demand, the units consumed are charged at a low price; when his demand is above the subscribed demand, the units of electricity in excess are charged for at a relatively high price. This form of tariff is used extensively in Norway.

Load, simultaneous maximum On the British electrical power system, the maximum load actually met on the national grid for any half-hour together with the load at that time on any stations not tied into the grid. This may differ from the simultaneous maximum potential demand by any load shed through voltage reduction, disconnection or any reduction in frequency.

Load suppression gear Automatic equipment for reducing the load on a turbine should the condenser vacuum fall to a predetermined level; also called 'vacuum deloading'.

Load, system peak On the British electrical power system, the annual maximum simultaneous load; it usually occurs in the late afternoon of some cold working day in December or January.

Load-on-top system A code of practice aimed at minimising pollution from oil-carrying ships; it involves passing the washings from tank-cleaning operations and residue from discharge of the original ballast water to an empty cargo tank nominated as the 'slop' tank. Fresh oil cargo is loaded on top of the final residue left after further discharges of water, the resulting mixture being acceptable to refineries despite some additional cost in removing the salt and water. Under the *International Convention for the Prevention of Pollution from Ships 1973*, all oil-carrying ships will be required to be capable of operating with this method of retention, or alternatively to discharge to reception facilities.

Lock hopper A vessel attached to a pressure gasifier to permit the intermittent passage of solid fuel without direct communication between the gasifier and the atmosphere.

Locomotive boiler A type of *boiler* used on steam locomotives; the boiler is fitted with an integral water-cooled firebox, the hot combustion gases passing from the firebox through a horizontal bank of fire-tubes situated within the shell of the boiler into a smoke box and out of a short chimney at the opposite end. The locomotive boiler is a relatively efficient steam generator. The blast pipe is situated in the smoke box below the chimney; through it passes the exhaust steam from the cylinders, the velocity of the steam inducing draught through the firebox. While the boiler is stationary, the creation of draught depends on the use of a 'blower', consisting of a ring of steam jets at the base of the chimney which can be operated by a valve in the cab. When the boiler is moving, the blast pipe takes over. Despite the high thermal efficiency of the boiler, the overall efficiency of the locomotive is quite low; only a small percentage of the heat passed

220

forward to the cylinders is converted into useful work. The general average efficiency of locomotives of modern design is about 7 per cent. A stationary locomotive boiler is a simple adaptation of the normal locomotive boiler; it is sometimes used where head room in a boiler house is limited, and appears to have been widely used to power oil-drilling machinery in various parts of the world.

Log sheet   A detailed record of instrument readings and plant operating details kept by a plant operator or supervisor.

London smog incidents   Acute episodes of heavy pollution associated with natural fog covering the Greater London area. The fog which covered the Greater London area during the four days 5–8 December 1952 was on a much different plane compared with those previously experienced. An *anticyclone* reached London from the north-west in the early hours of 5 December and then became stationary. On 6–7 December London Airport had a minimum air temperature of − 5 and − 6° C and a maximum just under 0° C. Soundings showed two *inversions*, one close to the surface and the other caused by descending (anticyclonic) air higher up. By 7 December, the two inversions were very close together. The final result was that London was at the bottom of a pool of cold stagnant air, with a very effective 'lid' overhead. The atmosphere contained a great deal of water in the form of very small droplets. Nearly 4000 people died.

Long flame coal   *Coal* of high *volatile matter*. *See short flame coal*.

Long-run marginal cost   *See marginal cost*.

Longwall mining   Historically the most important method of *mining* coal in Great Britain. In this system the whole of the seam within a specified area or panel is extracted in one continuous operation by taking successive slices over the entire length of a long working face. In a simple case, two parallel tunnels are driven from the main haulage road; these tunnels are called gate roads. A longwall face is opened up by driving a further tunnel between the remote ends of the gate roads.

Looping   The behaviour of a chimney plume when the environmental eddies so influence the plume that the *efflux velocity* or momentum and thermal buoyancy of the gases are ineffective and the plume zig-zags up and down. *See coning*.

Los Angeles smog   Smog of a photochemical nature, largely attributable to the effect of sunlight on *motor vehicle exhaust gases*. Basic to the Los Angeles control system is a programme of monitoring both weather conditions and levels of air contaminants. Three stages of alert related to the severity of exposure have been adopted, and appropriate control measures are instituted whenever necessary. In the first stage alert, all unnecessary activity which might pollute the air must be avoided. A second stage alert indicates a health menace, and the County Air Pollution Control Director may impose limitations on the general operation of vehicles, and may restrict the operations of public utilities and other industries to those essential to continued operation of the industrial complex. The third stage alert is evidence of a dangerous health menace, and appropriate measures may be taken under the California Disaster Act to limit activities to emergency needs.

Loudness The intensity of sound waves combined with the reception characteristics of the ear. Annoyance results from both the loudness and the frequency of a noise. Tones which stand out above the background noise, particularly those of high frequency, are usually the most annoying. Loudness depends on the response of the ear to sound, and the ear is not equally sensitive at all frequencies.

Low Btu gasification A process which employs steam–air gasification in a fluidised bed to produce low-thermal-value fuel gas from coal for use in high-efficiency combined-cycle electrical power generation. The gas has an energy value of about 4 MJ/m^3.

Low-heat value (LHV) Synonymous with *net calorific value*; a term most often used in the United States of America.

Low-level waste Part of the waste from various stages of the *nuclear fuel cycle*, typically containing only a few curies per cubic metre.

Low-pressure air atomiser *See blast atomiser.*

Low ram cooking stoker A *coking stoker* in which the ram or coal feeder is set low in the flue relative to the grate, thus enabling a wide ram to be fitted; this ensures a more even distribution of fuel across the grate than was possible with earlier high ram designs. A wide range of coals can be burned on the low ram stoker; non-coking to medium coking coals are suitable.

Low-temperature deposits Deposits on the cooler parts of a boiler system such as economisers and air-heaters consisting of alkali chlorides, sulphur compounds and phosphates. Dry-bottom pulvcriscd fucl-fircd boilcrs arc practically immune from 'back-end' corrosion, except when the chlorine and sulphur in the coal are very high (chlorine exceeding 0.6 per cent). *See high-temperature deposits.*

Lubricating oil A high-boiling paraffin hydrocarbon. The properties usually required are: (a) stability at high temperatures; (b) fluidity at low temperatures; (c) moderate change only in viscosity over a broad temperature range; (d) sufficient adhesiveness to keep it in place under high-shear forces. *See lubrication.*

Lubrication *See boundary lubrication; hydrodynamic lubrication; lubricating oil; solid lubricants.*

Lumen Unit of quantity or total flux of light; the total amount of light necessary to illuminate 1 square foot of surface to the value

222

of 1 *foot-candle*. In the **International System of Units** it is the unit of luminous flux; the flux emitted within a unit solid angle of 1 *steradian* by a point source having a uniform intensity of 1 *candela*.

Lurgi process A process originally developed in Germany to gasify brown coal; it has subsequently been applied in Britain for the total gasification of poor-quality coal. The process is carried out under conditions of high temperature and pressure in the presence of steam and oxygen. The gas produced is largely hydrogen and carbon monoxide; it is then purified by passing it with steam over a catalyst to convert the carbon monoxide and steam into carbon dioxide and hydrogen by the **water gas shift reaction**; the carbon dioxide is then removed in a Benfield plant. The final gas has a calorific value of about $16 \, MJ/m^3$. One plant is at Westfield, Scotland; another at Coleshill, Birmingham, England. *See Figure L.4.*

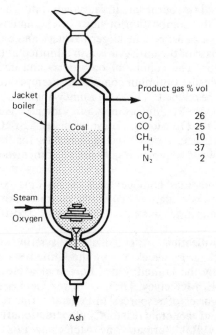

Product gas % vol	
CO_2	26
CO	25
CH_4	10
H_2	37
N_2	2

Jacket boiler

Coal

Steam

Oxygen

Ash

Figure L.4 Cross-section of Lurgi generator

Lux Unit of illumination; an illumination of 1 *lumen* per square metre.

L_x noise levels Noise levels in dBA which are exceeded for a specific percentage of the measurement period. For example, L_{10}

and L_{90} noise levels mean that the noise level in dBA was exceeded for 10 per cent and 90 per cent of the measurement period, respectively. The **Noise Advisory Council** of the United Kingdom has recommended the adoption of the L_{10} index for measuring disturbance by traffic noise. It has recommended also that existing residential development should in no circumstances be subjected, as an act of conscious public policy to more than 70 dBA on the L_{10} index unless some form of remedial or compensatory action is taken by the responsible authority.

M

McCashney incinerator A wood-waste incinerator consisting of a 'bottle'-shaped brick or steel shell, lined with a standard siliceous firebrick having an alumina content of about 22 per cent. The waste material is burnt as it is blown into the incinerator tangentially, the combustion of most of the material occurring while it is in suspension. The larger particles and any solid waste burn on the base of the unit. Air is also admitted at the bottom of the incinerator. The volume of both over- and under-fire air is regulated to ensure optimum combustion; temperatures of 1000–1400° C are developed. The McCashney incinerator, originally evolved by Mr R. McCashney of Victoria, Australia, was developed by the Division of Forest Products, CSIRO, for general industrial use; it is widely used in Australia for the disposal of sawdust and wood shavings at sawmills and other wood-processing plants.

Macerals Elementary homogeneous microscopic constituents of coal; they include alginite, collinite, cutinite, exinite, fusinite, inertinite, micrinite, resinite, sclerotinite, sporinite, telinite, vitrinite.

McKelvey classification A classification system for solid non-renewable resources developed by the United States Geological Survey and by the United States Bureau of Mines. It is named after Dr V. E. McKelvey, Director of the Geological Survey. It defines categories of resources in general terms of certainty of occurrence and economic feasibility of extraction. Its use involves a 'McKelvey Box' format. The McKelvey Box illustrates in summary form the current status of our knowledge of a particular resource, and can serve as a general guide in determining which categories of resource are most likely to satisfy future requirements. The system is currently used by the Australian Bureau of Mineral Resources. *See Figure M.1.* A major deficiency of the classification is that it conveys no impression of the availability or likely

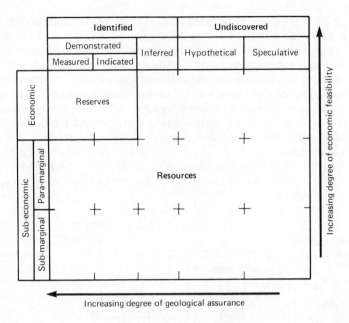

	Identified			Undiscovered	
	Demonstrated		Inferred	Hypothetical	Speculative
	Measured	Indicated			

Figure M.1 The McKelvey classification

availability of resources through time; the figures represent a judgement at a particular time and in terms of current economics, of the total quantity of a *resource* or a *reserve*.

Magazine boiler A hot water boiler equipped with a built-in hopper to hold fuel for up to 24 hours' continuous operation; from the hopper, fuel gravitates on to the fire and the rate of burning is thermostatically controlled. Automatic magazine boilers will operate at an efficiency of 80–85 per cent and can be supplied as a single boiler up to a rating of 10^7 kJ/h for central heating purposes. *See boiler.*

Magnesia, 85 per cent An insulating material, composed of 85 per cent of hydrated light basic magnesium carbonate and 15 per cent of long-fibred asbestos as a binding agent; it is suitable for temperatures up to about 300° C.

Magnesite brick A *refractory* containing 84–92 per cent of magnesium oxide.

Magnetic separator A magnetic device installed in a conveyor system to attack and remove any undesirable tramp iron in the coal.

Magnetohydrodynamic generation Direct electricity generation utilising hot flue gases and a magnetic field. In the simplest form of MHD generation a hot, electrically conducting fluid at about

2500° C is expanded through a nozzle and passed along a duct at high velocity. A magnetic field acts in a direction at right angles to the axis of the duct, and the electric currents induced in the flowing gas are collected on electrodes. Seeding of the gas with a trace of readily ionising material such as caesium or potassium vapour enhances the electrical conductivity by many orders of magnitude; cost necessitates a high rate of seed recovery. if such a technique could be combined with conventional plant it would serve as a 'thermodynamic topper', helping to improve power station efficiency. In 1977 a US-built 40 tonne superconducting magnet, the largest of its kind, arrived in Moscow for joint US–Soviet Union experiments in relation to magnetohydrodynamics. *See Figure M.2.*

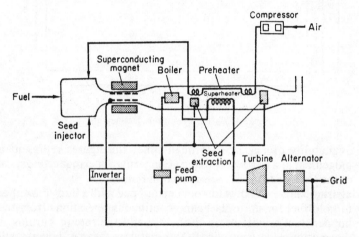

Figure M.2 Possible arrangement of fossil fuel power station of the future, equipped with an MHD 'topper' in addition to the normal steam turbo-generators

Magnox A magnesium alloy containing small quantities of beryllium, calcium and aluminium to reduce the possibility of rapid oxidation and fire; used in British nuclear reactors as a canning material for uranium fuel elements. Magnox melts at about 640° C. *See* **Magnox reactor**.

Magnox reactor A *nuclear reactor*, first designed and built at Calder Hall in the United Kingdom, which became the basis of a 'generation' of commercial nuclear power stations. The design uses uranium fuel with magnesium alloy (Magnox) as a canning material; graphite as a moderator and carbon dioxide as a coolant gas. *See* **Magnox**.

Main stop valve A valve which enables a boiler to be isolated from the steam pipe main and permits regulation of the flow of steam from a boiler. It is fitted to the highest part of the boiler, as far as practicable from the water surface. An anti-pumping pipe is usually fitted inside the boiler to the steam outlet to reduce the risks of *priming*, or *carry-over*. For pressures up to 10 bar the body of the stop valve is cast-iron; above this pressure, steel is used. The main stop valve incorporates a non-return valve if the boiler supplies steam to a common main.

Make-up water Water added to a boiler to compensate for the water lost as steam.

Malleable cast-iron *Cast-iron* made from constituents which solidify 'white', i.e. with all the carbon combined and none present as graphite; the 'white' castings are annealed to break down the hard brittle cementite and produce malleable castings.

Manchester process A wet scrubbing process for the removal of *hydrogen sulphide* from refinery and petroleum oil gas streams. The scrubbing medium is a suspension of ferric hydroxide, $Fe(OH)_2$, in a dilute solution of ammonia or sodium carbonate. Sulphur is recovered by blowing air through the solution. *See hydrogen sulphide removal.*

Manchester steam generator Introduced in 1925, a water-tube steam generator fired by pulverised coal; this proved a flexible, efficient and easily operated system of boiler firing. By 1952, 34 per cent of the coal burnt by the British Electricity Authority was in pulverised form; save in the oldest power stations, it is today the normal method of firing coal. The Manchester steam generator represented the first application of pulverised coal to steam raising, although it had been used towards the end of the nineteenth century for the firing of cement kilns and metallurgical furnaces.

Manganese fuel additive MMT Or methylcyclopentadienyl manganese tricarbonyl; an additive first sold in the United States of America in 1958 as an octane improver for gasoline in a mixture with tetraethyl lead (TEL) called 'Motor 33 Mix'. In low concentrations, it is claimed that MMT significantly improves the effectiveness of TEL, giving a 2–4 octane number increase compared with TEL alone. In recent years MMT has been used alone as an octane improver in unleaded gasoline; however, in October 1978 this practice was prohibited by the *Environmental Protection Agency*, owing to the deleterious effects of manganese on catalytic converter emission control systems. The addition of MMT is still allowed in leaded fuels. *See gasoline additives.*

Manifold A piping arrangement which allows one stream of liquid or gas to be divided into two or more streams, or several streams to be collected into one stream, as in the case of the inlet and exhaust manifolds of an automobile engine.

227

Manometer A *draught gauge* utilising a U-tube containing water.

Manufactured gas Gas produced as a product or by-product of a manufacturing process. This includes: gasworks gas or *town gas*, including gas produced by carbonisation and by the cracking of natural gas; *coke-oven* gas, obtained as a by-product of the manufacture of hard coke or metallurgical coke; *blast-furnace gas*, obtained as a by-product from blast furnaces; refinery gas, being non-condensable gas collected in oil refinery processes; *liquefied petroleum gas* other than obtained from natural sources; and other hydrocarbon gases obtained in the course of processing crude petroleum or its derivatives.

Marginal cost The increase in total cost of producing each successive increment of an output; the cost of producing $M + 1$ units, minus the cost of producing M units. As a concept, it includes social costs, although often construed more narrowly in accountancy terms. The short-run marginal cost is the cost incurred in making marginal or small changes, say in the energy output of a system, within existing capacity; it is also known in this context as 'marginal energy cost' or 'marginal running cost'. In thermal generating plant such cost is mainly that of fuel. It is important to distinguish sharply between the short-run marginal cost of changes in the energy output of the system and the long-run marginal costs of changes in the capacity of the system. Long-run marginal cost includes capacity costs, as well as energy costs. It is relevant to any consideration of the desirability of selling more electricity, which can only be supplied through the construction of new generating plant. *See Figure M.3.*

Marine boiler A cylindrical shell boiler with flat ends, equipped with several furnace tubes; the gases flow through these furnace tubes to the rear of the boiler, returning to a 'smoke box' at the boiler front through banks of small-diameter fire-tubes. The boiler may be single- or double-ended, the latter presenting a 'back-to-back' arrangement. Known as the Scotch marine boiler. *See boiler*.

Marine diesel oil A heavier type of *gas oil*, suitable for heavy industrial and marine compression-ignition engines.

Mass A basic property of matter, characterised by inertia and momentum and, within the influence of gravitation, weight. Even when objects become weightless, they lose none of their mass, continuing to display inertia and momentum. For the nuclear physicist, mass and energy are interchangeable.

Mass–energy equation *See Einstein's equation*.

Mass number The total number of protons and neutrons in a *nucleus*, equal to the integer nearest in value to the atomic mass when the latter is expressed in atomic mass units. In the symbol

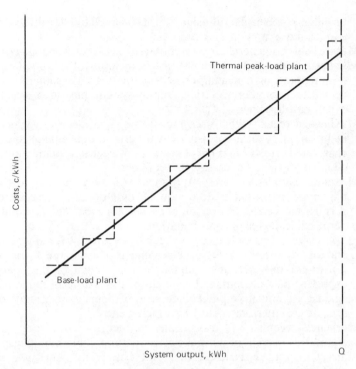

Figure M.3 The Short-Run Marginal Cost Curve (within system capacity) for a System up to Output Q. The horizontal short-run marginal cost curves of individual generating sets are shown by a broken stepped line. In a real system, the short-run marginal cost curve for the system may take a form departing from the linear, but always conforming to the same characteristics of increasing marginal running costs per kWh with increasing system output

for the nuclide, the mass number is indicated by a number following the element symbol; for example, U-238.

Mass spectrometry A technique for obtaining the mass spectrum of a beam of ions by means of suitably disposed magnetic and electric fields. The deflection of any individual ion in these fields depends on the ratio of its mass to its electric charge. Such a spectrum will appear as a number of lines on a photographic plate, each corresponding to a definite ratio value. *Isotopes* were first discovered in this way.

Mass transfer In respect of a fluid consisting of two or more components, in which a concentration gradient exists, the movement or flow of the components so as to reduce the concentration gradient. In a still fluid, or a fluid flowing streamline in a direction normal to the concentration gradient, mass transfer is effected by molecular diffusion, a slow process; in conditions of turbulence, molecular diffusion is supplemented by eddy diffusion, a much more rapid mixing process.

Mature coal   A general term for high-rank *coal*. *See immature coal.*

Maximum demand   In respect of electricity supply, the highest rate of simultaneous consumption of electricity by consumers connected to a common electricity supply system, measured usually over a half-hour period.

Maximum demand tariff   A *tariff* for the supply of electricity which, in its simplest form consists of two parts, a unit charge and a charge levied on the number of kilowatts of electricity of maximum demand registered over, say, half-an-hour.

Maximum permissible level   A term used loosely to refer to the maximum permissible dose rate of radiation, the maximum permissible concentration of a radioisotope or the maximum permissible degree of contamination of a surface.

Mayer curve   A graphical curve which gives directly, for any point and for any specific gravity, the percentage of ash in the 'floats' of a *float-and-sink test*; the cumulative ash content of the floats, expressed as a percentage by weight of the coal feed, is plotted against a cumulative percentage yield of floats over a range of specific gravity from 1.3 to 1.8 by increments of 0.1.

Mechanical rectifier   A transformer-rectifier for converting the alternating current of normal electricity supply into high-tension direct current electricity; this unit has found wide use in electrostatic precipitators. It takes the form of an insulated disc with contacts on its periphery, or an insulated arm with two contact blades with slip rings, rotated by a *synchronous motor*, so arranged that the high-voltage peaks of the AC wave form are selectively commutated. The high-tension current is picked up by jumping across an air gap from the AC contacts arranged around the insulated disc or insulated arms. Mechanical rectifiers are less costly than static types, essentially rugged, and the performance is largely unaffected by fluctuations in process conditions. They are, however, noisy, require radio and television suppressors and emit nitrous oxide fumes. *See electrostatic preciptator; rectifier; static rectifier.*

Mechanical stoker   A device which feeds coal or other solid fuel to a furnace automatically, displacing hand-firing methods. *See chain grate; chain grate stoker; coking stoker; low ram coking stoker; sprinkler stoker; travelling grate; trickle feed stoker; underfeed stoker.*

Medium-pressure air atomiser   *See blast atomiser.*

Medium voltage   Means a voltage exceeding 250 volts, but not exceeding 650 volts under normal conditions.

Mega-electron-volt (MeV)   One million electron-volts.

Megajoule A unit of 1000 kilojoules, or 10^6 J. One kilowatt-hour (kWh) = 3.6 megajoules. One gigajoule (GJ) = 1000 megajoules.

Megawatt One thousand kilowatts; a large unit of electrical power suitable for describing the generating capacity of power stations, or the capacity of an electricity supply system.

Megawatt-day The unit used to express the burn-up achieved or energy extracted from a fuel element; 1 megawatt-day = 24 000 kilowatt-hours.

Melt-down A description of the possible effects of a major breakdown in a nuclear reactor when the reactor becomes uncontrolled, the fuel elements fuse, and the molten mass burns through the base of the reactor and foundations and into the ground. The Rasmussen report (1974) on accident risks in United States commercial nuclear power plants estimated that with a power network of 100 nuclear reactors, a core melt-down could be expected to occur on average once every 175 years, although only about 1 in every 10 potential core melt accidents would produce measurable health effects. *See Harrisburg nuclear power plant incident.*

Mercaptans *See alkane thiols.*

Mercury A heavy liquid metal obtained generally from the roasting of mercuric sulphide ore (cinnabar). Mercury is used in older chlor-alkali plants, mercurial catalysts, the pulp and paper industry, and seed treatment, and is released to the environment also in the burning of fossil fuels, and in mining and refining processes. Mercury compounds may function as a cumulative poison affecting the nervous system (particularly in the form of methyl mercury, produced by the microbial conversion of inorganic mercury) and some mercury-containing organic compounds.

Merit goods and services Goods and services distributed by the state free (or almost free) to its citizens on the basis of merit (or need), e.g. hospital services, education, personal social services, employment services, and the arts. Merit goods differ from pure public goods in that they provide specific personal benefits for which the individual could pay if they were supplied through the market; however, the distinction is not clear-cut. *See public goods and services.*

Merry-go-round system A transport system between coal mines and power stations; as developed in Britain, a train has up to 43 hopper bottom wagons, each holding 32 tonnes of coal, the wagons remaining permanently coupled. At a colliery, with rapid loading equipment, the train moves slowly beneath the loading hopper which delivers the measured load of coal evenly into each wagon, distributing the weight equally across the axles. Trains of 1000

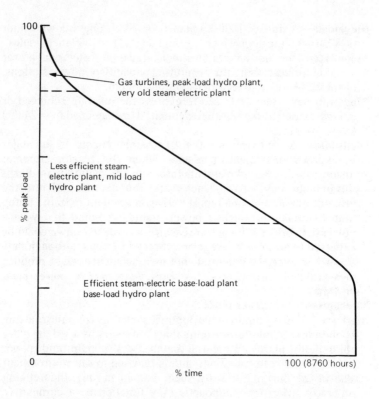

Figure M.4 Typical Annual System Load Duration Curve and Order-of-Merit
N.B. When water for a hydro plant is limited the plant will be operated at those times
which will effect maximum net system savings

tonnes or more can be filled in less than half-an-hour. On arrival at the power station, the hopper bottom floor of each wagon is unlatched and the coal flows into storage bunkers ready for use; when a wagon is empty, an automatic arm replaces the hopper bottom floor. The train is discharged in less than half-an-hour, never having stopped, and is ready to return to the colliery for another load. Power stations of 2000 MW capacity require some 70–100 train loads a week.

Mesons   Unstable particles having masses intermediate between those of the *electron* and the *proton*.

Metabolism   The processes in the body which involve the breaking down and building up of compounds, converting one to another in a great variety of chemical reactions. As the various processes of metabolism take place, energy changes occur and there is an overall release of some of the energy as heat.

Metallurgical coke   A hard coke produced in a *coke oven* for use in blast furnaces.

Methanation   A method of enriching a gas containing carbon monoxide and hydrogen by causing these two constituents to react over a catalyst to produce *methane*. The methanation reaction is:

$$CO + 3H_2 \rightarrow CH_4 + H_2O$$

The reaction is exothermic and is carried out between 300 and 400° C. The catalyst is usually based on nickel.

Methane CH_4. A non-toxic gas having a *calorific value* of approximately 40 MJ/m^3. Methane is the principal constituent of *natural gas*, being found in the petroleum districts of the world and other important areas such as the North Sea. Methane may be liquefied by refrigeration to a temperature of $-161°$ C; in this state it is reduced to $\frac{1}{600}$ of its volume as a gas. Methane is also produced during the process of digestion in sludge digestion tanks at sewage works, and occurs in coal mines as 'firedamp'. *See alkane series; sewage gas.*

Methane tankers   Ships specially equipped to carry liquefied natural gas from Algeria to Britain (Canvey Island). The storage tank insulation consists primarily of prefabricated balsa panels with plywood facing; insulation of the sides is augmented by a layer of glass fibre attached to the cold face. The temperature of the liquefied gas is $-161°$ C; in this state it is reduced to $\frac{1}{600}$ of its volume as a gas and requires no special pressure to keep it liquid. For maximum operating economy, the cargo 'boil-off' should not exceed 0.3 per cent per day of tank content; this is utilised to supplement the main boiler fuel. Two ships, *Methane Princess* and *Methane Progress*, alone deliver annually to Britain 700×10^3 tonnes of LNG or some 354×10^6 therms, the equivalent of 10 per cent of the gas load of England, Wales and Scotland. *See Canvey Island; natural gas pipelines.*

Methanol   Or methyl alcohol; derived from the fermentation of fibrous and woody plant materials, including cereal straw, *bagasse* logging and timber mill residues. *See ethanol.*

Metre   The SI unit of length, defined in terms of a number of wavelengths of a standard line of krypton-86; this is approximately the same as the original metre, which was about one forty-millionth of the circumference of the earth measured through the poles. From this unit is derived the unit of area, the square metre (m^2); and the unit of volume, the cubic metre (m^3).

Metric system   A system of measurement based on the centimetre, the gram and the second. It is often referred to as the 'CGS system'. The smaller size of the units, compared with the *British* *system*, commends them to the scientist. Some of the relationships between the units of the two systems are:

$$2.54 \text{ cm} = 1 \text{ inch}$$
$$453.6 \text{ g} = 1 \text{ lb}$$
$$39.37 \text{ in} = 1 \text{ metre}$$
$$2.2 \text{ lb} = 1 \text{ kg}$$

See dyne; erg; fluidity; International System of Units; MKS system.

Meuse Valley incident An air pollution incident of the 1930s involving respiratory illness and death. The topography of the Meuse Valley between Huy and Liége, a distance of some 24 kilometres, is that of a steep-sided trench about 1 kilometre wide and 120 metres deep. The valley is densely populated and is a centre of heavy industry. From 1 to 5 December 1930, prolonged stable weather conditions accompanied by the drainage of cold air (katabatic winds) from the uplands into the valley produced a severe inversion with heavy fog. Several hundred people became ill, and some 60 died as a result of respiratory illnesses. Many cattle had to be slaughtered.

MeV The symbol for one million electron-volts (1 MeV = 10^6 eV).

Micrinite A substance in coal derived from finely comminuted plant debris; a minor constituent of *clarain*, but a principal component of *durain*. It is a dull black material.

Microclimate A local climatic effect on a small scale from a metre to a kilometre horizontally and up to tree or house height. Vegetation, soil conditions, small-scale topography, structures and industrial activities may create pronounced microclimatic differences.

Microcrystalline wax Wax recovered from heavier lubrication oil fractions with much smaller crystals than *paraffin wax*; it is used for waterproofing.

Microcurie One-millionth of a *curie* (10^{-6} curie).

Microlithotype Microscopic banded coal type with a minimum width of 50 µm; includes carbargilite, clarite, clarodurite, durite, duroclarite, fusite, sporite, vitrinerite and vitrite. *See lithotype.*

Micromerograph A fast accurate instrument for determining the particle size distribution of powdered materials in the sub-sieve size range. It uses the principle of sedimentation in still air at atmospheric pressure. A cloud of particles is introduced at the top of a sedimentation column. These particles all fall the same distance at their terminal velocities on to the pan of a recording balance; a cumulative weight curve plotted against particle diameter is obtained directly from the recorder chart. It is claimed that particle size distributions of dry powders with particle size ranges from 1 to 250 μm are obtained easily from the micromerograph by this technique.

Micrometeorology The detailed study of physical phenomena in the lowest layers of the atmosphere, perhaps over a restricted area such as the site of a new town.

Micrometre A unit of length, being one-millionth of a metre (10^{-6} m). It is still commonly referred to as a *micron*. Symbol, μm.

Micromho The electrical unit of *conductance*.

Micron A unit of length. It equals 1×10^{-3} mm; 1×10^{-4} cm; 3.9×10^{-5} inch. Symbol, μm.

Micronutrients Mineral nutrients utilised by organisms only in minute amounts, e.g. iron, boron, copper, manganese, zinc, molybdenum and chlorine.

Micro-organisms Microscopic plants (bacteria, fungi and algae) or animals (protozoa, rotifers, crustaceans and nematodes) found in liquid wastes representing the active agents in biological treatment processes or the participants in the reduction activity.

Microprocessor Application Project (MAP) A programme launched by the British Government in 1979 to promote the use of microelectronics in industry.

Microprocessors Or microchips; the utilisation of microelectronics to achieve a more sophisticated control of industrial processes, particularly in respect of small batch processes in the chemical and food-processing industries. The concept is also relevant to fuel metering, fuel injection, ignition, and safety and emission control measures in the automotive industry.

Middlings A mixed material of adhering coal and dirt.

Mill differential pressure The difference in air pressure in inches water gauge between the inlet and the outlet of a pulverised fuel mill.

Millicurie 1 mc = 10^{-3} *curie*.

Milliroentgen 1 mr = 10^{-3} *roentgen*.

Mills See *ball mill; ball-race mill; high-speed mill; pulveriser; Raymond bowl mill*.

Mineral matter In coal, clay or shale mixed with varying proportions of free silica, silicates and other compounds of iron, calcium, magnesium, titanium and alkali materials. See *King–Maries–Crossley formula*.

Mineral wool An insulating material prepared from molten slag, glass or rock, blown into fibres by steam or air jet or spun by high-speed wheels. It may be supplied as a mineral wool base block, consisting of mineral wool fibres and clay, moulded under heat and pressure; or as mineral wool blanket, the fibres being compressed into blanket form held in shape between wire mesh or expanded metal lath. It is claimed that the temperature limit for block is 950° C, and for blanket 500° C.

Mining   See *auger mining; bord and pillar mining; longwall mining; opencast mining; opencut mining; prop-free front; strip mining*.

Mist   A suspension or dispersion of liquid droplets in an atmosphere.

Mixing height   The depth of atmosphere within which pollutants are dispersed and mixed; the depth available is often determined by the height of the base of an inversion layer.

MKS system   A system of measurement based on the metre, kilogramme and second. it is an attempt to create units of a useful size while retaining the advantages of the *metric system*. See *British system; International System of Units*.

Moderator   A material used in a *nuclear reactor* to slow down the neutrons by means of scattering collisions with the nuclei of the moderator. Moderators include *hydrogen* (in the form of light water), *deuterium* (in the form of heavy water), *beryllium* and *carbon*, in the allotropic form of graphite. See *neutron*.

Moisture   See *bed moisture; free moisture; inherent moisture*.

Mole   An SI unit (symbol, mol), being an amount of substance of a system which contains as many elementary units as there are carbon atoms in 0.012 kilogram of the carbon-12 atom. The elementary unit must be specified and may be an atom, a molecule, an ion, an electron, a photon, etc., or a specified group of such entities.

Molecular formula   A chemical formula which indicates the actual numbers of the constituent atoms in the molecule, as distinct from an *empirical formula*, which indicates the proportions only in which the constituent atoms are present in the molecule. For example, normal butane with an empirical formula of C_2H_5 has a molecular formula of C_4H_{10}. See *constitutional formula; graphical formula*.

Mollier chart   A diagram in which *total heat*, or *enthalpy*, is plotted against *entropy*.

Monazite   A thorium-bearing mineral, being a complex silicate. Most of the world's thorium resources occur in monazite.

Mond gas   A gas obtained by passing air and a large excess of steam over small coal at about 650° C.

Monitoring programme   The systematic deployment of measuring equipment for the purpose of detecting or measuring quantitatively or qualitatively the presence, effect or level of any polluting substance. For example, the scientific measurement of pollution has gradually become recognised as an essential prerequisite to remedial action. The measurement of the general atmospheric pollution to which people are constantly exposed is desirable for the following reasons:

(a) It enables an objective assessment to be made. It is, however, important that measurements should be made in more than one type of district in an area, or a false impression may be obtained as to the amount of pollution to which the inhabitants are exposed.

(b) It enables comparisons to be made with similar areas in other parts of the country.

(c) The information gained assists in making an intelligent decision as to the remedial measures necessary and their order of priority.

(d) It provides a means of judging the success of remedial measures once applied.

(e) The information is useful to an air pollution control agency considering plans for the industrial, commercial and residential development of an area.

(f) Knowledge of existing background levels of pollution is indispensable to the making of decisions in relation to chimney height proposals.

(g) Facts concerning existing pollution are an essential ingredient in an agency's campaign to convince the public of the need for a remedial programme.

(h) A quantitative picture of the pattern of air pollution throughout the country is a valuable aid to medical investigations into the relationship between air pollution and disease.

(i) It enables increases in air pollution that will result from industrial development in a given area to be predicted.

Motor Fuel (Lead Content of Petrol) Regulations 1976–79 Regulations which prescribe the maximum permissible lead content of petrol (in grams per litre) in the United Kingdom, and require petrol pumps in retail garages to be marked with star markings as specified in the British Standard (BS 4040) for petrol for motor vehicles. The star markings indicate the octane rating of petrol and maximum lead content. The maximum level of lead in petrol was reduced from 0.45 to 0.40 grames per litre from 1 January 1981. *See motor spirit*.

Motor method A test to determine the *octane number* of a *gasoline*, utilising an engine speed of 900 rev/min and a mixture temperature of 149° C (300° F). The conditions correspond fairly closely to the operation of car engines at high speeds and under heavy loads. It is a more servere test than the *research method*.

Motor spirit Blended light petroleum distillates used as a fuel for spark-ignition internal combustion engines, other than aircraft

237

engines. In the United Kingdom, the following system of classification of motor spirit has been adopted for finished motor spirit (the *octane number* being shown alongside):

2 star	under 94
3 star	94 and under 97
4 star	97 and under 100
5 star	100 and over

The United Kingdom average for gross energy value is 46.9 kJ/g (20 200 Btu/lb). *See Motor Fuel (Lead Content of Petrol) Regulations 1976–79.*

Motor 33 mix   *See manganese fuel additive MMT.*

Motor vehicle exhaust gases   The complex gases emitted from the exhausts of motor vehicle engines. In the *gasoline engine* some 15 kilograms of air are required per kilogram of fuel to ensure complete combustion; however, maximum power is achieved at a lower air fuel ratio of 12.5 to 1 and complete combustion does not take place. The result is the emission of substantial quantities of carbon monoxide together with other products of incomplete combustion such as alcohols, aldehydes, ketones, phenols, acids, esters, epoxides, peroxides and other oxygenates. Factors other than the air/fuel ratio also influence the composition and volume of exhaust emissions. The *diesel engine* operates with an excess of air and very little unburned fuel is normally exhausted; smoke emission is associated with engine overloading and poor engine maintenance. It emits much less carbon monoxide and hydrocarbons than the gasoline engine, but somewhat more *nitrogen oxides* and aldehydes. Motor vehicle exhaust controls have been introduced in many countries to curb this source of pollution. *See Los Angeles smog.*

Mott's classification   A system of coal classification using *calorific value* and *volatile matter* as the principal co-ordinates, on a *dry-mineral-matter-free* basis. *See coal classification systems.*

Mouthpiece   A cast-iron unit bolted to the back of the front plate of a boiler furnace tube; it gives access to the furnace for fuelling and other operations, but is normally closed by a fire-door. The space at the back of the mouthpiece is fitted with a firebrick liner to prevent the front and the boiler seams from becoming overheated.

Mud   In oil drilling, a mixture of clay, water and chemicals, that lubricates and cools the drilling bit during oil well drilling, carries cuttings to the top, and prevents oil and gas from gushing upwards. When the bit cuts into a porous oil-bearing formation, the fluid may escape and the circulation is lost; in this event, standard procedure calls for heavier mud to be pumped immediately into the shaft to counter the pressure of any oil present.

Multi-cellular collector   A *dust arrester* consisting of a large number of very small cyclones arranged in parallel; the individual cells may be vertical, horizontal or sloping. High collecting efficiencies are claimed over a wide range of particle sizes, although below those obtainable with the *electrostatic precipitator. See cyclone.*

Multi-fuel burner   Burners designed for the combustion of liquid, solid or gaseous fuels in the same unit, e.g. fuel oil, pulverised coal and natural gas. When a burner is equipped to handle gas and fuel oil, the gas may be fed through an annular tube with jets, surrounding the oil burner and igniter assembly.

Multi-gas burner   A gas burner invented by L. T. Minchin, based on the principle that very stable flames are formed when gas burns at a row of orifices lying along the ridge of a wedge-shaped cavity in the underside of the burner head; the flames are fan-shaped. In addition, the burner is constructed in such a way that a slow-moving stream emerges at certain critical points at the base of the

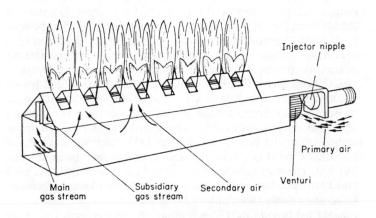

Figure M.5   The Tekni multi-gas burner (Source: Tekni Gas Ltd.)

flame, giving even greater stability for the flame formed by the main stream of fast-moving gases. Thus modified, it is claimed that the burner will give good stable flames even using Schlochteren methane, which contains 10 per cent nitrogen, and will not light back even on gases containing a high percentage of hydrogen. To switch from one gas to another it is only necessary to change the injector nipple, so that the rate of heat input is maintained constant. *See Figure M.5.*

N

Naphtha   A fraction obtained from the distillation of crude oil with a boiling point range of 65–170° C. Fractions in the lower boiling range 0–65° C are the 'light straight-run gasolines' and the 'straight-run gasolines' (although the latter are known in some UK refineries as 'straight-run benzine'). In the United States both the straight-run gasoline categories may be described as 'light naphtha'. Naphtha is needed both for gasoline production, through up-grading, and as a feedstock for the petrochemical industry for the production of plastics.

National Air Sampling Network (NASN)   An air pollution sampling network established in the United States in 1953. The NASN is operated in co-operation with health departments, air pollution control agencies and other local organisations. The basic network consists of about 225 sampling stations. Some of these are in operation every year, about 80 being situated in large cities and 35 at non-urban sites. The remaining stations, divided into two groups, sample during alternate years. Thus, about 175 stations are active in the network in any given year. Samples of suspended particulate pollutants are taken from the air every 2 weeks at each operating station of the network. Several common gaseous pollutants are sampled at more than 50 stations of the network.

National Coal Board (NCB)   A body created by the Coal Mines Nationalisation Act 1946 to take control of some 950 coal mines in Britain. The Board's statutory duty is to develop the industry efficiently, and provide coal in such quantities and at such prices as to further the public interest. In 1975 there were some 250 National Coal Board collieries and a colliery workforce of some 240 000. The coal industry lost 389 000 employees between 1960 and 1975. The largest consumer of coal is the *Central Electricity Generating Board*. *See Table N.1.*

National Coal Board Coal Classification   A classification system for coals prepared by the *National Coal Board* of the United Kingdom. The classification is based on the *volatile matter* content of coals on a *dry, mineral-matter-free* basis, and the caking power of 'clean coal'. A clean coal is defined as that with not more than 10 per cent ash. There are four main groups: 100, 200, 300 and 400–900. Each group is divided into classes and sub-classes. Coals of groups 100 and 200 are classified by using the parameter of volatile matter alone. Those in group 100 have a volatile matter content of under 9.1 per cent and consist of anthracites. Those in group 200 have a volatile matter content of 9.1–19.5 per cent. In this group fall the low-volatile steam coals, dry steam coals and coking steam coals. In group 300 the volatile matter content ranges from 19.6 to 32.0

NCB coal sales by markets, million tons	1977/8	1976/7
Power stations	75.7	75.1
Coke ovens	14.5	17.8
Industry	10.8	10.8
Domestic	7.7	7.6
Manufactured fuel plants	3.2	3.2
Other inland markets and own use	4.4	4.2
Exports	1.8	1.5
Total	118.1	120.2

Coal supply and demand, million tons	1977/8	1976/7
NCB production		
deep mines (including capital coal)	104.5	106.6
opencast, tip coal and licensed mines	14.5	12.3
Total NCB production	119.0	118.9
non-vested coal production	1.7	1.4
imports	2.6	2.4
Total coal supply	123.3	122.7
Demand		
inland consumption	119.7	122.6
exports	1.8	1.5
Total demand	121.5	124.1

Source: NCB, UK

per cent. In this group fall the medium-volatile coals and prime coking coals. In these first three groups (100–300) there is, in general, a close relationship between volatile matter content and caking properties. In the fourth main group (400–900), with over 32 per cent volatile matter content, is a wide range of caking properties at any given volatile matter content, and sub-divisions are made on the basis of the *Gray–King Assay*. See *coal classification systems*.

National Economic Development Council A Council set up by the British Government in 1961 with the following objectives: (a) to examine the economic performance of the nation with particular concern for plans for the future in both the private and the public sectors of industry; (b) to consider what the obstacles to quicker growth are, what can be done to improve efficiency and whether the best use is being made of the nation's resources; and (c) to seek agreement upon ways of improving economic performance, competitive power and efficiency, and increase the rate of sound growth. In 1975 the Council had set up some 37 working-parties to report on the situation in the various sectors of industry. In the same year the Council published a review entitled *Energy Conservation in the United Kingdom* (HMSO, London).

National Energy Advisory Committee (NEAC) An advisory committee established in 1977 to advise the Australian Government on energy matters. The Committee consists of 19 persons and represents a wide background of experience in Australian affairs. Standing groups have been formed to consider such matters as: energy conservation; energy resources, economics and modelling; energy research and development; oil and gas production, utilisation and transport. A number of reports have been published. *See National Energy Research Development and Demonstration Council.*

National Energy Research Development and Demonstration Council (NERDDC) A body set up by the Australian Government in 1978 to advise the federal Minister for National Development on the development and co-ordination of a national programme of energy research, development and demonstration. The Council consists of 12 persons of broad experience; several technical standing committees have been established. Funds have been made available to the Council by the federal government in order to promote research. In 1978/9 some $A 15 million became available to NERDDC for allocation. The Council is closely linked with the *National Energy Advisory Committee.*

National Industrial Fuel Efficiency Service (NIFES) Formed in May 1954, a non-profit British organisation which offers industry, local authorities and commercial property owners practical help in the use of heat and power, whatever type of fuel be used. NIFES provides heat and power surveys which show by precise measurement where waste occurs in factory premises and how it can be prevented; boiler house efficiency tests; advice on improving the efficiency of heat-using process plants, furnaces and power plants; checks on space-heating systems and advice on insulation; advice on the requirements of the Clean Air Act relating to smoke and grit emissions; and other services.

National Petroleum Advisory Committee A committee created by the Australian Government in 1979 to advise on appropriate arrangements for the equitable allocation of liquid fuels during any period of supply shortage; virtually every major fuel-using industry is represented on the Committee.

National Radiological Protection Board A body appointed by the British Government to be responsible for advising on possible dangers from all forms of radiation from whatever source.

National Smokeless Fuels Limited A subsidiary of the *National Coal Board* which manufactures a range of solid smokeless fuels for the domestic market, and coke for the steel and foundry industries.

National Society for Clean Air (NSCA) A voluntary society aiming to promote a cleaner atmosphere in Britain; it emerged from an earlier society, the National Smoke Abatement Society. The Society publishes a monthly journal *Clean Air*; conducts short courses and arranges lectures; and holds well-attended Annual Conferences. The headquarters of the Society are in Brighton, where it maintains an environmental library.

National Survey of Air Pollution, British An air pollution sampling network established in Britain in the early 1960s. Measurements of air pollution had been made since 1914 by local authorities and other interested bodies, in co-operation with the Department of Scientific and Industrial Research. However, the original system of measurements, while providing a rough indication of local conditions, was not sufficiently comprehensive to provide an accurate picture of the distribution of air pollution in different types of area throughout the country. Since information could only be obtained from a survey planned on proper statistical lines, and as the Clean Air Act became fully operative and more local authorities showed interest, the need for a more systematic national approach sharpened. As a result, a working-party consisting of representatives of local authorities, the Medical Research Council, the Meteorological Office, the Ministry of Housing and Local Government and DSIR was set up to devise a new survey. The recommended scheme was accepted by the Standing Conference of Co-operating Bodies for the Investigation of Atmospheric Pollution at their meeting on 14 November 1960. It was recognised that the pollution caused by dust was very localised and that it would be inappropriate to include dust deposition measurements in the new survey. The National Survey itself is confined to measurements of SO_2 and smoke. It was recommended that SO_2 in the atmosphere should be measured by the daily volumetric hydrogen peroxide method and smoke by means of a filter. Both were conveniently combined in the same piece of apparatus developed by DSIR. *See* **volumetric SO_2 apparatus**.

Natural circulation Circulation due to thermal and density effects, in contrast to forced circulation. The heating of water in a boiler results in some of the water becoming hotter than the rest; the hotter part, being less dense, rises, cooler water taking its place.

Natural coal dust The fraction of raw coal which passes a test sieve of 60 BS mesh (0.25 mm aperture); it differs fundamentally from the other coal constituents in that it consists largely of fusain concentrated in the dust during mining and subsequent opera-

243

tions—the volatile content is low, it is feebly caking or non-caking, and the ash content lies between 12 and 20 per cent. The majority of run-of-mine coals contain from 2 to 6 per cent of natural dust; it may rise to 15 per cent in untreated smalls. This dust tends to clog a fuel bed, and when dislodged by raking or forced draught, aggravates the grit emission problem.

Natural draught Draught produced by a chimney and the relative densities of the flue gases and the ambient air, without the assistance of fan power.

Natural gas A hydrocarbon gas obtained from underground sources, often in association with petroleum deposits. It generally contains a high percentage of methane, CH_4, with varying amounts of ethane, C_2H_6, and inert gases such as carbon dioxide, nitrogen and helium. The gross energy value is usually about 39 MJ/m^3, or 1050 Btu/ft^3. Most of the natural gas delivered by US utilities contains less than 5 per cent inerts. There are three main types of reservoirs: (a) reservoirs in which only gas can be produced economically, such gas being referred to as non-associated gas or unassociated gas; (b) condensate reservoirs, which yield relatively large amounts of gas per barrel of light liquid hydrocarbons, this gas also being called non-associated gas; and (c) reservoirs where gas is found dissolved in crude oil (solution gas) and in some cases also in contact with underlying gas-saturated crude (gas cap gas), both being called associated gas. *See methane*.

Natural gas liquids (NGL) A mixture of propane and heavier hydrocarbon fractions extracted from natural gas, usually at or near the point of gas production.

Natural gasoline *Gasoline*, which accompanies 'wet' *natural gas*, in many regions; it is removed from the gas by compression or absorption, or both. The gasoline is 'low-boiling', consisting mainly of hexanes, heptanes, octanes and pentanes. Also known as *casinghead gasoline*.

Natural gas pipelines Distribution mains carrying natural gas from its source to its market. In Britain a pipeline was constructed to carry imported natural gas from Canvey Island, Essex, to a terminal near Leeds, with suitable branches; the system was some 520 kilometres in length. Since the discovery of North Sea gas, further networks have been established to serve all the principal areas of Britain. In 1977 a 440 kilometre undersea gas transmission line began delivering gas from Norway's Ekofisk field to Emden, West Germany; this was the longest undersea line in the world. Pipelines serve most of the industrial areas of the United States, utilising the vast natural gas resources of that country. During 1966 a 720 kilometre pipeline between the Tyumen gas field in western Siberia and the industrial complex in the Urals was completed. The discovery of the world's second largest natural gas field in the Netherlands in 1959 has led to extensive pipeline developments in Europe.

Natural pollutants Substances of natural origin present in the earth's atmosphere which may, when present in excess, be regarded as air pollutants. These include: (a) ozone, formed photochemically and by electrical discharge; (b) sodium chloride, or sea salt; (c) nitrogen dioxide, formed by electrical discharge in the atmosphere; (d) dust and gases of volcanic origin; (e) soil dust from dust storms; (f) bacteria, spores and pollens; (g) products of forest fires. *See air pollution.*

Nautical mile Defined internationally as 1852 metres; equal to 1.15078 miles.

Neat gas burner A simple jet burner through which gas is ejected to burn with a luminous flame; a neat gas burner could be a simple drilled pipe without provision for premixing with air.

Neighbourhood noise A term used by the British *Noise Advisory Council* to cover a great variety of sources of noise which may, and frequently do, cause disturbance and annoyance to the general public in their homes and going about their lawful occasions. They include, for example: factory noise; noise from demolition, construction and road works; noise from ventilation and air-conditioning plant in buildings of all kinds; noise from sports, entertainment and advertising; and human noise arising from lack of consideration for others (loudspeakers, noisy parties, slamming of car doors, farewell hooting, and the like). The term does not embrace industrial noise as it affects workers, aircraft noise and traffic noise.

Neon A rare gas. When an electric current of sufficient intensity is passed through a glass tube containing neon gas under atmospheric pressure, a red glow is produced. Other rare gases produce different colours.

Nephelometer An instrument for measuring the amount of light scattered or absorbed by a suspension of particles; it is employed to determine suspended particle concentrations.

Neptunium Np. An *element*, at. no. 93. It is produced in a *nuclear reactor* by the absorption of a *neutron* by U^{238} to give U^{239}, the latter subsequently emits a beta particle, becoming Np^{239}. In turn this disintegrates into plutonium, Pu^{239}, by the emission of a further beta particle.

Neritic zone The relatively warm, nutrient-rich, shallow-water zone overlying the continental shelf; the marine counterpart of the littoral zone of a lake. Terminating at the edge of the continental shelf, sunlight normally penetrates to the ocean bottom, permitting photosynthetic activity and promoting the growth of a vast population of floating and anchored plants. The total amount of *biomass* supported by the neritic zone is greater per unit volume of water than in any other part of the ocean.

Net energy (or calorific) value *See energy value.*

Net reproduction rate The average number of female babies that will be born to a representative newly born female in her lifetime, if existing reproduction and mortality rates continue. If, for example, 1000 girls born in 1978 ultimately produce 1600 girl babies, then the net reproduction rate is 1.6. A net reproduction rate permanently greater than 1 means ultimate population growth, although this may be deferred if the existing age structure is unfavourable to growth. A net reproduction rate of less than 1 means an ultimate decline in population. In the United Kingdom the rate during the period 1935 – 39 was 0.78, compared with 0.98 in Australia and the United States. In the post World War II period all three countries have exhibited rates well above 1.

Neutral atmosphere In meteorology, a term usually applied to conditions in the lowest layers of the atmosphere when the *lapse rate* lies between zero and 10° C per 1000 m. *See inversion; superadiabatic lapse rate.*

Neutral flame In welding, a *flame* produced by an acetylene torch from a mixture of *acetylene* and oxygen in equal volumes.

Neutron An uncharged particle of mass 1.675×10^{-27} kg, which is a constituent of all nuclei except hydrogen. Outside a *nucleus*. A neutron is radioactive, decaying with a half-life of about 12 minutes into a *proton* and an *electron*. It is the agent which promotes the fission of ^{235}U in a *nuclear reactor*.

Neutron source Any substance that emits neutrons; for example, a mixture of radium and beryllium emits neutrons by a nuclear reaction between the beryllium nuclei and the alpha particles emitted by the radium.

Newton Unit of force; that force which, applied to a mass of 1 kg, gives it an acceleration of 1 m/s^2.

New York power pool A combination of seven investor-owned electricity supply undertakings, together with the Power Authority of the State of New York; one of the largest single energy-control centres in the United States of America has been constructed by the power pool. The control centre directs the generation and transmission of energy throughout New York State. There are interconnections with neighbouring power pools in Pennsylvania, New England and Ontario, Canada.

Niobium Nb. A rare metal and *element*; at. no. 41, at. wt. 92.906; possessing a low neutron-capture cross-section; it has been used as a canning material for nuclear fuel elements operating at high temperatures, e.g. when the *coolant* is liquid sodium.

Nitrides Compounds of metal and nitrogen. *See nitriding*.

Nitriding A case-hardening *heat treatment* process for producing a hard surface on certain types of steel by heating in gaseous ammonia. The ammonia cracks at the metal surface, causing the formation of *nitrides*.

Nitrogen N. A non-metallic *element*; at. no. 7, at. wt. 14; an odourless, colourless and, in the general atmosphere, chemically inert gas. The predominant constituent in air, it serves only as a diluent in combustion processes.

Nitrogen blanketing The use of nitrogen to displace air from above a liquid chemical in storage, which would otherwise react with the oxygen in the air.

Nitrogen cycle The circulation of nitrogen atoms brought about mainly by living things. An essential ingredient of proteins required by all living organisms, nitrogen enters the living part of the biogeochemical cycle in two different ways:

(a) Atmospheric nitrogen is converted into nitrogen compounds which can be utilised by plants and animals through nitrogen fixation effected by certain nitrogen-fixing bacteria and certain blue-green algae.

(b) The excrements of animals and dead bodies of animals and plants contain very complex nitrogen compounds which are broken down by ammonifying bacteria in the soil and converted into ammonia.

Some bacteria ('nitrifying bacteria') convert ammonia by oxidation into nitrates, while other bacteria oxidise the nitrites into nitrates. Among the factors considered essential for nitrification to proceed are the presence of phosphates, oxygen and a base (e.g. sodium or calcium) to neutralise the nitrous and nitric acids. The final oxidation product, nitrate, is utilised by plants for building up plant proteins. Both ammonia and nitrates can be easily assimilated by plants. Finally, a further group of denitrifying bacteria break up nitrogen compounds and release free nitrogen into the atmosphere, completing the nitrogen cycle.

Nitrogen oxides Oxides which include nitric oxide, NO, nitrogen dioxide, NO_2, and nitrous oxide, N_2O. They are produced in the combustion of organic matter and are thus introduced into the atmosphere from automobile exhausts, furnace chimneys, incinerators and other similar sources. Oxides of nitrogen undergo many reactions in the atmosphere. Nitric oxide reacts with oxygen to form nitrogen dioxide, although there is some controversy as to the extent and speed of this oxidation process. When NO and NO_2 are present, the following reaction proceeds rapidly:

$$NO + NO_2 + H_2O \rightarrow 2HNO_2 \text{ (nitrous acid)}$$

Nitrogen dioxide reacts with vapour and oxygen to give nitric acid vapour:

$$H_2O + 2NO_2 + \tfrac{1}{2}O_2 \rightarrow 2HNO_3 \text{ (nitric acid)}$$

Nitrogen dioxide can be injurious to health when inhaled, nitric acid being produced on the lung tissue with the moisture in the lungs. In the US Cleveland Clinic fire in 1929 many deaths were due to NO_2 from burning X-ray film. Nitric oxide can combine with the haemoglobulin of the blood to form an addition complex. Oxides of nitrogen from motor vehicle exhausts are important in the atmosphere of Los Angeles, California, where in the presence of sunlight they catalyse the formation of ozone.

Noise Advisory Council A body set up by the United Kingdom Secretary of State for the Environment in 1970 to keep under review the progress made generally in preventing and abating noise, and to make recommendations to government. The Control of Pollution Act 1974 significantly extended the powers of local authorities to deal with noise problems.

Noise exposure forecast (NEF) A technique for predicting the subjective effect of aircraft noise on the average person, exposure levels being expressed in NEF units. Factors which are taken into consideration are the frequency of aircraft movements and their distribution by day and night; the magnitude and duration of aircraft noise as determined by type, weight and flight profile; and the distribution of the noise energy over the spectrum of audible frequencies. In applying the NEF technique, a pattern of contour lines is drawn on the map of the area surrounding the airport.

Noise and number index (NNI) An index for the measurement of disturbance from aircraft noise; developed in the United kingdom in 1961 and since used extensively. The index was based on a social survey carried out around Heathrow Airport for the Wilson Committee, and takes into account the average peak noise level at the ground due to passing aircraft and also the number of aircraft which fly past. Local planning authorities have been advised to take aircraft noise into account in considering planning applications. The Surrey County Council has evolved a land-use zoning scheme, based on NNI contours, to control development around Gatwick Airport.

Noise rating A system whereby manufacturers indicate on their prod**u**cts the noise level emitted at a fixed distance from the appliance.

Noise zoning The statutory classification of areas according to usage; this method allows higher noise levels in areas where their effect will not be noticed, but maintains lower levels in more sensitive areas such as suburban residential sites and rural areas. An inherent problem in zoning lies in the difficulty of accurately defining the zones and of setting the boundaries between adjacent zones. Noise zones are not necessarily compatible with established land-use zones.

Non-associated gas *Natural gas* that occurs in an underground reservoir but not in association with economically recoverable oil deposits, other than light hydrocarbons or condensates.

Non-caking coal A coal which does not 'cake' when heated, leaving, after the *volatile matter* has been driven off, a dustlike coke. *See caking coal.*

Non-Ferrous metals Metals other than iron. *See brass; gunmetal; phosphor bronze.*

Non-renewable resources Natural resources which, once consumed, cannot be replaced, e.g. a ton of coal once consumed is gone forever in that form. Mineral resources generally are regarded as wasting assets of this kind. However, it is difficult to predict what the consequences of exhausting particular resources would be. For any particular mineral the exhaustion process could be gradual, accompanied, *ceteris paribus*, by a steady rise in its price. A rising relative price intensifies exploration and ensures treatment of lower grades of ore; recycling and reclamation of scrap and residues are also encouraged. Meanwhile, developments in substitute materials and processes and in the pattern of demand could mean that a mineral considered 'indispensable' at one stage could become redundant at another.

Normal temperature and pressure (NTP) A conventional reference standard of 0° C and 760 mmHg. *See standard temperature and pressure (STP).*

North Sea bonanza The benefits which Britain expects to receive from the successful exploitation of North Sea oil and gas reserves, both directly and through substantial relief to the balance of payments, in which imported oil has been an important item. Seismic prospecting for hydrocarbons under the North Sea has been in progress since 1962. Full-scale exploration started in 1964, following the enactment in Britain of the Continental Shelf Act 1964 and the subsequent ratification by Britain of the 1958 United Nations Continental Shelf Convention, covering the rights of coastal states to explore for and exploit the natural resources of the Shelf. Early emphasis was on the search for natural gas, mainly in

249

the Southern Basin of the North Sea, off the coasts of East Anglia and Yorkshire. The first gas came ashore in 1967. Supplies from several fields come ashore through marine pipelines to terminals at Bacton in Norfolk, Easington in Yorkshire, and Theddlethorpe in Lincolnshire, being distributed across the country by an extensive network of land pipelines. Offshore natural gas provides over 90 per cent of Britain's needs. The first important find of oil in the British sector of the North Sea was made in 1970; known as the Forties field, it is believed to be one of the largest in the world. The oil is brought ashore by an undersea pipeline some 180 kilometres long to Cruden Bay, near Peterhead in Aberdeenshire. A number of other important discoveries have also been made. These include the Brent and Ninian fields. By the end of the 1980s it is expected that Britain will be capable of producing enough oil to meet at least its own requirements. In 1973, 112 million tons of crude oil was imported. In 1976 the output of North Sea oil began to produce appreciable savings on imported oil. The output of North Sea oil was 37.3 million tonnes in 1977 and 51.3 million tonnes in 1978. The saving in the visible trade balance in 1978 was £2707 million. It is anticipated that by 1982 net imports of oil from elsewhere will be nil with a saving in the visible trade balance of over £7000 million.

Nozzle-mix burner   A *gas burner* in which all the combustion air is provided by high-velocity mixing at the base of the flame. Also known as a high-intensity or high-velocity burner. *See Figure N.1.*

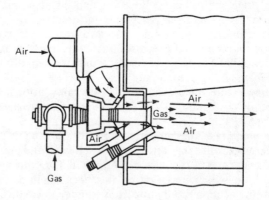

Fig. N.1   Cross section of a typical nozzle mix burner (Source: Urquhart Engineering Co Ltd).

Nuclear energy   *See energy.*
Nuclear Energy Agency, European   *See European Nuclear Energy Agency.*

Nuclear fission The splitting of a *nucleus* into two approximately equal fragments. The process is accompanied by the emission of neutrons and the release of energy. While neutron-induced fission is the most common, it can also be brought about by bombardment with heavy particles and by the absorption of photons. It may also occur spontaneously at a very slow rate. *See neutron; photon.*

Nuclear fuel Substances capable of producing heat as the result of the splitting or fussion of their atomic nuclei, and not through the chemical process of *combustion.* Fissionable *isotopes* are uranium-235 (from ore), uranium-233 (bred in reactor) and plutonium-239 (bred in reactor).

Nuclear fuel cycle A cycle comprising all of the operations from the mining and milling of uranium, through conversion, enrichment and fuel fabrication, to spent fuel storage and ultimate disposal, or reprocessing and recycling of valuable materials and disposal of the radioactive wastes.

Nuclear fusion Or thermonuclear fusion; a type of nuclear reaction in which two light nuclei come together to form one heavier nucleus; the reverse of nuclear fission. However, as in fission, very large amounts of energy are given off. Fusion reactions only take place at staggering temperatures—100 000 000° C or above. Fusion reactions are the source of energy given off by the Sun; they occur also when a hydrogen bomb explodes. A controlled fusion reaction has yet to be achieved. One avenue of research has been in respect of reactions between two hydrogen isotopes, deuterium and tritium, to form the heavier nucleus of helium and a neutron, accompanied by the release of energy. In 1978, at Princeton University, USA, research workers achieved a temperature of 60 million degrees Celsius, more than twice as high as any temperature previously achieved. *See Tokamak magnetic field configuration.*

Nuclear power programmes The gradual introduction of nuclear power stations into electricity supply systems in many countries; by 1979, 208 nuclear power stations were in operation, 209 under construction and 106 on order. Objections on environmental grounds have slowed down, and even halted in some instances, the development of programmes. In 1979 the United States of America had more than 70 operating reactors (over 50 000 MW), representing about 13 per cent of the total electricity supply. In Japan 13 nuclear power stations with a total capacity of 11 000 MW were under construction. Britain had 10 operational nuclear power stations.

Nuclear reactor A device in which nuclear *fission* takes place as a self-supporting chain reaction. A typical reactor comprises: (a) fissile material such as uranium or plutonium; (b) *moderator*; (c) reflector; (d) control elements; (e) provision for the removal of heat by means of a *coolant*. *See advanced gas-cooled reactor (AGR); boiling water reactor; Canadian pressurised heavy water reactor (CANDU); converter reactor; Dounreay fast reactor; Dragon high-temperature reactor; Dungeness 'B' nuclear power station; fast breeder reactor; heavy water-moderated nuclear reactors; heterogeneous reactor; homogeneous reactor; light water-moderated nuclear reactors; Magnox reactor; pressurised water reactor; pressurised water thorium–uranium converter reactor; savannah, N.S.; steam generating heavy water reactor (SGHWR); swimming pool reactor; zero energy reactor*.

Nuclear Regulatory Commission (NRC) A United States federal commission which is responsible for the licensing of nuclear plant and their inspection. It has the power to close down plants which appear to be unsatisfactory in operation, and has done so on several occasions.

Nucleon A constituent of the atomic *nucleus*; a *proton*, or a *neutron*.

Nucleonics The practical application of nuclear science, and the techniques associated with these applications. The term is also commonly used to mean the electronic techniques used in nuclear instrumentation and measurements.

Nucleus The positively charged core of an *atom*, which contains practically the whole mass of the atom while possessing only a minute fraction of its volume. It has a diameter of between 10^{-12} and 10^{-13} cm, and contains neutrons and protons. The charge equals the *atomic number*.

Nuclide A type of atom characterised by a given number of protons and neutrons in its nucleus.

Nutrient stripping A tertiary treatment of waste waters, either to reduce the rate of *eutrophication* of receiving waters or to permit the re-use of water for domestic purposes. Constituents of interest are those responsible for stimulating excessive growth of algae, namely compounds of phosphorous and nitrogen. Methods range from chemical coagulation to advanced treatment processes such as those developed for the desalination of sea-water and brackish waters. The disposal of the concentrates arising in all advanced waste water treatment processes presents an additional problem.

O

Oberhausen rotor furnace A *steelmaking furnace* of German origin, comprising a very slowly rotating barrel-shaped vessel. A primary oxygen lance projects oxygen into the bath of molten pig iron, while a secondary lance blows oxygen over the bath. This secondary oxygen supply burns the carbon monoxide released from the boiling metal.

OCCR gasifier A gasification unit in which fuel oil is gasified; partial combustion and cracking occurs on the base of a refractory lined vertical chamber. The product gas then passes to the furnace, where secondary air is admitted to complete combustion. The calorific value of the hot raw gas is about 8 MJ/m^3.

Ocean thermal energy conversion (OTEC) The concept of utilising the temperature differences of 30° C or more which occur between the surface of an ocean and its depths to achieve a continuous supply of power. One scheme employs ammonia as a working fluid; this is boiled in a heat exchanger by the relatively warm ocean surface water, the resulting vapour being used to drive a turbine generator. The ammonia is returned to a liquid state in a closed cycle by condensing it in another heat exchanger cooled with cold water from a depth of about 700 metres, where the temperature difference is some 22° C. In 1979 the US Department of Energy converted a former T-2 tanker ship into an ocean-going test platform to explore the viability of OTEC systems.

Octane number A method of ranking gasolines according to their resistance to detonating explosions when used as fuels in internal combustion engines. An octane number is numerically equal to the percentage by volume of iso-octane (2,2,4-trimethyl pentane) in a mixture of normal heptane and iso-octane having the same knocking tendency as the fuel being tested. This mixture is used because *n*-heptane has bad anti-knock properties and iso-octane has extremely good anti-knock properties. The scale is based on 0 for *n*-heptane and 100 for iso-octane. Gasolines may also be rated above 100 octane by use of extrapolation formulae. Octane number is the universal scale used to define the anti-knock or anti-pinking qualities of a motor spirit. *See aviation mixture methods; gasoline additives; motor method; research method; tetraethyl lead.*

92-Octane petrol A new grade of petrol introduced into New South Wales, Australia, by Caltex in August 1979 as a conservation measure, superseding 89-octane petrol. It was claimed that the new petrol, coloured purple, would be suitable for use in two of every five cars on the road. It was intended that this would reduce the demand for super-grade petrol while offering a slightly cheaper fuel of similar effectiveness. At about the same time, the octane rating of super petrol was cut from 98 to 97 as part of the Australian Federal Government conservation package.

Octopus system A system of firing pulverised coal; the airborne fuel is distributed to several firing points by supply lines from a single source. It is in use, for example, for top-fired annular brick kilns, the coal being delivered under control to several rows of feedholes on the kilns so as to give a curtain of flame across the kiln at the different rows.

Odoriser A chemical substance added to a gas to give it a distinctive smell. *Town gas* produced from oil feedstocks requires an odoriser to be added as a statutory safety precaution. The most commonly used substance is tetrahydrothiophene.

Ohm The unit of electrical resistance; the electrical resistance between two points of a conductor when a constant difference of potential of 1 volt, applied between these two points, produces in the conductor a current of 1 *ampere*, this conductor not being the source of any electromotive force. This fundamental relationship between voltage and current is known as Ohm's Law; named after G. S. Ohm, a German physicist (1787–1854).

Oil *See Petroleum*.

Oil burner A piece of equipment whose function is to deliver fuel into a *combustion chamber* in a form suitable for combustion; this is achieved by vaporising or 'atomising' the fuel. *See air-assisted pressure jet burner; blast atomiser; gun-type burner; inside mix burner; outside mix burner; pressure jet burner; rotary burner; self-proportioning burner; spill-type burner; steam blast burner; vaporising burner; wide-range pressure jet burner*.

Oil Companies International Marine Forum (OCIMF) An organisation created by the oil industry to relate to the shipping world and provide data to the *Intergovernmental Maritime Consultative Organisation (IMCO)*.

Oil gas Gas manufactured by the gasification of oil with steam in a chamber containing hot chequer-brickwork; the process is a cyclic one, as the reactions are endothermic. *See endothermic reaction*.

Oil market information system An information system on the international oil market, developed by the *International Energy Agency* to enable governments to play a more effective role in relation to the oil industry. The agreement on the International Energy Programme which emerged from the Washington Energy Conference held in November 1974 specified that one of the main areas of activity of the International Energy Agency should be the setting up and operation of an information system on the international oil market. The participating countries are asked to make available to the Agency on a regular basis information relating to the oil companies operating within their jurisdiction. By 1979 some 200 oil companies, producers and importers, refiners and traders, were submitting information. The result is much published information on the situation in the international oil market and the activities of oil markets. The information also provides a basis for the operation of the Agency's emergency oil allocation system. *See Figure O.1.*

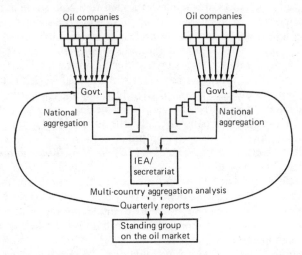

Figure O.1

Oil preheating   *See primary heating; secondary heating.*

Oil producer cartel   *See* **Organisation of Petroleum Exporting Countries (OPEC).**

Oil refinery capacity   The capacity of oil refineries to receive and process crude oil, the gross capacity being about 5 per cent greater than net output capacity, owing to process losses and wastes. By 1976 world refinery gross capacity reached 75 million barrels a day. Western Europe had a capacity of 21 million barrels a day, surpassing that of the United States of America for the first time. The communist world had a refining capacity of some 13 million barrels a day. Japan's capacity exceeded 5 million barrels a day.

Oil refining   The process by which crude oil is divided up and made into marketable products.

Oil reserves   Oil accumulations economically recoverable at current prices and costs; the reserves are usually classified in the petroleum industry as 'proved', 'probable' or 'possible'. The 'proved' and 'probable' reserves are essentially similar to the 'demonstrated' reserves of coal, while the 'possible' reserves are similar to 'inferred' reserves.

Oil shale   A fine-grained sedimentary rock containing organic matter that yields substantial amounts of oil when heated in a closed retort; the organic matter is mostly insoluble in ordinary petroleum solvents. Oil shales have been deposited in the sea, in lakes and in coal swamps. Australia has all three types of occurrence. The largest deposits in the United States of America lie under a broad swathe of land overlapping Utah, Wyoming and Colorado; schemes are being developed to 'cook' shale *in situ* to melt out the oil and enable it to be pumped to the surface.

Oil slug injection system   A system for firing top-fired continuous brick kilns; oil is circulated through a *ring main*, each separate row of feed holes being supplied by a branch line connected between the main flow and return lines. The branch line incorporates off-take points at each feed hole; a regulating valve at each feed hole controls the size of the oil slug. At the end of each branch line a control panel regulates the number of slug injections, normally ranging from 1 to 12/min.

Oil spills   *See Amoco Cadiz incident; IXTOC 1 incident; oil tankers; Torrey Canyon incident.*

Oil stability   A property of importance in the storage of oil; the term means the retention of initial properties, but also refers to the avoidance of sludge or sediment formation from the oil substance itself. This property is related to the probability of blocked supply lines and filtering equipment.

Oil synthesis   A process for the production of oil from hydrogen and carbon monoxide.

Oil tankers   As distinct from *methane tankers*, ships designed to carry crude petroleum or petroleum products; the world tanker fleet in 1976 was 321 million tonnes deadweight. Of the world tonnage, over 40 per cent was employed between the Middle East and Western Europe, and 11 per cent between the Middle East and Japan. The world tanker fleet doubled in capacity during the five years 1971–76. Tankers account for the largest ships afloat, transport economics favouring vessels of up to 500 000 tonnes. In July 1979 a collision occurred between two supertankers offshore from Tobago in the Caribbean; the ships were the *Aegean Captain*, of 210 275 tonnes, and the *Atlantic Empress*, of 292 666 tonnes; the two ships were thus carrying between them some 500 000 tonnes of crude oil. The burning ships were towed away, but not without leaving an oil slick covering some 130 square kilometres. This illustrates the scale of tanker shipping. *See Figure O.2; Amoco Cadiz incident; Torrey Canyon incident.*

Oil traps   Geological configurations which permit the trapping of oil. The two main types of trap are structural and stratigraphic. Structural traps include domal traps, in which the rock beds have been formed into an inverted dome; the anticlinal trap, a kind of elongated dome; the fault trap, formed by a fracture plane running down through layers of rock, making it possible for the sediments on each side of the fracture to slip out of alignment with an impervious layer being brought opposite a tilted reservoir rock.

256

Stratigraphic traps can be formed in several ways. One example is that of the up-dip edge of a wedging-out sand layer grading into an impermeable clay or shale. Another type of stratigraphic trap is sometimes provided by an ancient reef buried by impervious sediments. An oil reservoir is not an underground lake; it is an accumulation of liquid hydrocarbons within the pores of a particular kind of rock.

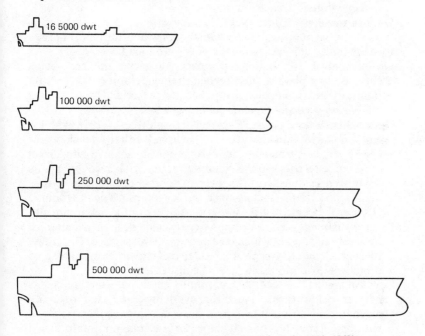

Figure O.2   The growth in the size of tankers since the 1939–45 War
(source: IMCO)

Olefin series C_nH_{2n}. Aliphatic *hydrocarbons or alkenes*, in which two carbon atoms may share more than one bond; these compounds are described as 'unsaturated', since by breaking one of the double links additional atoms of hydrogen or other elements may be added. They can also combine with each other by forming cross-linkages; that is, they polymerise. Examples of olefins are: (a) ethylene, C_2H_4; (b) propylene, C_3H_6; (c) butylene, C_4H_8. This type of hydrocarbon is formed when oil is subjected to high temperatures in cracking processes.

Oleum spirits   A petroleum 'cut' between 150 and 200° C, meeting certain other specifications.

Once-through boiler   A steam boiler in which no water recirculation takes place and which possesses no steam or water drum.

Onia–Gegi process A cyclic gasification process based originally on heavy oil and later adapted to accept light distillates. The process is one of partly thermal and partly catalytic cracking in the presence of an excess of steam and a nickel-bearing catalyst, with a process temperature of between 760 and 800° C. The calorific value of the gas produced is about 20 MJ/m^3.

On-load refuelling The replacement of the fuel elements of a *nuclear reactor* while the reactor is on-load.

On-load washing A method of cleaning gas-side boiler surfaces by applying a carefully controlled amount of water from spray nozzles to the fouled surfaces while the boiler is on-load.

On–off control An automatic device by which the heat being supplied to a plant is cut off when the temperature exceeds a set point; in this system heat is supplied at a prescribed rate, or not at all. *See proportional response control.*

Ontario Centre for Resource Recovery Established in 1978 at North York, Ontario, Canada, a centre concerned with the development of a 15 year waste management programme to achieve a high level of resource recovery in the community; materials recovery plants have been established to recover paper, tin cans, glass, ferrous metals, non-ferrous metals and organic material. The programme is under the control of the Ontario Ministry of the Environment.

Opencast mining. The working of coal seams near their outcrops, i.e. near the point at which coal appears at the surface. Generally, the actual outcrop edge of a seam is very inferior, owing to prolonged oxidation, and has to be discarded.

Open cut mining A technique of mining employed when the coal is not far below the surface; the overlying earth and rock are mechanically stripped to expose the coal, which is then removed with or without blasting. Includes *strip mining* and multi-bench mining.

Open cycle A mode of operation of a *heat engine* in which the working fluid is used only once.

Open flash point A *flash point* determined in an 'open' apparatus, the sampling cup having no cover. The open flash point of a liquid is a few degrees higher than the closed flash point. *See Abel flash point apparatus; Pensky–Martens flash point apparatus.*

Open-hearth furnace A *steelmaking furnace* consisting of a shallow bath capable of holding 60–400 tonnes of metal at a time; the smaller furnaces are static, the larger tilting. Construction is of special bricks inside a steel casing. In the end wall are situated the oil or creosote-pitch burners and ports for the admission of preheated combustion air; in earlier designs using coke-oven or producer gas, provision was made for this to be preheated also. The preheating of air (and fuel gas, if used) takes place in regenerative heat exchangers situated below the furnace; waste gases from the furnace flow through the regenerators before passing to a *waste-heat boiler*, and hence to the chimney. The flow of the gases is reversed every 20 minutes or so, the combustion air flowing alternately through the regenerators at each end of the furnace. A typical charge consists of 60 per cent steel scrap and 40 per cent pig iron; the three stages of charging, melting and refining take about 10–14 hours.

Open-loop control system A control system in which a controlled variable is allowed to vary in accordance with the inherent characteristics of the control system and the controlled power apparatus for any given adjustment of the conroller. *See closed-loop control system; feedback controller*.

Operational research The application of scientific and mathematical techniques to complex problems arising in the direction and control of systems of men, machines, materials and money, to improve the speed, quality and cost of performance. The approach normally involves the development of a symbolic model of the system, incorporating measurements of factors such as chance and risk, in an attempt to predict and compare the outcomes of alternative decisions, strategies or controls.

Opportunity cost The real cost of satisfying a want, expressed in terms of the cost of the sacrifice of alternative activities. For example, if capital funds could earn 7 per cent elsewhere, then that is their cost in present use. If machinery has no alternative use whatsoever, the opportunity cost is zero; historical cost in this instance is irrelevant. The opportunity cost of factors currently obtained from outside the firm is measured by the price currently paid for their services. In the case of factors already owned by the firm, it is measured by the amount for which factors could be hired out or sold to another firm. If a government chooses to build more schools and finds it must cut down on its road construction programme, then the cost of the schools programme can be represented as so many kilometres of road.

Optical pyrometer A device for measuring temperatures, the hot object being observed through a telescope inside which a standardised electric lamp is fitted. The current passing through the lamp is adjusted until the brightness of the filament matches the brightness of the object. The current is measured with an ammeter whose scale is calibrated in units of temperature. A colour match may be made against a standardised tungsten filament. The instrument is suitable for temperatures ranging from about 700° C up to about 3500° C. *See radiation pyrometer*.

Organic A general term used for all chemicals based on carbon.

Organisation for Economic Co-operation and Development (OECD) An organisation which came into being in September 1961 to promote policies designed: (a) to achieve the highest sustainable economic growth and employment and a rising standard of living in member countries; (b) to contribute to sound economic expansion in member as well as non-member countries in the process of economic development; and (c) to contribute to the expansion of world trade on a multilateral, non-discriminatory basis in accordance with international obligations. The members of OECD are Australia, Austria, Belgium, Canada, Denmark, Finland, France, the Federal Republic of Germany, Greece, Iceland, Ireland, Italy, Japan, Luxembourg, the Netherlands, Norway, Portugal, Spain, Sweden, Switzerland, Turkey, the United Kingdom and the United States of America. In relation to the energy situation, the Ministerial Council of OECD in 1979 concluded: 'There is now a real danger that, without responsible policies by oil consumers and producers alike, the energy situation will seriously damage the world economy.'

Organisation of Petroleum Exporting Countries (OPEC) An association of countries formed in 1960 for the purpose of promoting the interests of the oil-exporting countries. The countries comprised Algeria, Bahrain, Brunei, Ecuador, Gabon, Indonesia, Iran, Iraq, Kuwait, Libya, Nigeria, Oman, Qatar, Saudi Arabia, Trinidad and Tobago, the United Arab Emirates, and Venezuela. Until the Arab–Israeli Yom Kippur war in October 1973 the aim of OPEC was essentially defensive; oil was more of a buyers' market and resources were effectively in the hands of the international oil companies. OPEC imposed oil sanctions against countries friendly to Israel, and in any event the market began to develop into a sellers' market. From $US 2 a barrel in mid-1973, the price of oil rose to $US 10 a barrel in 1974. It created an oil crisis, and a balance-of-payments crisis for many countries, and helped precipitate the worst post-war recession. A unitary pricing system was adopted by OPEC, although in the late 1970s this developed into a two-tier pricing arrangement. In June 1979 Saudi Arabian light crude was offered at $US 18 a barrel, while other prices ranged between $US 20 and $US 23.50 a barrel. In 1979 the members of OPEC held about 80 per cent of the non-communist world's proven oil reserves and more than two-thirds of total world reserves. They supplied about 30 million barrels of oil daily, being 63 per cent of the supply of crude oil to non-communist countries. About 75 per cent of this came from the six member countries on the Persian Gulf: Saudi Arabia, Iran, Iraq, Kuwait, the United Arab Emirates and Qatar. Saudi Arabia is by far the largest exporter. In December 1978 political unrest caused a total halt to Iran's crude oil output; this had only partly recovered by mid-1979. About 80 per cent of the supplies to Europe come from OPEC countries. However, the United States is the largest single petroleum importer, the second largest being Japan.

Orifice meter A flowmeter employing as a measure of flow the pressure differential across an orifice, i.e. the pressure measured on the upstream and downstream sides of the orifice, as fitted in a pipe or duct.

Orsat apparatus Instrument for the determination of *carbon dioxide, oxygen* and *carbon monoxide* in flue gases. It consists essentially of three glass vessels which are partly filled with absorbent liquids through which a measured sample of flue gas may be passed. A solution of potassium hydroxide (caustic potash) is used to absorb carbon dioxide; a mixture of pyrogallic acid in caustic potash is used to absorb oxygen; finally, a solution of ammoniacal cuprous chloride is used to absorb carbon monoxide.

Orthohydrous coal *Coal* of normal or typical *hydrogen* content for the type species. *See per-hydrous coal; sub-hydrous coal.*

Ostwald diagram A graph showing the relationship between carbon dioxide and oxygen for a particular fuel; the oxygen in the air is graphically connected to the maximum possible carbon dioxide in the flue gases for the fuel concerned. The oxygen and carbon dioxide readings should fall along this line. *See Figure O.3.*

Ostwald viscometer A *viscometer* for measuring the *viscosity* of kerosine, gas oils and diesel oils. The viscosities are expressed in *centistokes*.

Outage The amount by which plant availability differs from the total capacity of the system through its being out of service owing to breakdown, essential maintenance, or other causes.

Outcrop coal Part of a coal seam visible at the surface of the ground.

Outlet flow heater A heater situated at the outlet of a liquid fuel storage tank; it has an immediate effect on the outlet temperature of the oil.

Outside mix burner An *oil burner* in which the oil is released into the atomising fluid at the outlet from the burner.

Over-deck superheater *See superheater.*

Overfeed combustion *Combustion* in which *ignition* takes place from the bottom of the charge of fuel, the *ignition plane* travelling upwards in the same direction as the air flow. The principle is illustrated in *Figure O.4.* This type of combustion tends to smoke formation, and if the coal has any caking properties, these are well developed during combustion. *See underfeed combustion.*

Overfire jets Jets of air or steam directed over a fire bed in order to promote turbulent mixing, thus helping to achieve smokeless

261

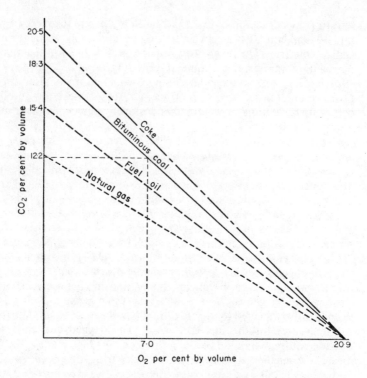

Figure O.3 Ostwald diagram

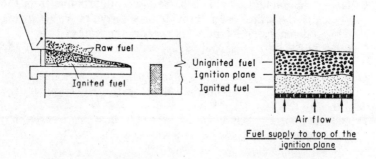

Figure O.4 Overfeed combustion as illustrated by the sprinkler stoker

combustion. There are three types of jet: (a) secondary air jet (separate fan); (b) steam-induced secondary air jets; (c) straight steam jet. Jets must be so designed as to penetrate the depth of the furnace, and sufficient in number to ensure turbulence over the whole area of the fuel bed.

Overhead In an oil refinery distilling operation, that portion of the charge which is vaporised.

Oxidant In the study of *air pollution* any chemical substance with an oxidation potential greater than that of oxygen, e.g. ozone, O_3.

Oxidation The addition of oxygen to a compound; the reverse of *reduction*. An oxidising atmosphere is one containing free oxygen.

Oxide An ash formed by the union of oxygen with, for example, a metal. The rusting of iron is the result of oxidation, the rust being described as iron oxide.

Oxides of Nitrogen *See nitrogen oxides.*

Oxo-process The reaction of *synthesis gas* with olefins to produce alcohols for use in solvents and plasticisers; the process operates at 170° C in the presence of a catalyst.

Oxy-fuel burner A *burner* in which a supply of gaseous or liquid fuel and a supply of oxygen are delivered to the same nozzle. The fuel may be natural or other gas, or oil. Oxy-fuel burners have been employed in the steel industry as an alternative to oxygen lancing, producing the same result without copious emissions of iron oxide fume. The simplest oxy-gas burner consists of two concentric tubes with oxygen supplied through the central tube and gas through the annulus. Burners usually incorporate an arrangement to produce sufficient turbulence to mix thoroughly the oxygen and fuel.

Oxygen A non-metallic *element*; symbol O, at. no. 8, at. wt. 16; an odourless, colourless gas which supports both combustion and life. Commercial or tonnage oxygen is produced from liquid air by fractional distillation. Oxygen as such is used in oxy-acetylene torches, oxy-fuel burners and in oxygen lances for steel-making. *See air.*

Oxygen recorder An instrument which uses the magnetic properties of oxygen for measurement purposes; oxygen is paramagnetic, while the other principal components of flue gases are diamagnetic. In one instrument a Wheatstone Bridge circuit is utilised, the 'out-of-balance' in the circuit being proportional to the oxygen present. *See Orsat apparatus.*

Ozone O_3. A gaseous oxidising agent which has been used in air conditioning; above minimal levels, however, it is an irritant to human beings and animals. It is a natural constituent of the atmosphere, occurring in concentrations of about 0.01 ppm; the toxicity threshold for workers is 0.1 ppm. Ozone is produced by certain high-voltage electrical equipment; it is also produced in certain circumstances when photochemical reactions occur in the atmosphere between ultra-violet light (sunlight) and the oxides of nitrogen and hydrocarbons emitted to the atmosphere by motor vehicles. *See Los Angeles smog; nitrogen oxides; oxidant.*

Ozonosphere A layer of the atmosphere between about 10 kilometres and 50 kilometres above the surface of the earth, containing *ozone*. The ozone is produced through the splitting of some oxygen molecules, the resulting atomic oxygen recombining with unaffected molecules to yield ozone (O_3).

P

Package boiler A compact steam generator supplied as a self-contained unit complete with all fittings, draught equipment pump, controls, etc., the whole being mounted on a bedplate. This type of boiler can be installed speedily, connections to fuel supply, water and electricity being all that is necessary. Package boilers operate up to 30 bar pressure and evaporative capacities range up to 5 kg/s steam. Thermal efficiency is in excess of 80 per cent. Strict attention must be paid to the quality of the feed water. *See boiler*.

Pair production The conversion of a gamma-ray *photon* into an *electron* and a *positron*.

Paraffin A refined petroleum distillate intermediate in volatility between motor spirit and gas oil; used for lighting and heating. A hydrocarbon belonging to the methane series. *See isoparaffin*.

Paraffin wax A wax removed from lighter lubricating oil stocks during refining; it has relatively large crystals. The wax is used for candles, tapers, food cartons, electrical insulation and polishers, and as a waterproofing agent. *See micro-crystalline wax*.

Paramarginal resources *Resources* that border on being economically recoverable under current prices and costs by current technology.

Parent A radionuclide which disintegrates to yield a given nuclide, known as 'the daughter', either directly or as a later member of a radioactive series. *See daughter*.

Partial combustion *Combustion* in which the supply of air (or oxygen) is so restricted that it is incomplete. For example, in the process for producing synthesis gas from methane by reaction with a limited amount of oxygen, the products are carbon monoxide and hydrogen; with sufficient oxygen to complete combustion the products would be simply carbon dioxide and water:

$$CH_4 + \tfrac{1}{2}O_2 \rightarrow CO + 2H_2$$
$$CH_4 + 2O_2 \rightarrow CO_2 + 2H_2O$$

Partial combustion reactor *See reactor*.

Partial pressure In connection with a mixture of gases, the pressure of one of the components of the mixture.

Participation crude Crude oil belonging to a host government within the *Organisation of Petroleum Exporting Countries (OPEC)*, in proportion to its stake in the producing company.

Particle size distribution The percentage by weight or number of each of the specified fractions into which a sample of particulate matter is divided

Partition curve A curve which indicates for each *specific gravity* or size fraction the percentage of it contained in one of the products of the separation. *See ecart probable.*

Pascal Unit of pressure; the pressure produced by a force of 1 newton applied, uniformly distributed, over an area of 1 square metre. *See Newton.*

Pass-out turbine A *steam turbine* from which steam is bled for process work or heating; the remaining steam passes through the low-pressure zone of the turbine to the condenser. *See back-pressure turbine.*

Peak load A transient maximum demand on a source of supply such as a steam generator, or gas or electricity undertaking. Peak loads tend to occur regularly at certain hours of each day; they also occur at certain times of the week and characterise certain seasons.

Peat The youngest member of the coal series, a *fuel* consisting of layers of dead vegetation in varying degrees of decomposition, occurring in swampy hollows in cold and intemperate regions. Fresh plant growth at the surface adds material to the decomposing debris; peat may be found in layers several metres thick. Light in colour near the surface, at deeper levels it is brown and even black. The amount of moisture in peat ranges up to 90 per cent, but more typically is 25–45 per cent. It is a bulky fuel, and even when well-dried its *calorific value* is only about 10 kJ/g. Its ash and sulphur content is very low. Extensively used in Scotland and Ireland, it is dug out and dried during the summer months. The Electricity Supply Board of Ireland has been using peat for electricity generation for the past 25 years. Peat is available in two forms: (a) sod peat, which consists mainly of large irregularly shaped sods; (b) milled peat, which consists of multi-sized fibrous particles, of which about 75 per cent pass through a 5 mm sieve. Sod peat is burned in boilers with chain grate stokers and a maximum capacity of 20 MW; milled peat is fired as pulverised fuel in boilers of up to 40 MW capacity. For the same heat content, sod peat has four times, and milled peat eight times, the bulk of normal coal.

Pebble stove A *heat exchanger* for providing preheated air or other gas for use in industrial processes. In its simplest form the stove consists of small refractory pebbles enclosed in a brick-lined steel shell, thus offering an alternative to conventional *chequer-brickwork* arrangements. The initial heating of the bed may be achieved by the combustion of fuel or the utilisation of waste gases; after a sufficient degree of heating has been achieved, the hot bed is available for the heating of cold air or other gas. In this form the process is cyclic, and more than one stove is necessary to achieve a constant supply of preheated air. Two-chamber arrangements are available, in which the pebbles are heated in an upper chamber and then pass through to the lower chamber, where they heat the incoming air; after cooling, the pebbles are returned to the first chamber. The process being continuous, one such unit only is required to supply a continuous stream of preheated air. *See blast-furnace stove; Howden-Ljungstrom air preheater.*

Peltier effect The evolution or absorption of a quantity of heat, depending on the direction of the current, proportional to the total electric charge, when an electric current flows across the junction between two different metals or semiconductors. This effect is due to the presence of an electromotive force at the junction. Named after Jean Peltier (1785–1845).

Penetration test A test to determine the degree of hardness of bituminous material; penetration is expressed as the distance that a standard needle vertically penetrates a sample of the material under prescribed conditions.

Pensky–Martens flash point apparatus Apparatus for determining the *flash point* of liquid fuels with flash points above 49° C (120°F). A sample of oil is heated in a closed vessel until a temperature is reached at which the vapours in the air space above are sufficient to form an inflammable mixture and ignite when a flame is applied. The flash point so determined is known as the 'closed' flash point; if the test is carried out with the sample cup uncovered, an *open flash point* is determined. *See Abel flash point apparatus.*

Pentanes C_5H_{12}. Low-boiling paraffin hydrocarbons. The *n*-pentane has a boiling point of 37.8° C (100° F) and a specific gravity of 0.63. *See liquefied petroleum gases.*

Percolation A situation in which a fuel–air mixture becomes over-rich during the evolution of vapour in a *carburettor* which causes fuel to percolate into the inlet manifold of the engine.

Perfect gas A gas which, at all temperatures and pressures, satisfies the relationship

$$pv = RT$$

where p = pressure;
  v = volume occupied by 1 mol of the gas;
  R = gas constant;
  T = absolute temperature

The gases hydrogen, helium, oxygen and nitrogen give the nearest approach to a 'perfect gas'. Also known as an 'ideal gas'.

Performance number The percentage gain in 'knock limited power' developed by a typical supercharged aviation gasoline engine when operating on leaded iso-octane, compared with that obtained when operating on unleaded iso-octane.

Per-hydrous coal *Coal* containing more *hydrogen* than is normal for the type species. *See orthohydrous coal; sub-hydrous coal.*

Permafrost Perennially frozen ground.

Peroxyacetyl nitrates (PAN) A component of photochemical smog; field levels of 0.01–0.05 ppm will injure sensitive plants. Causes eye irritations and other health effects. *See Los Angeles smog.*

Persistence An important characteristic of a pollutant in an environmental medium (air, water or soil) or in living tissue; substances that persist (remain active for a long time) in a toxic form, such as certain heavy-metal metabolites, organochlorine compounds and polychlorinated biphenyls (PCBs), are particularly hazardous.

Petit Badon solar heating installation A solar heating system developed at the Mas du Petit Badon in the Camargue, France, to provide heating for all the buildings of the Tour du Valat Foundation, an organisation for the study and conservation of nature. The plant consists of 50 flat solar panels and 36 parabolic collectors with 3 batteries producing a total output of 160 kW. The collectors are set in the roof of one of the buildings and cover a surface area of 310 square metres. Heat obtained by the collectors can be released directly to the hot water or heating systems or to one of three storage units. These storage units can be charged to 110°C. The excess energy available is used for the desalination of sea-water. The plant is situated in a nature reserve where the use of polluting fuels is forbidden. The plant was fully operative in 1976.

Petro-Canada The Canadian federal government oil company established in 1975. It has been undertaking high-cost, high-risk, exploration work in the Arctic islands and offshore eastern Canada and Labrador. Essentially the company has engaged in 'frontier exploration'. It is likely to become involved in the development of tar sands.

Petrochemicals Chemicals manufactured from the products of oil refineries; mainly a post-war development based largely on ethylene, propylene and butylene produced in the cracking of gasoline fractions. It is the unsaturated olefine compounds which are used to produce chemicals; the saturated paraffin compounds (methane, ethane, propane and butane) are used as refinery fuel or sold as bottled gas or supplied for *town gas* enrichment.

Petrodollars Currency (often US dollars) earned by the sale of crude oil and petroleum products, particularly by Middle Eastern countries, notably members of the *Organisation of Petroleum Producing Countries (OPEC)*.

Petrol Essentially a British, Australian and New Zealand term applied to fuel available at car service stations, and referred to in North America as 'gasoline'. *See gasoline.*

Petrol engine *See Gasoline engine.*

Petroleos Mexicanos The Mexican government agency exercising monopoly control over oil exploration and drilling, production and marketing, in Mexico. Mexico's proven hydrocarbon reserves in 1979 were estimated at 40 billion barrels, ensuring an important future as an oil-producing country. In addition to onshore production, exploration and development has taken place offshore in the Bay of Campache. *See Ixtoc 1 incident.*

Table P.1 TWELVE OF THE LARGEST US OILFIELDS

Field	Discovery year	State	1976 production (in 1000 bbls)
Proudhoe	1968	Alaska	—
East Texas	1930	Texas	67 094
Wilmington	1932	California	60 200
Panhandle	1918	Texas	11 163
Yates	1926	Texas	27 476
Midway Sunset	1894	California	38 317
Kelly Snyder	1948	Texas	67 844
Huntington Beach	1920	California	14 991
So-Vel-Tum	1955	Oklahoma	31 465
Long Beach	1921	California	2 506
Ventura	1916	California	10 841
Oklahoma City	1917	Oklahoma	1 780

Petroleum According to generally accepted theory, a substance derived from the remains of plant and marine life that existed on the earth millions, in some cases hundreds of millions, of years ago. A complex and variable substance, petroleum ranges from solid bitumen, through liquid oils, to highly volatile natural gases such as *methane*; all are essentially mixtures of compounds made up from the elements *hydrogen* and *carbon*, being known as 'natural hydrocarbon'. *See crude oil; hydrocarbons; petroleum products (see Table P.1).*

Petroleum coke A residue remaining after the complete distillation of oil; the two principal types are known as 'delayed process coke' and 'fluid process coke', both being produced in a fine granular form. They are usually high in sulphur content, up to 10 per cent. *See delayed-coking process; fluid coke process.*

Petroleum Institute Environmental Conservation Executive (PIECE) A body created under the auspices of the Australian Institute of Petroleum to provide a forum for the review of environmental matters of concern to the oil-refining industry. The executive maintains contact with government, convening discussions from time to time on broad-ranging matters. Although based in Melbourne, meetings are held in most of the capital cities of Australia.

Petroleum products Refinery products which include motor spirits, aviation turbine fuels, kerosines, diesel oils, furnace oils, liquefied petroleum gas, bitumens, lubricating oils and greases, and all other products derived directly or indirectly from crude oil through a refining process.

Petroleum revenue tax (PRT) A tax levied in the United Kingdom in addition to corporation tax; petroleum revenue tax is assessed on the profits of each oilfield for periods of 6 months; PRT has been levied at a flat rate of 45 per cent. Deductible costs for assessment purposes include royalties, exploration costs for the field, operating costs, capital expenditure and transportation costs.

Petroleum spirits *See white spirit.*

Petrology A branch of geology concerned with the study of the individual mineral components, structure and history of rock masses, including coal.

Petromin The General Petroleum and Mineral Organisation of Saudi Arabia, being a government-owned agency created in 1962 for the development of petroleum and mining resources in that country. The headquarters of the Organisation are in the capital, Riyadh. *See Petronal.*

Petronal A London-based affiliate of the Saudi oil company *Petromin*, created to market oil in Europe.

pH A term used to express the degree of acidity or alkalinity of a solution; a pH value is the logarithm, to the base 10, of the reciprocal of the concentration of hydrogen ions in an aqueous solution. Thus, hydrogen ion concentration $= 1 \times 10^{-x}$, where x equals the pH. A neutral solution contains 1×10^{-7} gram-equivalents of hydrogen ions per litre; the pH value is therefore 7. Acid solutions contain more than 1×10^{-7} grams-equivalents of hydrogen ion per litre and consequently the pH values are less than 7; conversely, alkaline solutions have pH values greater than 7. The range within which pH values are expressed covers the scale 0–14. Acidities or alkalinities greater than those represented by this scale are expressed as 'per cent concentrations'.

'Phimax' A *gas coke* of a highly reactive nature produced in continuous vertical retorts from specially selected coals; it is a smokeless fuel, very free-burning, and suitable for any domestic appliance *See **authorised fuels; smoke control area.***

Phosphor bronze An alloy of copper, tin and phosphorus, with or without the addition of other elements; these alloys are used where resistance to corrosion or wear is required. e.g. as in bearing or steam pipe fittings.

Photochemical reaction A chemical reaction which may occur when a substance is exposed to radiation, mainly visible and ultra-violet radiation. *See **Los Angeles smog; smog.***

Photoelectric cell A device in which electrons are stimulated by the action of light energy to create an electromotive force. This principle is put to practical use in measuring the density of smoke in a chimney, the smoke passing between a lamp and the photoelectric cell upon which its rays have been focused. *See **smoke density indicator.***

Photon A quantum of electromagnetic radiation. It has an energy hv, where h is **Planck's constant** and v is the frequency of the radiation, or hc/λ where λ is the wavelength of the radiation and c is the velocity of light.

Photosynthesis The formation in chlorophyll-containing (green) plants of carbohydrates from atmospheric carbon dioxide and water, the energy to affect photosynthesis being supplied by the Sun. As a result of the process, free oxygen is released into the atmosphere. An equation for photosynthesis may be presented:

$$\text{carbon dioxide} + \text{water} \xrightarrow[\text{green plant}]{\text{light energy}} \text{carbohydrate} + \text{oxygen}$$

In photosynthesis carbon becomes incorporated in the processes of the cell; it becomes fixed. The green pigments, or chlorophyll, appear to be the pigments that convert light energy to chemical energy in the process. In most plants chlorophyll is found in structures called chloroplasts within the cells. The leaf appears to crack moisture into the component parts, oxygen and hydrogen, for hydrogen is needed to transform the carbon dioxide into carbohydrates. This step appears to take place within the chloroplasts.

Photovoltaic conversion A technique for generating electricity directly from solar energy, utilising silicon cells. The direct conversion of radiation into electrical energy in this way was discovered in the nineteenth century, but it was not until the mid-1950s that these photovoltaic effects in germanium and silicon positive–negative junctions were demonstrated with efficiencies of around 10 per cent. Efficiency is defined as the ratio of electrical output power to the incident solar power. Silicon cells, the most popular, have a virtually unlimited life; the process is silent, static and without waste discharge. High costs have restricted their use to satellite communication systems and other special applications. Spacecraft are normally powered by arrays of silicon cells that produce direct current. The use of photovoltaic cells for the large-scale generation of electricity is dependent on research and development, which could lead to a substantial reduction in the costs per watt generated. Exxon has been working on the technology since 1970, calculating the cost of power produced at $1–3 per kilowatt-hour; the pathway to electricity generation from this source has been described as 'uncertain, risky and far from clear.'

'Phurnacite' A solid smokeless fuel in the form of carbonised ovoids; the briquettes are made from fine *Welsh dry steam coal* and pitch. The ovoids burn for long periods without attention, and are suitable for use in room heaters, boilers and cookers. *See authorised fuels; smoke control area.*

Phytotoxicant A chemical agent that produces a toxic effect in vegetation.

Pick breaker A device for reducing the size of large *run-of-mine* coal. Strong pick blades mounted rigidly on a solid steel frame move slowly up and down, the coal passing under the picks on a slowly moving horizontal plate conveyor belt. Several machines may be placed in series, with screens in between to remove fines. *See Bradford breaker.*

Pickling The use of hot or cold acid solutions to remove oxides and scale from metal surfaces.

Picocurie One-millionth of a *microcurie* (10^{-12} curie).

Pile An obsolescent term for a *nuclear reactor*, derived from the first reactor, which consisted of a pile of graphite blocks containing pieces of uranium.

Pinking *See knocking.*

Piston-type gas holder A vertical steel plate tower, with a moving piston floating on stored gas. *See wet gas holder.*

Pitch *See coal tar fuels.*

Pitot tube A device for the measurement of fluid flow in ducts. In one form it consists of two concentric tubes, the centre one terminating at the tip of the pitot tube in a pitot head hole. The outer tube has a series of holes drilled in it, the pitot static holes. When the pitot tube is placed in a flowing gas with the nose pointing upstream, the static holes, because there is no gas flowing into or out of them, measure the static pressure of the gas. The pitot head hole measures the combination of static pressure and velocity pressure due to the movement of the gas. The difference between the two pressures can be used as a measure of gas velocity.

For this purpose the chambers at the back of the holes are connected to two tappings at the other end of the pitot tube. The pressure difference is measured by an inclined gauge manometer.

Planck's constant h. The ratio of a quantum of radiant energy of a particular frequency to the frequency; the value is 6.6253×10^{-27} erg per second.

Plan for Coal A plan for the development of the nationalised coal industry endorsed in 1974 and 1977 by the Coal Industry Tripartite Group, involving the British Government, the mining unions and the *National Coal Board*. The Plan for Coal covers developments to the mid-1980s.

Plant availability *See availability*.

Plant load factors *See load factor, plant*.

Plant mix The proportions of different types of plant in a system, or contemplated for the future. In respect of electricity generating plant, the different types are usually classified as coal, oil, natural gas, nuclear, hydro, pumped storage and gas turbine. The first four are simply labelled by the fuels they use and are all steam-electric plants. Hydro plant implies hydro-electric plant, in which the water used for electricity generation flows entirely via dams from natural catchment areas. In pumped storage schemes water is pumped from a lower reservoir to a higher reservoir, base-load electricity at off-peak times being used, and released from the upper reservoir to generate electricity at peak times. Gas turbine plant is used for peak-load generation purposes.

Plant tissues Complex compounds comprising carbohydrates such as cellulose, lignin, starches and sugars; and proteins such as fats, oils, resins and waxes. The carbohydrates, which form a high proportion of most plant tissues, consist of carbon, hydrogen and oxygen; the proteins are compounds of carbon and hydrogen, with little or no oxygen. Plants synthesize their component tissues from atmospheric carbon dioxide and mineral-charged waters from the soil.

Plasma An ionised gas possessing properties different from those of a normal gas; plasma is a very good conductor of electricity, and is strongly influenced by magnetic fields. A plasma arc is a high-temperature ionised gas stream obtained by passing gas through an electric arc; it is used for cutting, welding and melting.

Plastic refractories Refractories composed of a refractory clay and grog, together with certain additions, to impart strength and volume stability at operating temperatures. Complete monolithic structures can be obtained without jointing, and linings can be made to any shape and thickness; plastic refractories are also resistant to spalling.

Plate precipitator   An *electrostatic precipitator* which comprises a number of vertical plates between which the gas passes horizontally. Plates are about 25 cm apart and from 6 to 9 m in height; discharge electrodes hang centrally between them at about 25–30 cm intervals. Plate precipitators are usually subdivided into a number of zones in series which may be energised and rapped separately.

Platformer   An oil refinery reforming process employing a platinum catalyst. Naphthas vaporised by heat and mixed with recycle hydrogen are passed through the catalyst beds. The catalyst promotes molecular rearrangement among the hydrocarbons to yield a reformed product: the platformate, containing a high percentage of aromatics which have great value in the compounding of high-grade *gasoline*.

Play   One or more accumulations of petroleum (typically either a cluster or a linear trend of such accumulations) that were formed under similar geological conditions, particularly regarding their source, migration paths and entrapment.

Plutonium   Pu. A product of the radioactive decay of *neptunium*; contains isotopes 238 and 239. Plutonium is an important by-product in nuclear reactions; it is recovered during the chemical processing of irradiated fuels. Used in the manufacture of weapons, it is hoped that it will ultimately be used as a fuel in fast reactors of the type being developed at Dounreay. *See Dounreay fast reactor; Windscale chemical processing plant.*

Plutonium economy   Descriptive of an economy in which the basis of electricity generation is the fast breeder reactor using *plutonium* as a fuel, the plutonium being recovered from the spent fuel elements of nuclear reactors using natural or enriched uranium. Hence, the economy hinges on the use of plutonium, and becomes increasingly dependent upon supplies of it; the risks of the diversion of plutonium into the manufacture of nuclear weapons are significantly increased. As a result, some people feel that the move to the *fast breeder reactor* should be discouraged, and that the world should go no further than the natural uranium and enriched uranium fuelled reactor.

Pneumatic controller   An automatic control system in which a Bourdon tube is used as a temperature measuring device; the movement of the Bourdon element is communicated to a fuel-controlling valve by air pressure on a diaphragm. By the introduction of a 'feedback' bellows into the system a proportional response control can be achieved. *See Bourdon gauge; proportional response controller.*

Pneumatic conveyor   A system by which loose material, e.g. fine coal, is conveyed through tubes by a stream of air. The air stream may be created by the expansion of compressed air through nozzles.

Pneumoconiosis Descriptive of various diseases of the lungs caused by the inhalation of dust particles in occupations such as coal mining, quarrying and the handling of asbestos.

Poise A unit of *dynamic viscosity* in the *metric system*; viscosities are often tabulated in centipoise, where 1 poise = 100 centipoise. *See viscosity.*

'Polluter pays' principle A principle which equates the price charged for the use of environmental resources with the cost of damage inflicted on society by using them. The price charged may be levied directly, e.g. as taxes on the process which generates pollution or as the purchase price of licences which entitle the holder to generate specific quantities of pollutants. If the producer or consumer can avoid the additional expense, he will attempt to do so; there is, therefore, an incentive to refrain from using the polluting item or to change consumption patterns or production processes in ways which mitigate pollution. Alternatively, in a reversal of principle, measures may take the form of direct payments from the public purse to polluters not to pollute. The difficulty with either procedure is to decide the 'right' price to charge or the 'right' subsidy to pay; both involve an assessment of the monetary value of a clean and unimpaired environment. For this reason, non-market techniques of pollution control are frequently preferred; under the 'polluter pays' principle, non-market measures encompass the promulgation of regulations governing in various ways the emission or effects of pollution. Regulations eliminate some of the uncertainty which is inherent in market approaches. The 'polluter pays' principle was affirmed at the United Nations Conference on the Human Environment held in Stockholm in 1972. However, an examination of the practices of members of the *Organisation for Economic Co-operation and Development (OECD)* reveals that a majority provide some incentive to industry to install pollution control equipment.

Pollution Any direct or indirect alteration of the physical, thermal, biological or radioactive properties of any part of the *environment* by discharging, emitting or depositing wastes or substances so as to affect any *beneficial use* adversely, to cause a condition which is hazardous or potentially hazardous to public health, safety or welfare, or to animals, birds, wildlife, fish or aquatic life, or to plants.

Polycyclic aromatic hydrocarbons (PAH) A variety of chemical compounds such as benzopyrene, dibenzopyrene, dibenzoacridine, occupational exposure to which can cause cancer in man.

Polymerisation An oil refinery process for producing high-octane gasoline components; it is a normal adjunct to a catalytic cracking unit. The process results in the combination of several molecules to form a more complex molecule, possessing the same *empirical formula* as the former.

Pony motor A motor provided for the purpose of promoting nuclear reactor coolant gas circulation, with the reactor shut down or at low power.

Population explosion A phenomenal increase in the rate of natural growth of the world population; while the time required for the population to double was about 1500 years during the period 8000 B.C. to A.D. 1650, it now takes only about 35–37 years. This rapid increase of population is due to a combination of the application of modern drugs and large-scale public health measures (control of specific diseases and improved sanitation) which have depressed death rates and extended the span of human life without any accompanying fall in birth rates. On the basis of present population trends, the United Nations Organisation forecasts for the year 2000 a world population of between 5.5 and 7 thousand million people. This prediction takes account of the fact that the world's annual rate of natural increase fell from 1.90 per cent in 1970 to 1.64 per cent in 1975; even those who consider zero population growth desirable tend to recognise that the great flywheel of human population will not be stopped overnight.

Porosity In respect of coke, the fraction of the total volume which is represented by the pores in the pieces; it is defined:

$$1 - \frac{apparent\ specific\ gravity}{true\ specific\ gravity}$$

See *voidage*.

Positive displacement meter A self-contained meter for measuring feed-water flow on the delivery side of a feed pump; a type in which a piston sweeps a definite volume of water along the feed pipe at every stroke. See *feed-water meter*.

Positron A particle identical to the *electron* except that its charge is positive. It is emitted in the beta decay of many radioactive atoms and is formed in the process of *pair production*.

Post-aerated burner A *gas burner* in which all the air required for combustion is introduced as secondary air.

Posted prices The published list prices of oil, although the *Organisation of Petroleum Exporting Countries (OPEC)* has adopted instead a *unitary pricing system*.

Potential energy See *Energy*.

Potential temperature, gradient of In the lowest layers of the atmosphere, the difference between the adiabatic lapse rate and the actual atmosphere lapse rate. When the gradient of potential temperature is zero, the buoyancy of a mass of rising gas is constant and there is no theoretical ceiling. If the gradient is negative, the buoyancy of the gas will increase as it rises, and if it is positive, it will decrease and there will be a height at which the buoyancy vanishes. *See adiabatic process; lapse rate.*

Poundal An older unit of force; the force needed to give a mass of 1 pound an acceleration of 1 ft/s^2.

Pour point That temperature which is 2.8° C (5°F) above the temperature at which the oil just fails to flow when cooled under prescribed test conditions. *See pour point depressant.*

Pour point depressant A lubricating oil additive intended to lower the temperature at which oil will just flow; typical compounds are wax-naphthalene condensates or polymeric compounds. These depressants are only suitable for waxy (paraffinic) oils.

Power Plant capacity or rating, expressed in kilowatts or megawatts.

Power alcohol Industrial *ethanol*, used as a fuel.

Power factor In its application to *alternating current*, a measure of the proportion of useful current employed to that expended in overcoming the effects of inductance and capacity.

Power forming An oil refinery unit in which low-octane naphtha is converted to a high-octane component of premium motor spirit; the naphtha is first hydrofined and then reformed over a platinum catalyst in an atmosphere of hydrogen at high pressure.

Power kerosine A volatile *kerosine*, with distillation limits essentially between 150 and 260°C and of good anti-knock value. It is used as a fuel for some spark-ignited engines, e.g. tractors. It is known alternatively as vaporising oil.

Power pool The interconnection of two or more electricity supply systems, combined with centralised control of the operation of all the power stations in the enlarged system. *See New York power pool.*

Poza Rica incident An air pollution incident involving respiratory illness and death. At Poza Rica, Mexico, on 24 November 1950, some 22 persons died and 320 were hospitalised as a result of the malfunction of an oil refinery sulphur recovery unit; large quantities of *hydrogen sulphide* were vented to the atmosphere.

Meteorological data indicated that a pronounced low-altitude temperature *inversion* prevailed at the time. Unlike other acute air pollution incidents, a single source and air pollutant were responsible.

PPHM Parts per hundred million. When used in relation to the concentration of a gaseous pollutant, it refers to the number of parts by volume of the polluting gas in one hundred million parts of the air in which it is diffused.

Prandtl number A dimensionless group of factors which serve as a criterion of temperature gradient similarity in fluids; it is expressed:

$$\frac{C_p\mu}{K}$$

where C_p = specific heat;
K = thermal conductivity;
μ = viscosity.

Pre-aerated burner A *gas burner* in which the primary air is mixed with gas before arrival at the burner.

Pre-boiler burner An arrangement whereby solid fuel is consumed in a separate furnace fitted to the front of the boiler, the hot gases produced in the burner passing through into the boiler proper; the furnace is circular in section, water-cooled and supplied with fuel from a hopper mounted on the top. A fan provides primary and secondary air to the furnace. It is claimed that the fitting of a pre-boiler unit to a *sectional boiler* can result in a fuel saving of 15–18 per cent.

Precipitation sampler Automatic sampler; samples are collected in receptacles designed to open only during periods of rainfall. Thus, these samples contain only the pollutants carried by precipitation and do not collect ordinary dust-fall.

Preferential combustion When a compound of two combustible elements burns in an insufficient amount of air, the taking of the limited amount of oxygen by the element with the greater affinity for oxygen, while the other element is liberated. For example, if hydrogen sulphide is burned in an insufficient supply of air, the combustion of hydrogen takes preference over the combustion of sulphur, and sulphur is deposited:

$$2H_2S + O_2 \rightarrow 2H_2O + 2S$$

This principle is adopted to achieve sulphur recovery in oil refineries.

Pre-ignition In relation to the spark-ignition engine, the ignition of the fuel–air mix prior to the firing of the spark, usually as a result of overheated surfaces. See *auto-ignition*.

Premix burner A *burner* in which the fuel and oxidiser are mixed prior to ignition. See *direct burner*.

Pre-mixed flame A *flame* arising from the combustion of gases, the gases already being intimately mixed prior to ignition; most and possibly all the necessary oxygen for combustion is contained in the mixture, and the rate of combustion depends on the rate of the chemical reactions alone and is not dependent on the rate of mixing with an external source of oxygen. *See diffusion flame.*

Prepared town gas (PTG) A specially prepared gas used in carburising processes. It is produced by the thermal decomposition of coal gas in the presence of iron turnings, which removes *oxygen* and gives a satisfactory $CO:CH_4$ ratio. The gas provides a 'balanced' furnace atmosphere in which the *heat treatment* of steel may be conducted without scaling, *carburising* or *decarburising*.

Present worth The capital sum, which, if invested at the time of the commissioning of a plant, would together with the interest earned suffice to pay the annual capital charges on the plant as they occur year by year. If it is necessary to calculate an average annual cost over the whole life of the plant, the total present worth may be converted to an 'equivalent average annual cost' by multiplying by the appropriate annuity rate. Present worth factors and annuity rates may be found from *Inwood's Tables*, Archer's *Tables of Repayments*, etc. For a 12 per cent interest rate, the present worth factor is given by:

$$\frac{1}{1.12^m}$$

where m = years of life of the plant (assumed for depreciation purposes). The present worth method offers a general and systematic approach to the problem of allocating a series of irregular costs to the units of a continuous output.

Pressure A measure of force per unit area; in the past it has been expressed in pounds per square inch (psi) or grams per square centimetre (g/cm^2). A current unit is the *pascal*.

Pressure distillation Petroleum distillation in stills under pressure to produce a cracked, light, gasoline-bearing distillate product.

Pressure gasification of coal *See Lurgi process.*

Pressure gauge An instrument fitted to every steam boiler to indicate steam pressure. It is usually of the Bourdon type. The dial should be not less than 15 cm in diameter, with graduations in pascals ranging from zero to twice the operating pressure. An exception to this is the *critical pressure gauge*. The dial must indicate: (a) the boiler operating pressure, in red; (b) the maximum permissible working pressure, in purple. The gauge requires a stop valve so that it may be removed while the boiler is under steam. *See Bourdon gauge.*

Pressure jet burner An *oil burner* in which oil is supplied under pressure to a special nozzle which converts the pressure energy of the oil into kinetic energy; the oil is forced through tangential slots in a sprayer plate to impart a high rotational speed to the oil, the oil then leaving the burner through a swirl chamber and precision orifice which controls the conical angle of the oil mist. *See Figure P.1*. The mist droplets vary in size between about 20 and 200μ. The oil is supplied to the burner at pressures of 7 bar upwards. A **turn-down ratio** of about 3 to 1 only is attainable unless special wide-range pressure jets are used. *See **atomisation; wide range pressure jet burner***.

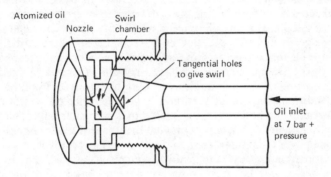

Figure P.1 Section of pressure-jet nozzle (Source: British Petroleum)

Pressure mill A mill in which the air carrying the coal within the mill is under pressure, and not under suction as in a suction-type mill.

Pressure vessel The vessel containing the fuel elements *moderator* and *coolant*, of a *nuclear reactor*. Its purpose is to enable the reactor to be operated at pressures above atmospheric in order to improve the heat transfer properties of the coolant. It is generally constructed of steel, or of prestressed concrete.

Pressurised water reactor A *nuclear reactor* using water as the *moderator* and *coolant*, in which the system is maintained at a high pressure to prevent boiling. The water is circulated through a heat exchanger to generate steam in a secondary circuit.

Pressurised water–thorium–uranium converter reactor A nuclear reactor installed at Indian Point Power Station, 24 miles north of New York City on the Hudson River, USA, by the Consolidated Edison Company of New York; the station has an electrical output of 275 MW. The reactor operates at 100 bar and is cooled by the circulation of water. The reactor core consists of 120 fuel elements; each of the fuel elements contains 195 fuel rods arranged in a square lattice. The rods are stainless steel tubes loaded with fuel pellets. Thorium-232 does not itself undergo fission but absorbs neutrons to form thorium-233; this isotope decays to protactinium-233, which in turn decays to the fissionable isotope uranium-233, which then fissions to produce heat energy.

Price The value of a commodity or service in terms of money; an exchange ratio. The term 'price' is used in relation to the exchange value of goods and some services; in respect of human services the equivalent terms 'wage', 'salary', 'commission' and 'fee' are used. Again the price of borrowing money is called 'interest' and the price of hiring land or equipment 'rent'.

Price discrimination The charging of different prices to different groups of individuals for the same goods or services for reasons not associated with difference in costs. This may occur when there is a geographical separation of markets, the structure of demand in each market being different. It may also occur in respect of railway freight charges. It has also been practised by medical practitioners in communities of wide disparity of wealth. The conditions for discrimination are: (a) the seller can control the amount of a product that is offered to a particular buyer; and (b) either the seller can legally prevent the resale of the commodity by one buyer to another, or such resale would be costly or physically impossible for buyers to arrange. The results of price discrimination appear to be mixed, promoting greater output and consumption in some instances and proving restrictive and anti-social in others.

Price elasticity of demand *See demand.*

Price mechanism System used in a competitive society for the distribution of scarce resources through the agency of price. If too little of commodity A and too much of commodity B is being produced, the demand price for A rises, and that for B falls. The signal is thus given to the productive system to make the necessary adjustments. If competition were perfect, these adjustments would continue until marginal price and cost were equal. The prices of all goods and services form an interrelated system; what we are prepared to pay for A depends on the terms which we can obtain for B, C, D, etc. In other words, all prices are mutually determined. Price tends to equate the amount sellers are prepared to offer for sale and the amount which buyers wish to buy.

The market system theoretically leads to a point of maximum efficiency through the marginal equivalence of costs and prices. Producers seek to minimise their costs, while consumers maximise their satisfaction. At the point of maximum efficiency a redistribution of output could not increase the satisfaction of anybody without reducing that of someone else.

The price mechanism is defective, however, in that it does not necessarily succeed in regulating supply as required; in that it does not necessarily operate without inflation; and in that it does not automatically adjust social and private marginal net products. There are areas in which the price mechanism not only fails, but could not operate. In these instances the state often finds it necessary to produce goods which cannot be provided by the market, e.g. low-income and welfare housing. In addition, in world commodity markets price has proved a singularly ineffective weapon for achieving reductions in output. Small diminutions in price of demand have produced gluts of astonishing proportions.

Prices Justification Tribunal (PJT) Established by the Australian Government in 1973 to assist in curbing inflationary trends in prices. The tribunal considers the justification for price increases proposed by companies; it is not in itself a price-regulating body. In considering proposed price increases, it has regard to increased costs which have actually been incurred, distinguishing between 'avoidable' and 'unavoidable' cost increases; wage increases which exceed those defined in wage agreements; productivity as an element offsetting some increases in costs; and the level of profitability of the company and its capacity to maintain a reasonable level of investment and growth. The tribunal has been involved, for example, in the question of increases in the prices of petroleum products.

Primary air Combustion air introduced into a furnace through the fuel bed by natural, induced or forced draught; or mixed with the fuel and delivered to the furnace under pressure. *See secondary air; tertiary air*.

Primary burner In respect of an incinerator, a burner installed in the primary combustion chamber to dry out and ignite the material to be burned. *See secondary burner*.

Primary heating The heating of heavy viscous grades of oil in the storage tank in order to keep the oil at a viscosity at which it may be pumped; this can be done by steam or hot-water coils placed in the tank, or by electric heating elements. The minimum storage temperatures vary with viscosities.

Priming (1) The entrainment of boiler water in the steam produced; priming may be due to excessively high water levels, or due to foaming as a result of excessive salts in the water. (2) In pumping, the replacement of residual air in a pump by the fluid being pumped.

Producer gas Gas manufactured by the action of air and steam on coke or coal. A *gas producer* comprises a cylindrical water-cooled shell fitted with a fuel-charging hopper at the top and a fire grate at the bottom equipped with air–steam blast tuyeres; this unit is frequently followed by gas-cleaning plant. The composition of the gas produced depends on the proportion of water vapour in the blast and on the type of fuel gasified. A typical analysis for a coke-produced gas, the constituents being expressed as percentages, is as follows: carbon monoxide, 29; hydrogen, 11; carbon dioxide, 5; nitrogen, 55. The calorific value is about 6 MJ/m^3.

Production, structure of time The time interval between a decision to undertake production and the beginning of the outflow of the product; the more capitalistic or 'capital-intensive' the method of production, the longer will be the time interval. In respect of a power station some 5–6 years or more may elapse between a decision to build and the initial flow of electricity into the grid.

Productivity The efficiency with which productive resources, i.e. labour, capital and land, are used; the motive for increasing productivity is to produce more goods at a lower cost per unit of output while maintaining quality. It is not easy to measure changes in productivity, but the relationship of output to man-hours expended is often adopted as a rough guide; this criterion relates to input of labour only, without reference to capital or land, and in certain circumstances may give a misleading picture.

Profit-taking Sales by 'bull' speculators of commodities or shares, following a rise in price, in order to profit.

Projected diameter The diameter of a circle which has the same area as the projected profile of the particle.

Project Independence Launched by President Richard M. Nixon in 1974, a plan to reduce United States dependence on imported foreign oil and promote the role of nuclear power. It was envisaged that 40 per cent of the electricity requirements of the United States would be generated by nuclear plant by the late 1980s. A vast increase in coal production was also planned.

Project Sunshine A Japanese national research programme involving the development of solar and geothermal energy, hydrogen fuel and synthetic natural gas. It is designed to cope with Japan's energy problems over the next 25 years. The programme has included the construction of four solar houses equipped with solar-type absorption chillers.

Propane A hydrocarbon gas, C_3H_8, being a member of the *alkane series*; it is used for heating, metal-cutting and flame-welding purposes. It can be stored under pressure as a liquid at atmospheric temperatures, but is more volatile than *butane*, and high pressures are needed to keep it in liquid form. By reason of its chemical composition it is classified as a C_3 hydrocarbon. Propane burns in air to carbon dioxide and water; it is available in cylinders for domestic and commercial use; hence the term 'bottled gas'. *See liquefied petroleum gases (LPG).*

Propane de-asphalter An oil-refining process which extracts feed for the fluid catalytic cracking unit from *residuum* by its solubility in liquid propane.

Prop-free front A technique used in relation to machine-mined coal in which the entire length of a coalface, perhaps 200 metres or more, is free of conventional props. Self-advancing power-operated steel roof supports provide this uncluttered working area, along which chain hauled power loaders operate, cutting coal and automatically loading it on to the face conveyor. There are over 200 000 powered support units in use in the British coal industry.

Proportional counter A device for detecting charged particles and photons by the *ionization* which they produce in a gas between two electrodes. *See radiation detector.*

Proportional response controller An automatic device by which a proportion of the heat being supplied to a plant is cut out when the temperature exceeds a set point. This type of control ensures smaller temperature variations than with an *on–off control*. A technique known as 'proportional response with reset' combines the two methods to give accurate control at the desired temperature.

Proportionate control mixing gas burner A gas burner system in which the two flows of air and gas are held in a constant ratio by means of an automatic flow ratio controller. *See gas burner.*

Proration A system used in the United States and some other oil-producing countries whereby production from every well (except new wells, or the largely exhausted 'strippers') is limited to a given percentage of its capacity, and buyers are usually obliged to take the same percentage of production from all the wells connected to their gathering facilities. In the United States the amount of production allowed each month is usually prorated to market demand, as estimated by the competent State authority.

Protein A group of complex nitrogenous organic compounds of high molecular weight; protein molecules consist of many hundreds of amino acids joined together by a peptide linkage into one or more interlinked polypeptide chains. *See energy value.*

Protium The hydrogen isotope with a mass number of 1.

Proton A fundamental particle which is a constituent of all atomic nuclei. It has a mass of $1.672\,48 \times 10^{-24}$g and a positive electric charge of 1.602×10^{-19} *coulomb*. The number of protons in a *nucleus* represents the *atomic number* of the *element*.

Protoplasm The matter of which biological cells consist; it is usually divided into two parts, the cytoplasm and the nucleus.

Proximate analysis The determination in a sample of fuel of: (a) moisture content; (b) volatile matter content; (c) ash content; (d) fixed carbon content. *See adventitious ash; ash; fixed carbon; free moisture; inherent ash; inherent moisture; ultimate analysis; volatile matter.*

PSI Pounds-force per square inch.

PSIA   Pounds-force per square inch absolute.

PSIG   Pounds force per square inch gauge.

Psychometric chart   A chart which defines the relationships between dry bulb temperature, wet bulb temperature, moisture content, relative humidity and total heat of air.

Public goods and services   Goods and services distributed by the state free to its citizens, being indivisible benefits shared by everyone equally. No private entrepreneur could provide and sell such goods and services efficiently on an individual basis. Examples are external defence, police, maintenance of law and order, regulation of industry, overseas representation, tax collection, civil defence. *See merit goods and services.*

Pulsation bottles   Pressure vessels inserted between compressor cylinders, or after compressors, to smooth out pressure pulses produced by compressors.

Pulse jet engine   A type of *ram-jet engine*, in which the combustion process is discontinuous, being arranged to occur at intervals between which the pressure in the combustion chamber is allowed to build up. German flying bombs of World War II were powered by pulse jets.

Pulverised fuel   Fuel, usually coal, finely ground; the desirable degree of fineness is governed by the use to which the pulverised fuel is to be put. A standard for fuel to be fired in large water-tube boilers is that 65 per cent of the pulverised fuel should pass through a 200 BS mesh with elimination of coarse particles above 30 mesh. The most suitable type of fuel for pulverising is bituminous coal containing over 20 per cent of volatile matter and of not more than medium coking power; the ash content may be as high as 20 per cent. *See pulveriser*.

Pulverised fuel ash (PFA)   Finely divided greyish ash resulting from the combustion of pulverised coal. It has a large surface area of 2000–5000 cm^2/g, and 75 per cent or more will pass through a 200 BS sieve. A typical percentage analysis for fly ash from Central Electricity Generating Board power stations is: silica, 42.6; alumina, 32.4; ferric oxide, 10.4; lime, 5.1; magnesia, 2.5; sulphate, 0.9; alkalis, etc., 1.1; loss on ignition, 5.0. The 'loss on ignition' is due to unburnt carbon, a variable percentage depending on combustion conditions which may be as low as 1 per cent and may rise to 12 per cent; the average for British base-load stations is about 4 per cent. Formerly regarded solely as a waste product, over one million tons a year of PFA is being used in Britain in building materials.

Pulveriser   A device for grinding fuel as finely as possible with a minimum expenditure of power. By reducing coal to a fine powder it can be carried into a furnace by an air blast. Pulverisers fall into three categories—slow, medium and high-speed types. *See **ball mill; ball-race mill; high-speed mill; Raymond bowl mill***.

Pumped storage A means of 'storing' electricity in the form of water by pumping it to a high level when there is electricity available to drive pumps, perhaps during the night when electricity demand is low, and then letting the water flow down through water turbines and so generating electricity the following day when demand is high. It is a technique for helping to reduce the effects of fluctuating demand for electricity on an electricity supply system; the addition of a pumped storage station to a system saves the installation of an equivalent capacity of conventional generating plant that would otherwise be required to meet the peak demand. A pumped storage scheme is in operation at Ffestiniog, Wales, the upper reservoir being about 300 m above the lower reservoir. At peak load periods Ffestiniog station generates 300 MW. *See Figure P.2*.

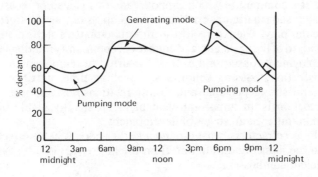

Figure P.2 Typical daily load curve for pumped storage scheme

Pure coal The '*dry, mineral-matter-free*' portion of coal; the basis to which all data and observations are referred when fundamental questions are being considered. The '*as received*' basis is adopted for practical and commercial consideration.

Pyrites A hard, yellow sulphide of iron which occurs as an impurity in coal; 'fool's gold'. Pyritic sulphur may occur in lenses, bands, veins, joints, balls or fossils; or it may occur as finely disseminated particles. It may vary in size from particles a few microns in diameter up to lumps several feet in diameter.

Pyrolysis The decomposition of a substance by heating.

Pyrolysis gasoline One of the liquid hydrocarbon fractions condensed from the cracked gas stream, following the cracking of *naphtha* under heat. It may be blended into gasoline.

285

Pyrometer A device for measuring high temperatures. *See light-sensitive pyrometer; optical pyrometer; radiation pyrometer; suction pyrometer; thermoelectric pyrometer; total radiation pyrometer; two-colour pyrometer; venturi pneumatic pyrometer. See also line reversal method; thermometer.*

Pyrometric cones Small pyramids about 3 cm high made of selected mixtures of oxides and glass which soften and melt at known temperatures; they are widely used in the ceramic industry as a method of measuring high temperatures in refractory heating furnaces.

Q

Quality of life In current usage, a phrase which appears to cover a miscellany of desirable things not necessarily recognised, or adequately recognised, in the market place. Some qualities of the life of a community which cannot readily be valued or measured include such matters as civil liberties, compassion, justice, freedom and fair play. Then there are the important matters of the material standard of living, health and education, community relationships, employment opportunities and security of employment, job satisfaction, housing conditions, travelling time, nutrition and general stress, personal and family relationships. Environment protection is an important element in the quality of life of all communities at all stages of development.

Quarl A refractory throat around a burner port; it helps direct air into the flame, and radiation from the hot refractory also assists efficient combustion.

R

Rad A unit of energy dissipation, or radiation dose, in any medium; it is applicable to any type of radiation, whereas the *roentgen* is restricted to X-rays and gamma rays. One rad is equal to an energy absorption of 100 ergs per gram, e.g. of human tissue. One millirad = 10^{-3} rad.

Radar Abbreviation of the description 'radio detection and ranging'; any locating, navigating or guiding system that employs microwaves. Radar operates independently of weather and visibility conditions because it depends on the ability of microwaves to bounce back from solid objects.

Radial-tip fan blading A design of blading usually preferred for centrifugal induced-draught fans; it has a greater resistance to erosion, and there is a lower tendency for grit and dust in the gas stream to build up on the back of the impeller blades compared with *backward-curved fan blading* . *See Figure R.1 ; fan.*

Figure R.1   Radial-tip fan blading.

Radiant energy   *See energy.*

Radiant superheater   *See superheater.*

Radiation   The transmission of heat from one point to another without affecting the temperature of the medium through which it passes. The heat dissipated from a surface is influenced by the difference in temperature between the radiating surface and the receiving surface, and by the coefficient of emission. The coefficient of emission is unity for a *black body*. Other examples are: red brick, 0.94; refractory brickwork, 0.75–0.8; polished metals, 0.04–0.09. *See heat transfer.*

Radiation absorber   A substance which absorbs atomic particles and radiation. The control of nuclear reactors is commonly effected by the use of control rods which contain materials, particularly boron and cadmium, which are good absorbers of thermal neutrons.

Radiation detector   A device for detecting the presence of *radioactivity*. *See Geiger–Müller counter; ionisation chamber; proportional counter; scintillation counter.*

Radiation pyrometer   A device for measuring temperatures, utilising the emission of radiant energy from a hot body. It is suitable for measuring temperatures higher than those measured by a *thermoelectric pyrometer*, or for measuring the temperatures of the inside of a furnace or kiln. There are three kinds of radiation pyrometer available: (a) *total radiation pyrometer;* (b) *optical pyrometer;* (c) *light-sensitive pyrometer. See pyrometer; thermometer.*

Radioactivity   The spontaneous emission of ionising particles and rays following the disintegration of certain atomic nuclei. When a large number of radioactive atoms are present, the random nature of individual emissions is obscured, and the radiation appears to be regular and uniform. It can be defined as the number of disintegrations per unit time taking place in a radioactive

287

specimen. A unit of radioactivity is the *curie*, which is 3.7×10^{10} (37 thousand million) disintegrations per second. The following sources of radioactive emissions to atmosphere are listed in *Meteorology and Atomic Energy* (United States Atomic Energy Commission, 1955):

Mixing and handling uranium ores.
Chemical production of brown oxide or uranium (UO_2).
Machining radioactive and toxic metals (e.g. uranium and beryllium).
Atomic laboratories.
Particle accelerators.
Nuclear reactors.
Chemical processing plants for reactor fuels.
Waste disposal.

Plants handling radioactive materials are designed to prevent emissions of radioactive material to the atmosphere. Methods used for the removal of radioactive particles from gas streams are the same as those used for non-radioactive particulate matter, except that the collection efficiency must be almost perfect.

Man is continuously exposed to ionising radiations from natural sources; radioactive substances are widely distributed in rocks, soils, water and air. Sedimentary rock areas contribute an average dose of about 45 millirems per year, while granite areas such as Aberdeen in Scotland contribute more than double this dose. Higher background levels are to be expected in areas containing deposits of uranium, thorium and radium, perhaps several times as much. Natural gamma radiation directly from the earth and from buildings represents usually about one-half of the total external radiation to which populations are exposed. All food and drinking-water contains minute quantities of radium which are ingested into the system; the variations in content depend upon the radioactivity of the rocks and soils from which they are derived. *Radon* in the atmosphere and cosmic rays also contribute to the general level of radiation.

The world average dose received from natural and unavoidable background radiation by the individual is estimated to be about 150 millirems per year. This may vary from 80 to 300 millirems, depending upon the composition of the ground, building materials, altitude, foodstuffs, and so on; the average for the United Kingdom has been estimated to be about 100 millirems per year. Sources of additional radiation to which a person may be exposed include the various forms of radiation used in medicine and industry, the radioactive isotopes produced in nuclear reactors, accidents in nuclear reactors, and fall-out resulting from the explosion of nuclear weapons.

Radiochemistry The branch of chemistry concerned with the chemical processing of irradiated material or naturally radioactive material in order to isolate the various radioactive isotopes, and with the use of radioisotopes in the study of chemical problems.

Radiography A method of detecting flaws, inclusions, lack of homogeneity, etc. in solid objects by making shadow images on photographic emulsions by the action of X-rays or gamma rays which have been differentially absorbed in their passage through the object.

Radiometer An instrument for measuring the radiant heat emitted by an open fire. A test grate is raised above the floor so as to be at the centre of an imaginary hemisphere; the radiation falling on the inside surface of this hemisphere is measured by thermopiles. The readings are integrated to a value for total radiant energy.

Radionuclide Any radioactive nuclide, a nuclide being an atom characterised by the constitution of its nucleus, i.e. by the atomic number and the mass number.

Radiopotassium The chief source of radioactivity inside the human body; about $\frac{1}{500}$ of body weight is due to potassium. About 1 potassium atom in every 8000 is K^{40}, which is a radioactive isotope.

Radiosonde A balloon-borne instrument which transmits measurements of pressure, temperature, etc., as it rises and moves with the wind.

Radon A radioactive gas which is released from the soil into the atmosphere. It is also released during the combustion of coal, and for this reason there is more radon in the air over cities than in country air.

Raffinate The refined product resulting from a solvent refining process in an oil refinery.

Rainout Or washout; a mechanism whereby small particles in the atmosphere are removed by being captured in raindrops as they form.

Ram jet engine A simple type of jet propulsion system; as with all jets, the air is compressed, while fuel is injected into it and burnt. The burnt gas at high pressure and temperature provides the energy source for movement. In the ram jet engine, air is admitted continuously at the forward end, being compressed only by the forward movement of the vehicle; the whole of the exhaust gas is available for propulsive thrust.

Ramsbottom test A test to determine the *carbon residue* of a liquid fuel, which is generally preferred to the *Conradson test*. A glass bulb containing a sample of oil is heated in a bath of molten metal maintained at 550° C. After cooling, the residue is weighed and expressed as a percentage of oil used.

Range plant   Or range system; turbine and boiler plant connected by a common steam main which enables a turbine to draw steam from two or more boilers, and for the steam from any boiler to be taken to two or more generating sets.

Rank   In respect of coal, the degree of metamorphosis or 'coalification' that the original plant debris has undergone during the geological ages since it was deposited. Lignites and brown coals are low-rank coals, while anthracite is of the highest rank.

Rankine cycle   An ideal cycle for a steam engine, proposed by Professor W. J. M. Rankine (1820–1872) as a standard of comparison with the performance of actual engines. The cycle comprises four stages: (a) steam passes from the boiler to the cylinder at constant temperature; (b) steam expands adiabatically to the condenser pressure; (c) heat is given to condenser at constant temperature; (d) condensation is completed and condensate returned to boiler. In the Rankine cycle the work done is equivalent to the total heat, H_2, in the steam at the end of the adiabatic expansion, subtracted from the total heat, H_1, in the steam at the beginning of the expansion. The heat supplied is equal to the sensible heat, h_2, in the condensed steam subtracted from H_1. The cycle efficiency of the engine is calculated as follows:

$$\text{efficiency} = \frac{\text{work done}}{\text{heat supplied}} = \frac{H_1 - H_2}{H_1 - h_2}$$

Rankine scale   A temperature scale based on three fixed points: (a) the lowest (0°) being abolute zero; (b) the second (491.67°) being the temperature of melting ice; (c) the highest (671.67°) being the boiling point of water at normal atmospheric pressure. The Rankine and Kelvin scales are absolute scales; 0° R = 0 K; 1.8 Rankine degrees = 1 Kelvin. *See absolute zero; temperature scales.*

Rasmussen report   *See melt-down.*

Rational analysis   The resolution into chemical types of a mass of rock or coal. In respect of coal the constituents determined are: (a) resins, waxes and extractable hydrocarbons; (b) cuticular plant remains; (c) humic substances; (d) opaque matter; (e) *fusain*.

Rational classification   A system of coal classification based primarily on the *ultimate analysis* of the different chemical entities present in coal. *See rational analysis.*

Raw gas   Gas before purification.

Raymond bowl mill   A medium-speed *pulveriser* for the production of *pulverised fuel*. Raw coal enters a bowl rotating at between 74 and 150 rev/min; there it is ground between a ball ring and rollers.

Hot air for drying and conveying the pulverised fuel enters the lower mill casing, and a rapid circulation of hot air is maintained through the system. The powdered coal is drawn through a classifier, where the oversized particles are returned to the inlet of the mill. The outlet temperature aimed at is usually about 80° C.

RBE *See relative biological effectiveness.*

Reactive fuel *See reactivity.*

Reactivity In relation to a fuel, e.g. coke, the measure of ease of ignition from cold and of the relative rate of combustion under specified conditions. British Standard 3142 defines standards for domestic cokes.

Reactor Combustion units in which a fuel and oxidiser are chemically combined to obtain a specific chemical compound, e.g. the combination of sulphur and oxygen to produce sulphur dioxide. In a 'partial combustion' reactor a less than stoichiometric supply of air is used, as in the reaction of carbon with oxygen to produce carbon monoxide.

Reactor kinetics The branch of reactor technology concerned with the study of the behaviour of a reactor when its conditions of operation are not steady.

Réaumur temperature scale A temperature scale in which the fixed points are 0° (freezing point of water) and 80° (boiling point of water); a little-used scale.

Reciprocal poise *See fluidity.*

Reciprocating pump A positive displacement pump consisting of a piston moving back and forth within a cylinder. With each stroke a definite volume of liquid is pushed out through the discharge valve.

Reclaimed coal *Coal* recovered from a stockpile, as distinct from being received direct from a supplier.

Recorder An instrument fitted with a chart, rotated by some form of clockwork or electric mechanism, upon which a pen draws a diagram indicating a measured change in an operating condition, e.g. temperature or smoke emission. The charts provide a permanent record.

Rectifier A device for the conversion of an alternating current into a direct current by the inversion or suppression of alternate half-waves. *See electronic valve rectifier; electrostatic precipitator; mechanical rectifier; silicon diode rectifier; static rectifier.*

Recuperative air preheater An *air preheater* constructed of tubes or plates, the gases passing on one side and the air on the other, usually in a contra-flow manner in order to ensure a maximum heat transfer through the plates.

Recycling The return of discarded waste materials to the production system for utilisation in the manufacture of goods, with a view to the conservation as far as practicable of non-renewable and scarce resources. Recycling goes beyond the reuse of a product (such as glass milk bottles) and involves the return of salvaged materials, such as paper or metals or broken glass, to an early stage (pulping or melting) of the manufacture process. Some recycling has always been profitable to certain industries, e.g. the return of steel scrap to the steel industry. Opinion now seems to favour an increased tempo and scale of recycling to conserve resources for the future of mankind beyond what may be profitable in the shorter term. The capacity of an industry to recycle is in many cases limited by technical as well as economic considerations. On the other hand, the glass industry is already recycling considerable quantities of cullet; the aluminium industry currently finds it economically acceptable to encourage the return of aluminium drink cans. *See Figure R.2; **non-renewable resources**.*

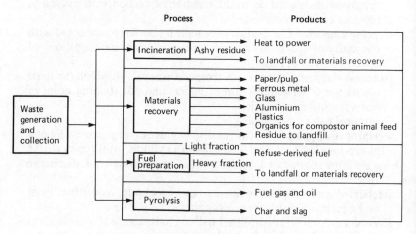

Figure R.2 Recycling or resource recovery process.

Redox Acronym for 'reduction–oxidation'.

Redox energy-storage system A system developed at the Lewis Research Centre of the National Aeronautics and Space Administration (NASA) for storing electrical energy; chemical energy is converted into electrical energy when two reactant fluids, chromium chloride and iron chloride, are pumped through a stack of flow cells. In each cell the fluids are kept separate by a special membrane; the fluids transfer electrical charge through the membrane as each fluid reacts at a separate inert electrode surface, but the chromium and iron remain in solution barred from passing through the membrane. To recharge, the fluids are pumped through the stack again but with electrical energy supplied from

292

an outside source. The basic reactants can be used indefinitely in this reversible reaction. The system is a substitute for conventional lead–acid storage battery systems, and claimed to be much superior. *See electric battery.*

Red top burning   The condition of a fuel bed which is uniformly incandescent at its surface. *See black centre burning.*

Reducing   In petroleum refining, the removal of light hydrocarbons by *fractional distillation.*

Reduction   The removal of oxygen from a compound; the reverse of *oxidation.* A reducing atmosphere is one devoid of free oxygen.

Redwood viscometer   An instrument for measuring the *viscosity* of fuel oils, expressed in Redwood seconds at 37.8° C (100°F). Ordinary fuels are measured in Redwood No. 1 seconds; more viscous oils in Redwood No. 2 seconds. The Redwood No. 2 cup has a larger orifice which allows the sample of oil to flow out in about one-tenth of the time taken in the Redwood No. 1 apparatus.

Reference method   A method of analysis against which others are compared. Usually a method of high accuracy, but time-consuming and less suitable as a routine method.

Refinery capacity   In respect of oil refining, the maximum through-put of crude oil for which the plant is designed. A refinery's output of refined products will usually be about 5 per cent smaller than this, allowing for waste and for the products burned to keep the refinery and its auxiliary plant in operation.

Refinery flares   A method of disposing of surplus gas at oil refineries by burning such releases at the top of a flare stack. Fuel gas cannot be released to the atmosphere unburnt, partly because of odour and partly because of explosion hazards. The simple burning of refinery fuel at the top of a stack involves the production of considerable smoke; flares are made to a large extent smokeless by either steam injection or water spray close to the point of ignition. The adjustment of the steam or water to meet the requirements of the flare is usually by manual control, but may be automatic.

Refinery gas   Any form or mixture of gas gathered in a refinery from the various stills.

Reflectance   A measure of the extent to which a surface will reflect light or other form of radiation. If smoke-contaminated air is drawn through a white filter paper, the reflectance will be reduced; this reduction is the basis of the British smoke stain methods for measuring the concentration of smoke in the atmosphere.

Reflector   A layer of material surrounding the core of a *nuclear reactor*, whose purpose is to reduce the escape of neutrons by means of scattering processes which result in the return of many of the neutrons to the core; a reflector material should have a high neutron-scattering cross-section and a low neutron-capture cross-section.

Refluxing In *fractional distillation* in oil refining, the return of part of the distillate to the *distillation tower* to assist in making a more complete separation into the required fractions; the material returned is the reflux.

Reformed gas Low thermal value gas obtained by the *pyrolysis* and steam decomposition of high thermal value gases, e.g. natural gas or oil-refinery gas.

Reformer A catalytic refining unit used in oil refineries to upgrade low-octane gasolines into high-octane gasolines by rearrangement of the molecular structure. The catalyst usually contains platinum. Reformers generally yield by-product hydrogen.

Reformer pretreater A hydrogenation unit used in oil refineries for purifying the feed to reformers. The pretreater effectively removes catalyst poisons and protects the platinum catalyst in the *reformer*.

Refractory A material used in lining furnaces; it must be capable of resisting a moderate load at high temperatures without distortion or collapse, changes in temperature, and the action of molten metals, slags and hot dusty gases. *See alumina; aluminous firebrick; chrome brick; chrome-magnesite brick; dolomite brick; firebrick; hot face insulation; magnesite brick; plastic refractories; semi-siliceous brick; silica; silica brick; siliceous brick; sillimanite brick.*

Refrigeration The cooling of air or liquid by passing it over coils containing a cooling medium, which can be refrigerant, chilled water or brine. Chilled water and brine are generated from a liquid-chilling machine which has a refrigerant cycle; the water or brine is pumped in a closed circuit to the air conditioning or process equipment, and then back to the chiller for recooling. Ammonia, carbon dioxide, sulphur dioxide, methyl chloride and *freons* are all possible refrigerants, although the freons have proved most popular, because of their special characteristics. *See refrigeration plant.*

Refrigeration plant Plant specifically designed to achieve *refrigeration* and comprising four basic components: (a) an evaporator, where heat is absorbed from the medium to be cooled (either air or liquid) by the evaporation of the liquid refrigerant; (b) a compressor, which raises the temperature and the pressure of the refrigerant gas from the evaporator by compressing it; (c) a condenser, which is a heat transfer coil, where the hot gas is cooled by the condensing medium; (d) an expansion valve, which reduces the pressure of the liquid refrigerant to a point where it will vaporise at low temperature in the evaporator. *Figure R.3* gives a diagrammatic presentation of the refrigeration cycle.

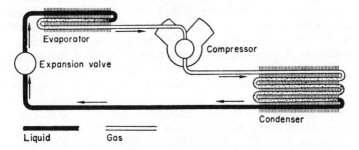

Figure R.3   Refrigeration cycle. (Source: Lion Group Ltd.)

Regeneration   In catalytic processes in oil refineries, the burning of the deposits on the catalyst with an oxygen-containing gas.

Regenerative condenser   A steam condenser which, in addition to condensing the steam, heats the condensate.

Regenerative feed-heating   The use of steam bled from a *steam turbine* to heat the feed-water intended for the boiler.

Regenerator   A *heat exchanger* consisting of a mass of *chequer-brickwork* or other heat-absorbing solid constructed or placed in a chamber; during the heating stage, hot waste gases from the furnace pass through the chamber and heat the chequers, escaping via a reversing valve to the chimney. After a prescribed interval, the direction of flow in the chamber is reversed and cold combustion air drawn in, acquiring the stored heat of the chequers and passing in a preheated condition to the furnace. In glass-melting and open-hearth steel furnaces, for example, regenerators of this type are employed, a reversal of gases taking place once every 20 minutes or so.

Reheat   The passing of the exhaust steam from the high-pressure cylinder of a turbine through a reheater section of the boiler; there it is reheated usually to its original temperature before passing to the intermediate-pressure cylinder of the turbine.

Reid vapour pressure (RVP)   An important specification for gasoline; it is a measure of the vapour pressure of a sample at 38° C (100° F). The test is usually made in a 'bomb' and the results are reported in pounds per square inch or grams per centimetre. Low-boiling hydrocarbons have a high vapour pressure.

Reinluft process   A process for removing sulphur dioxide from flue gases, involving the use of a falling bed of activated carbon through which the flue gases are passed, SO_2 and SO_3 being adsorbed. The sulphur dioxide gas is subsequently recovered by desorption at a higher temperature, when the adsorbent is regenerated without loss of activity. *See **Battersea gas washing process; Howden–ICI process; Fulham–Simon-Carves process.***

Relative biological effectiveness (RBE) A dimensionless unit standing for the relative biological effectiveness of various radiations with respect to X-rays. The RBE for X-, gamma, and electron radiation is approximately 1; the RBE for protons, fast neutrons and alpha particles is, typically, 10. That is, fast neutrons can be ten times more effective in producing biological change than gamma rays.

Relative Humidity The ratio of the amount of water vapour in the atmosphere to the amount which would saturate it at the same temperature; or the ratio of the actual vapour pressure of the water vapour to the maximum or saturated vapour pressure of the water vapour at the same temperature. At *dew point* the relative humidity is 100 per cent; a rise of temperature without the addition of more vapour lowers the relative humidity, the *absolute humidity* remaining the same, while a fall in temperature raises it. A hygrometer is used to measure relative humidity.

REM *See **Roentgen equivalent man.***

Reprocessing The stage of the nuclear fuel cycle at which plutonium and uranium in spent fuel elements are separated from the other actinides and fission products which constitute waste.

Rerun oil In a petroleum refinery, oil which has been redistilled.

Research method A test to determine the *octane number* of a *gasoline.* utilising an engine speed of 600 rev/min and a mixture temperature of 49° C (120° F). The result gives good agreement with engines operating under mild conditions with a low engine speed; it is a less severe test than that offered by the *motor method.*

Research octane number (RON) The *octane number* of a fuel established by the *research method.*

Reserve The quantities of a mineral ore or fuel that may be extracted profitably under existing conditions; a reserve is usually a fraction of a *resource*, and in changing conditions relating to technology, costs and market conditions requires constant revision. Reserves may be subdivided into proved, probable, possible and potential. Proved reserves are those which can be recovered almost entirely with the facilities available within the forecast market and other conditions. Probable reserves are those with a fair or good probability of being recovered with available facilities within the forecast conditions. Possible reserves are those which will prove recoverable only under improved circumstances relating to knowledge, facilities or conditions. Potential reserves are those which might be anticipated from the discoveries of new deposits in promising areas. The term 'measured reserve' is employed in relation to a mass of ore that has been exposed and sampled on three sides and is ready to mine; an 'indicated reserve' has been exposed on two sides and all evidence supports the conclusion that the mass outlined is ore; an 'inferred reserve' is a reserve which has been identified through a reasonable amount of geological data, but which has not been explored in the manner of measured or indicated reserves. *See McKelvey classification.*

Reserve capacity Or back-up capacity; the electricity generating capacity held in reserve to meet any unusually high demand for electricity, or to compensate for the outage of regular generating capacity. *See spinning reserve.*

Residence-time The time taken by an element of fuel to pass through a combustion chamber; the time depends on the particular path that the element has followed in its passage through the chamber, the possible paths depending on the flow pattern within the chamber.

Residual A quantity of material or energy which is left over or 'wasted' when, in the course of human activity, inputs are converted into outputs in production and consumption. Examples are waste-heat and gaseous pollutants from thermal electricity generation, slag from metal ore refining. Outputs have prices in normally existing markets; residuals do not. Through a change in economic circumstances or technological advance, a residual may become an output. Scrap metal or glass recovery at a refuse plant converts a waste into an output. *See recycling.*

Residual fuel Any liquid fuel containing the *residuum* from the crude distillation or thermal cracking of petroleum. *See fuel oil.*

Residuum The most non-volatile portion of petroleum; residua are sometimes described as long or short residua. A long residuum is the non-volatile residue from an atmospheric pressure distillation; a short residuum is obtained from vacuum distillation.

Residuum hydro desulphuriser An oil refinery unit utilising a special catalyst and a fairly high hydrogen partial pressure. Hydrogen is consumed in the process and sulphur is released as H_2S.

Resistance The resistance which an electrical conductor or appliance offers to the passage of an electric current. It is measured in ohms. *See ohm.*

Resistance heating The production of heat by direct application of a voltage to a resistor; the efficiency of conversion of electrical energy into heat by this method is 100 per cent. Resistor material may be an alloy of nickel and chromium, graphite, molybdenum or silicon carbide.

Resistance thermometer A device for measuring temperature. It depends upon the known variation of electrical resistance of a metal wire as the temperature changes. The detection element is a wire of platinum or nickel; the measuring element is an electrical instrument for measuring the electrical resistance of the detecting element and registering this in terms of temperatures. The resistance thermometer is capable of measuring temperatures from -240 to $+1000°$ C but, for industrial use, it is not recommended for temperatures above 600° C. *See* **pyrometer; thermometer**.

Resource All valuable minerals identified by exploration, including those classified as reserves and those which are marginal or unprofitable. Development depends on the amount thought available, the percentage physically extractable with known technology, the extraction costs and the prices attainable. Thus, an oil shale field may be identified containing crude oil of the order of 1000 billion barrels; however, of this, perhaps only 50 billion barrels or less may be profitably extracted, under existing conditions. If prices increase, then some resources become recoverable reserves; the reserves increase while the resource may remain unchanged unless further exploration reveals more. Resources may be subdivided into identified or potential resources, depending on the extent of exploration and the degree of confidence in the findings. In a larger context, the term 'resource' is applied also to environmental resources such as air, water, soil and heritage values in the widest sense. *See McKelvey classification*.

Resource rent tax A tax based on profits in excess of a level necessary for the commencement or continuation of a project. It may combine the functions of royalties and income taxes, but it could also be superimposed on an existing system of taxes and royalties.

Restricted hour tariff A *tariff* offering low rates for supplies which are restricted automatically by time switches to certain off-peak hours of the day.

Reversion pressure test burner A test burner which assesses the tendency of a non-aerated flame to 'lift' or move away from the burner head. The gas supply pressure to the burner is increased until the flame lifts, and is then gradually reduced until the flame returns to the burner head; this pressure is the reversion pressure. A high reversion pressure ensures flame stability.

'Rexco'. A reactive coke produced under 'low-temperature' carbonisation conditions, when coal is heated in retorts to a temperature of the order of 600° C; it is produced by the National Carbonising Co., Ltd. It ignites readily, burns with very little smoke emission, and is well suited for all domestic appliances, including old-fashioned stool-bottom grates. The volatile content is about 7 per cent.

Reynolds number Symbol Re; a non-dimensional ratio which defines the type of flow occurring in a pipe. It is calculated:

$$Re = \frac{vdp}{n}$$

where v = velocity of fluid;
  d = pipe diameter;
  p = density of fluid;
  n = absolute viscosity.

Any self-consistent units may be used. Generally, streamline flow = $< Re2000$; transitional flow = $Re2000$ to 4000; turbulent flow = $Re > 4000$.

Rich gas Gas of high calorific value, exceeding 35 MJ/m³. *See **lean gas***.

Rich mixture *See **lean mixture***.

Rich oil *Absorption oil* containing dissolved natural gasoline fractions.

Richter energy magnitude scale A scale which measures energy impact; the magnitude of impact ranges from 8×10^2 J (scale 0) to 1.2×10^{18} J (scale 9). Scale 0 is about equal to the shock caused by an average man jumping from a table; general destruction can be caused from a magnitude of 6 on the scale, depending on the circumstances. Earthquake tremors are invariably expressed in the Richter scale.

Riddlings Unburnt pieces of solid fuel which fall between the firebars of a furnace.

Ridley report The Report of the UK Committee on National Policy for the use of Fuel and Power Resources (Cmnd 8647, 1952).

Ring Circuit An electrical circuit or main arranged in the form of a ring or loop, as distinct from a single, open-ended length; it is used in domestic situations to provide an adequate number of socket outlets of a 13 amp universal type at a lower cost than the traditional system of radial wiring to each point.

Ring main (1) A system for supplying *pulverised fuel* to boilers or furnaces. It consists of a pulveriser of the slow, medium or high-speed type, from which pulverised coal is extracted and delivered by an exhauster fan to a cyclone situated above a bin; this cyclone extracts the coal dust, which is deposited in the bin. Feeders at the base of the bin supply coal to ring mains which encircle the shop in which the furnaces or boilers to be fired are situated. Pulverised fuel not consumed is returned to the cyclone for redistribution. Ring mains vary in length up to 300 m; they enable batteries of medium-sized furnaces to be serviced. (2) A system for supplying oil to a number of burners fitted to boilers or furnaces; oil is pumped from a storage tank, preheated if necessary, and passed through an oil heater in which it is raised to the correct atomising temperature. From the heater the oil is circulated through a ring main to which branch pipes serving the individual burners are connected. Through continuous circulation the oil is kept at the correct temperature.

Ringelmann Chart   A shade chart, devised over 70 years ago by Professor Ringelmann of France, used for the observation from a distance of the density of smoke issuing from a chimney. The estimation is made by comparing the shade of the smoke against shade cards held some 15 metres from the observer and in line with the chimney; the shade cards consist of grills of black lines, each grill being 10 centimetres square. The shades are:

Number	Description	Approx. % black on shade card
1	Light grey	20
2	Darker grey	40
3	Very dark grey	60
4	Black	80

Clean air legislation frequently prohibits the emission of 'dark smoke' from chimneys, save for short periods permitted by Regulation. Dark smoke is usually defined as 'smoke which, if compared in the appropriate manner with a chart of the type known as the Ringelmann Chart, appears to be as dark as or darker than shade 2 on the chart'. British Standard 2742:1958 describes the use of the Chart in detail; BS 2742M describes the use of a miniature smoke chart.

Ritchie Boiler   *See thermal storage boiler.*

Rittinger's law   A statement that the work required to produce material of a given size from a larger size is proportional to the area of the new surface produced.

Roentgen   Symbol r; a unit of radiological dose. It is the quantity of X- or gamma radiation capable of liberating ions carrying 1 electrostatic unit of charge of each sign in 0.001 293 grams of air (equivalent to 1 cm^3 of dry air at 0° C and 760 mmHg). This is equivalent to the release of 83.8 ergs of energy in 1 gram of air. The megaroentgen (Mr) = 10^6 roentgens; the kiloroentgen (kr) = 10^3 roentgens; and the milliroentgen (mr) = 10^{-3} roentgen.

Roentgen equivalent man   Symbol rem; the quantity of any ionising radiation such that the energy imparted to 1 gram of tissue has the same biological effect as an absorbed dose of 1 *rad* of X-radiation. It follows that:

$$\text{dose in rems} = (\text{dose in rads}) \times (\text{relative biological effectiveness})$$

Roentgen equivalent, physical Symbol rep; a unit of absorbed dose which has now been replaced by the *rad*.

Roga index An index of the caking properties of a *coal*; it is determined in a laboratory tumbler test of a coke button which has been made by heating a mixture of the coal under consideration, and anthracite.

Rotary burner Or spinning cup burner; a low-pressure air burner in which oil is distributed over the air blast from the lip of a rapidly rotating cup. The cup is rotated by an air turbine or an electric motor. *See oil burner.*

Rotary drilling An oil-well drilling system in which the rock formation is penetrated by a rotating bit connected to a hollow drill pipe through which fluid is pumped to convey the rock cuttings to the surface.

Rotary table A heavy geared circular steel body having a square hole cut at its centre, for engaging and rotating the drilling string; the table rotates in the horizontal plane and is normally driven by chains from a draw-works or winch; used in the drilling of oil wells.

Rotometer An instrument for measuring the rate of fluid flow. It consists of a tapered vertical tube with circular cross-section containing a float which is free to move in a vertical path to a height dependent upon the rate of fluid flow upward through the tube. It is based on the principle of *Stokes' law*.

Royal Commission on Environmental Pollution A Commission set up in 1970 as a standing body to advise the British Government as a whole. The terms of reference have been: 'To advise on matters, both national and international, concerning the pollution of the environment; on the adequacy of research in this field; and the future possibilities of danger to the evnironment.' The Royal Commission is intended to serve as a general watchdog on pollution and environment protection. A number of substantial reports have been published. In 1975, in its Fifth Report, *Air Pollution Control: An Integrated Approach*, the Commission urged the creation of a central inspectorate to control all forms of pollution. It recommended also that the Council of Engineering Institutions (CEI) and the Council of Science and Technology Institutes (CSTI) should consider establishing a joint diploma in pollution control. This would be awarded to engineers and scientists suitably qualified and experienced in all aspects of pollution control to meet the needs of policy-makers. The matter was referred by CEI/CSTI to the Council for Environmental Science and Engineering, which reported with recommendations in 1978.

Royal Commission on the Great Barrier Reef A Commission set up by the Australian federal and Queensland state governments to examine the implications of oil drilling on the Great Barrier Reef. The Reef lies off the Queensland coast and is widely regarded as the largest assemblage of living corals in the world. The Commission concluded in 1975 that if petroleum drilling were permitted, some spills, ranging from small to substantial, would occur. Two commissioners found there was a risk of uncertain magnitude to marine life and recommended that drilling should not be permitted in large areas of the Reef. The chairman, the remaining member of the Commission, was not convinced that drilling could be conducted anywhere on the Reef unless more research provided some clear evidence that no harm would result. Following the report of the Royal Commission, the Australian federal government decided that no further exploratory drilling would be permitted until research into the effects of oil on coral reefs had been undertaken.

Ruhrgas gasifier A *cyclone gasifier* for gasifying coal fines to make producer gas. Developed by Ruhrgas A.G.

Rundle oil shale deposits *Oil shale* deposits in Queensland, Australia; demonstrated reserves of recoverable oil shale at Rundle represent 2000 million barrels of crude shale oil and gas. The average grade of these substantial deposits is 77 litres/tonne. In all, in 1979 Australia had about 3600 million barrels of *in situ* demonstrated resources of shale oil.

Run-of-mine coal Coal as raised from the mine or pit, before screening.

Run-of-retort coke Or run-of-oven coke; coke from a retort or oven before undergoing screening or other process.

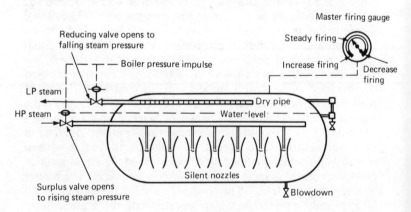

Figure R.4 Ruths' accumulator.

Rutherford Symbol rd; a unit of radioactivity. It is the quantity of any radioactive nuclide in which the number of disintegrations per second is 10^6.

Ruths accumulator A steam accumulator; it operates by accepting steam through a surplus valve from a high-pressure main and condensing it in water, while on demand it releases steam through a reducing valve into a low-pressure main. It is particularly useful in a system with peaky loads, and where there is a demand also for low-pressure steam. An old boiler shell may be used as the pressure vessel. *See Figure R.4.*

S

Safety rod A neutron absorbing rod which can be inserted rapidly into the core of a *nuclear reactor* in the event of an emergency, thus enabling the reactor to be shut down

Safety valve A device to prevent a boiler from working in excess of the 'maximum permissible working pressure' by automatically discharging to the atmosphere the excess steam generated in the boiler. Each valve should be large enough for its purpose, and all industrial boilers, with the exception perhaps of the small vertical boiler, should be fitted with at least two safety valves. Thus, if one valve fails, the other will function. Each safety valve should be tested daily to ensure that it will blow at the correct pressure. *See dead-weight safety valve; lever or steelyard safety valve; spring-loaded safety valve.*

Sales gas Natural gas that has been processed to remove condensate, liquefied petroleum gas and any carbon dioxide in excess of 3 per cent; it consists of a mixture of methane, ethane, minor amounts of other hydrocarbon gases and some carbon dioxide.

Samarium Sm. An *element*, at. no. 62, at. wt. 150.35. Samarium is produced in a *nuclear reactor*; samarium-149 has a high neutron-capture cross-section. *See fission products.*

Sampling The taking or withdrawing of a quantity or fraction of something, for analysis, in such a way that it may be assumed that the sample is representative of the whole.

Sampling probe A tube inserted in a chimney or duct in order to draw off into measuring equipment a sample of gas.

Sankey diagram A heat flow diagram for an industrial process in which the quantity of heat in the various items of the *heat balance* for the plant is represented by the width of a band. *Figure S.1* shows a Sankey diagram for an oil-fired *economic boiler*, indicating the ingoing and outgoing flows of heat; the widths of the bands are proportional to the amounts of heat represented.

303

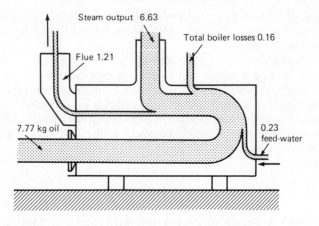

Fig. S.1 Simple diagrammatic heat-flow diagram or Sankey diagram

Sanitary land-fill *See land-fill.*

Santa Barbara 'blow-out' An incident which occurred on 28 January 1969, while the Union Oil Company was drilling its fourth well from an offshore rig located about 13 kilometres from the shore of the Santa Barbara Channel, off the coast of California. Pressure in the hole rose suddenly and blew out all the drilling mud, and crude oil spurted from the ocean floor. The blow-out continued for 10 days. The first slick of oil reached Rincon, 24 kilometres from the drilling platform, some 5 days after the blow-out; 3 days later it arrived at Santa Barbara, surging on to the beaches, promenades and jetties. Seagulls avoided the oil, but many grebes dived into it and died slowly. Santa Barbara was declared a Federal disaster area. The Union Oil Company spent close on $5 million making good the damage, collecting blackened birds, and washing and caring for them. Dispersives were sprayed on the oil. Tons of straw were blasted at high pressure on to water, beaches, rocks and docks, functioning as 'blotting paper'. The straw was later collected and removed. Blackened sand was removed from the beaches. The oil well was sealed with mud and cement, but cracks continued to occur in the ocean floor around the platform, causing seepage; special measures were adopted to deal with this seepage.

Sapropel Initially a slimy, putrefying, sediment formed in deep, stagnant waters where the remains of minute organisms, algae, spores, pollen and plant fragments accumulate in virtually complete absence of air; on consolidation sapropel contains a higher proportion of hydrocarbon-rich compounds. The sapropelic coals used to be a valuable source of gas. *See **boghead coal; cannel coal**.*

Sasol A coal-to-oil project undergoing development by the South African Coal, Oil and Gas Corporation. The objective is to reduce the country's dependence on imported oil. Sasol mines its own coal. The project has reached a fully commercial stage, a second plant being commissioned in 1981. Details are given in *Table S.1.* The Sasol plant is based on *Fischer–Tropsch synthesis*.

Table S.1 SASOL II COAL CONVERSION PLANT, REPUBLIC OF SOUTH AFRICA

Plant cost	$1.5 billion
Plant on stream	1981
Coal feed	40 000 tons per day
Products	
motor fuel	30 000 barrels per day
tar products	200 000 tons per year
ethylene	100 000 tons per year
ammonia	100 000 tons per year
sulphur	90 000 tons per year
other chemicals	50 000 tons per year

Saturated steam Steam which has taken up its full quota of latent heat, containing no moisture or suspended unevaporated water; synonymous with *dry saturated steam*.

Saturation The condition in which a substance has taken up the greatest possible amount of another substance, e.g. air which is incapable of taking up more moisture at that temperature.

Savannah, N.S. The world's first nuclear-powered merchant ship, launched on 21 July 1959; a United States test ship built to prove the feasibility of nuclear power for merchant ships. The reactor system is a pressurised water type consisting of a single reactor with two main coolant loops and two steam boilers. The uranium oxide fuel, in the form of pressed sintered pellets, is contained in stainless steel tubes; each of the 32 fuel elements contains 164 of these tubes. The reactor is controlled by stainless steel rods containing enriched boron.

Saybolt system A system for measuring *viscosity*, based on the time taken for the flow of a liquid from one vessel to another. The Saybolt Universal Second (SUS) is the time in seconds for 60 ml of oil to flow out of the cup in a Saybolt viscometer through a specified aperture. Saybolt Furol Seconds (SFS) are used for viscous oils. Roughly, 1 Saybolt Furol Second = 10 Saybolt Universal Seconds.

SCA Steel-cored-aluminium conductor used in electricity transmission systems.

Scaling A condition in which the thickness of oxides and corrosion products reaches a certain value, the layer cracking and falling away from the metal, often leaving it exposed to fresh attack.

Scarfing The use of oxygen flame jets to remove surface defects from steel.

Scintillation counter A device for detecting charged particles or photons by the flashes of light (scintillations) which they produce in certain materials known as phosphors, e.g. sodium iodide and anthracene. The light falls on the cathode of a photomultiplier, which converts each scintillation into a pulse of current and amplifies the latter so that it can operate associated counting equipment. *See radiation detector*.

Scotch marine boiler *See marine boiler*.

Scram rod Or safety rod; a neutron-absorbing rod which can be inserted rapidly into the core of a nuclear reactor in the event of an emergency, thus shutting it down. In normal operation it is usually suspended magnetically above and outside the core so that it falls under gravity in the event of a power failure.

Scraper conveyor A conveyor which removes ash from the back of a furnace, bringing it to the front, where it is discharged into an ash shute.

Screening The passing of coal over bars, perforated plates or wire mesh screens, so that sizes smaller than the openings fall through. *See cylindrical screen; shaking screen; trommel; vibrating screen*.

scrubbing An absorption system in which gaseous or particulate pollutants are removed from a stream of air or gas by contact with a liquid. The scrubbing liquid may be in the form of a spray, or a bath over which the gas passes, or may pass over packing in a tower.

Seasonal tariff A *tariff* for the supply of electricity, in which a higher price per unit of electricity applies during the winter months than during the summer months.

Second *See atomic time (AT)*.

Secondary air Combustion air introduced above or beyond a fuel bed by natural, induced or forced draught; it includes all 'overfire' air which may be introduced through the front or side walls, or through the bridge wall, of a furnace. Fuels containing a very high proportion of *volatile matter* require the greater part of the total weight of air required for combustion to be supplied over the grate. The burning of *wood* is an example of this. On the other hand, *coke*, with a very low volatile content, requires only enough secondary air to burn the *carbon monoxide* to *carbon dioxide*. *See overfire jets; primary air; tertiary air*.

Secondary burner In respect of an incinerator, a burner installed in the secondary combustion chamber to maintain a minimum temperature of about 760° C *See primary burner*.

Secondary heating The heating of oil to bring it to the correct temperature for efficient *atomisation*. This is achieved by heating the oil with steam or electricity, with thermostatic control. With all plants and grades of oil the temperature of the oil is raised to that at which its viscosity is between 80 and 100 seconds Redwood No. 1 at the burner. Examples of heating temperatures in relation to the desired oil viscosities are: (a) light fuel oil, 50–60° C); (b) medium fuel oil, 80–90° C; (c) heavy fuel oil, 120° C. *See primary heating; viscosity.*

Second Baku Name for the very large oil and gas fields developed in the Soviet Union between the Volga and the Urals, matching the original oil resources of Baku in Caucasia. Developments in Soviet Central Asia and western Siberia have suggested a 'Third Baku'. A national pipeline system ensures that most parts of the Soviet Union are supplied from the producing areas. Natural gas is supplied to the Moscow conurbation.

Sectional boiler Cast-iron or steel hot-water boiler, made up in sections which enable it to be increased or decreased in size; most widely used for central heating. Cast-iron boilers are most common, and can be used in buildings where the working head does not exceed 38 m; above this, steel boilers are necessary. Sectional boilers have a life of about 20 years, and can be operated with efficiencies of up to 70 per cent with mechanical firing. *See boiler.*

Sedimentation The determination of the terminal velocities of particles by introducing a known weight of dust into a liquid, generally water or alcohol, and measuring the amounts of dust falling out at predetermined times.

Seebeck effect The generation of an electric current, when two wires of different metals are joined at their ends to form a circuit, the two junctions being maintained at different temperatures. Named after T. J. Seebeck (1770–1831). *See thermo-electric pyrometer.*

Seeboard process A wet scrubbing process for the removal of *hydrogen sulphide* from refinery and petroleum oil gas streams. The gas is scrubbed with a solution of sodium carbonate, Na_2CO_3. The dissolved hydrogen sulphide is subsequently removed by blowing air through the solution. *See hydrogen sulphide removal.*

Seagas process A cyclic process developed by Britain's South Eastern Gas Board for the catalytic manufacture of gas from oil; a lime catalyst is used to facilitate the reactions. Feedstock varying from liquid butane to residual oils can be used. In the blow stage, counterflowing air is blown through the air preheater, burning off any carbon; the preheated air brings the catalyst to a working temperature of 788–982° C and also heats up the vaporiser and steam preheater, leaving the system via a waste heat boiler. A steam purge follows, in the same direction. In the make stage, steam is admitted in the opposite direction to the blow. It passes through the preheater and then meets a counter-current oil spray; the steam–vaporised oil mixture then passes through the catalyst, where the desired reactions take place. The constituents of a typical gas produced, expressed in percentages, are: hydrogen, 50; methane, 16; carbon monoxide, 15; carbon dioxide, 9; hydrocarbons, 7; nitrogen, 3. The calorific value is about 20 MJ/m^3.

Seger cones Small cones of clay and oxide mixtures, calibrated within defined temperature ranges at which the cones soften and bend over; they are used in furnaces to indicate, within fairly close limits, the temperature reached where the cones are placed.

Segregation The tendency of coal, when poured, for the larger pieces to separate themselves; thus, when non-graded coal is poured on to a cone-shaped heap, the larger pieces run down the sides of the cone. Segregation may occur when there is transfer of coal from one conveyor to another and in bunkers. It can be reduced by fitting a flat plate above the outlet from bunkers, and by using travelling chutes to feed coal into hoppers.

Seismic reflection method A method used in the exploration for oil. A charge of dynamite is exploded in the ground, usually within about 30 m of the surface; the shock waves travel through the ground and bounce back from underground layers of rock. Seismometers placed at intervals on the surface receive the shock waves and transmit them to a recording instrument, where they are recorded on a seismogram. From these records, geophysicists determine the position and shape of anticlines and other formations lying deep in the earth. *See petroleum.*

Selenium **Se.** An *element* used in the making of photoelectric cells; at. no. 34; at. wt. 78.96. *See photoelectric cell.*

Self-proportioning burner A low-pressure air blast burner, supplying all the air required for combustion through the burner, and maintaining a correct air/fuel ratio over its complete range of fuel output; the oil and air flow rates are adjusted simultaneously by the movement of a single lever. *See oil burner.*

Semet–Solvay process A cyclic oil gasification process using thermal cracking in steam. The plant consists of two carburettors or gas generators connected at the top. During the 'blow' period *primary air* passes through one generator, burning off carbon previously deposited on the *chequer-brickwork*, and enters the second chamber, in which oil is burned in the presence of

secondary air. During the 'make' period cracking oil is injected into both generators from the top, with steam passing through both chambers in the same direction as in the 'blow' period. The process is then repeated with reverse directions for air and steam. Heavy oil is used in the process with a cracking temperature of 650–900° C. The energy value of the gas produced is 39–45 MJ/m³.

Semi-siliceous brick A *refractory* containing 78–85 per cent of silica, the balance being mainly alumina.

Sensible heat Heat, the intensity of which can be measured by thermometric techniques; as distinct from latent (or hidden) heat, which produces no change in temperature. The heat required to raise the temperature of water from freezing point to boiling point is sensible heat. *See latent heat; total heat.*

Sensitive flame A gas *flame* which changes in shape when sound waves fall upon it.

Separation nozzle process A process for *uranium enrichment* using a succession of nozzles of decreasing size to separate U^{235} from U^{238}; four or five hundred nozzles are needed in this process.

Servomechanism A device for amplifying by electronic or other means a small impulse from a measuring instrument; the larger force is then capable of operating a valve or other mechanism.

Settlement chamber A dust-arresting device for use in conjunction with boilers and furnaces; it consists of a rectangular chamber or an enlargement of a flue or duct, the effect being to reduce the velocity of the gases, allowing grit to settle out. The chamber may contain a system of baffles to deflect the gases and assist in the removal of grit. Devices of his nature are only effective in respect of particles larger than, say, 100μm.

Severity A term which describes the manner in which an engine rates a fuel on the road. A 'severe' engine will rate a fuel near to its motor method octane number, while a 'mild' engine will give a road octane number near to its research method octane number; modern overhead-valve engines are in the latter category, while side-valve engines tend to be severe. *See octane number.*

Sewage gas *Methane* produced naturally during the process of digestion in sludge tanks at sewage works. In large works it is utilised for heating the tanks to accelerate the sludge digestion process, power generation, space heating and lighting, hot water supplies and laboratory work. *See natural gas.*

Seyler's classification A classification system for coals; the positions of all coals are plotted in a diagram using the percentages of (a) total carbon, (b) hydrogen, as the main parameters and co-ordinates. As all coals with a specified *volatile matter* content occupy positions along a straight line, this led to the construction of a series of nearly parallel 'isovols'. Similarly, all coals having any specified *energy value* occupy positions along a straight line, and thus a series of roughly parallel 'isocoals' was constructed. The isovols and isocals are approximately at right angles to each other and inclined to the carbon and hydrogen co-ordinates. *See coal classification systems.*

Shadow price   A price or value imputed to unpriced social benefits or losses, or to resources which are not satisfactorily priced in commercial markets. The concept is much used in cost–benefit analysis. In evaluating any project, the economist may adjust a number of market prices, and attribute prices to unpriced gains and losses likely to arise.

Shaft mine   A mine in which the seams are reached by a vertical shaft from the surface.

Shaking screen   A horizontal rectangular screen given a reciprocating motion in a lengthwise direction by means of an eccentric crank operating at 80–120 rev/min.

Shale oil   An oil obtained by the *destructive distillation* of oil shale; crushed shale is placed in retorts and the organic material of shale, known as kerogen, is cracked with gas or steam at 350–500° C. The crude oil produced is similar to crude petroleum but contains higher percentages of sulphur, nitrogen and oxygen; the refining of shale oil is similar to that of crude petroleum. The yield of a tonne of shale oil can vary considerably, although an average might be of the order of 100 litres, and some 450 litres of ammonia liquor. Oil shale resembles low-grade black coal; there are large deposits in various parts of the world. For example, there are large deposits in Queensland, Australia, at Rundle. Generally, oil from this source has been economically unattractive, although with the oil crisis of the 1970s fresh interest has been taken. In June 1979 the US House of Representatives approved legislation to enable the President to promote synthetic fuel production from coal, shale rock and other materials.

Shatter test   Test for the impact hardness of coke; 22 kg of coke over 5 cm size is dropped four times from a height of 2 metres on to a metal plate. A sieve analysis follows. A good metallurgical coke should show 75 per cent over 5 cm in size; 85 per cent over 3.5 cm.

Shell gasification process   A process consisting of the non-catalytic, partial oxidation of hydrocarbon feedstocks using oxygen and steam under pressure of 20–60 bars, at temperatures of about 1200° C. A water gas is produced which can be used for a variety of purposes. *See Figure S.2.*

Shell phosphate process   A wet scrubbing process for the removal of *hydrogen sulphide* from refinery and petroleum oil gas streams.

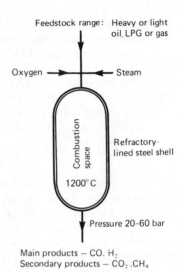

Feedstock range: Heavy or light oil, LPG or gas

Oxygen → ← Steam

Combustion space

1200°C

Refractory-lined steel shell

Pressure 20–60 bar

Main products — CO. H$_2$
Secondary products — CO$_2$ .CH$_4$

Figure S.2   Shell gasification process:

The scrubbing medium is a solution of tripotassium phosphate which removes hydrogen sulphide in the following reaction:

$$K_3PO_4 + H_2S \rightarrow K_2HPO_4 + KHS$$

The hydrogen sulphide is regenerated by boiling; the sulphur may be recovered in a *Claus kiln*. *See hydrogen sulphide removal.*

Shell smoke meter   An instrument for measuring stack solids developed by the Shell Oil Company for use with oil-fired installations. It comprises a small motor-drive vacuum pump which draws a sample of the flue gases through a filter paper for a period of 1 minute, the pressure drop across the paper being maintained constant at 3 inHg; the timing is automatic. The stain on the filter paper is then compared with a scale of shades graduated from 0 (white) to 9 (black); the nearest matching shade is known as the Shell Smoke Number. Shade 9 is the equivalent to chimney emissions of less than shade 1 on the *Ringelmann Chart*, while shade 6 represents the lowest concentration of stack solids which is just visible to human eye. The results obtained by this instrument are influenced by the type of fuel and size of smoke particle; a given weight of very fine particles has a much greater blackening effect than a similar weight of relatively coarse particles.

Shim rod   A name sometimes given to control rods used for making coarse adjustments in the reactivity of a nuclear reactor. Such rods can often only be moved slowly, but are capable of making relatively large changes in the reactivity.

311

Short-circuit An electric current taking a shorter path than intended; as the resistance of the unintended path is usually low, a rush of current takes place and blows the *fuse*.

Short flame coal *Coal* of low *volatile matter*. See *long flame coal*.

Short-run marginal cost See *marginal cost*.

Short ton, US A customary United States unit of weight of 2000 pounds; comprising 20 short hundredweights, each of 100 pounds. The following relationships apply:

$$\text{one US short ton} = 0.893 \text{ UK long ton (approx.)}$$
$$= 0.907 \text{ metric tonne (approx.)}$$

The long or 'gross' ton of 2240 pounds is little used in the United States, while the metric tonne is being used to an increasing extent.

Shot-rain Cleaning A method of cleaning fouled furnace tube surfaces by discharging or 'raining' a large quantity of steel shot over the surfaces periodically.

Shuffling The process of rearranging the positions of the fuel rods in the channels of a nuclear reactor, in order to equalise burn-up.

Shuttle kiln A thin-walled steel-encased rectangular intermittent kiln in use in the United States ceramic industry. There are two types: (a) the envelope kiln, which has two fixed bases on which the ware is set and a movable cover on wheels which is pushed on rails over the base being fired; (b) the car shuttle kiln, which has fixed walls and crown of thin wall construction, and cars set with ware are pushed into the kiln for firing.

SI *See International System of Units (Système International d'Unités)*.

Siegert formula A formula for calculating the heat loss in flue gases passing through a chimney:

$$S = \frac{K(t_2 - t_1)}{L}$$

where S = percentage of the net *energy value* lost in the flue gases;
t_1 = ambient air temperature, °C;
t_2 = flue gas temperature, °C;
L = percentage CO_2 content of the flue gases;
K = a constant equal to: anthracite 0.67; bituminous coal 0.63; coke 0.70; and oil 0.56.

Siemens Symbol S; a unit of electrical conductance, equivalent to 1 reciprocal ohm, or mho. The unit is named after Sir William Siemens (1823–83).

Sieving A relatively simple and quick way of obtaining a sizing or grading analysis of a sample of dust or solid fuel; it is only practicable down to a particle size of about 50 μm.

Sigma calorimeter A continuous-recording *gas calorimeter* based upon the differential expansion of two concentric steel tubes. One of the tubes is heated by the combustion of gas under test; the other is cooled by the flow of combustion air. The differential expansion of the tubes is transmitted to a pen recorder, calibrated in heat units.

Silica SiO_2. The dioxide of silicon; it occurs in crystalline form as quartz, cristobalite and tridymite, and is an essential constituent of the silicate groups of minerals. Silica and silicates constitute an important part of the mineral impurities in coal. *See **acid refractories**.*

Silica brick A *refractory* containing at least 92 per cent of silica.

Siliceous brick A *refractory* containing from 85 to 92 per cent of silica, the balance being mainly alumina.

Silicon diode rectifier A transformer-rectifier for converting the alternating current of normal electricity supply into high-tension direct current electricity; this design has found use in electrostatic precipitators. It is claimed that these rectifiers operate at 98 per cent efficiency. They have no moving parts. *See **electrostatic precipitator; rectifier**.*

Sillimanite brick A *refractory* containing 55–65 per cent of alumina and 25–35 per cent of silica.

Single-phase power *See **three-phase power**.*

Single pole A description applied when a switch or fuse is inserted in only one of a pair of wires in an electrical circuit; the switch or fuse is described as a 'single-pole' type. *See **double pole**.*

Sinking fund A fund built up by equal periodic instalments in order to accumulate a certain sum by a given date for some specific purpose, e.g. the replacement of a physical asset. The interest is itself periodically reinvested and allowed to accumulate with the periodic contribution of principal. The sinking fund instalment, consisting of an annual payment, is equal to the original capital cost of the asset multiplied by a sinking fund factor equal to:

$$\frac{i}{(1 + i)^n - 1}$$

where i is the rate of interest and n the plant life expressed in years. *See **depreciation; depreciation fund method**.*

Sintering The fritting together of small particles to form larger particles; the conversion of fine dust into hard agglomerates. Sintering is employed in the steel industry to effect a chemical and physical improvement of ores being charged to the blast furnaces; it is a process in which fines are mixed with coke breeze and passed through a sintering furnace. A sintering furnace consists of an endless strand of travelling perforated pallets on which the mixture is spread; this moving grate may be some 3 metres in width and up to 40 metres in length. The mixture ignites as it passes under a firebrick arch, jets of burning gas being directed upon it. Air passes downward through the strand bed, and the sinter mix is completely burned through by the time it reaches the end of the strand; upon cooling, the sinter is broken and graded. The gases from the strand, discharge end, breakers and screen require de-dusting before discharge to atmosphere. Large quantities of sulphur dioxide are emitted, and chimneys of up to 150 m or more are often essential.

Sized coals *See graded coals.*

Size distribution *See size fraction.*

Size fraction A portion of a powder, dust or fuel sample composed of particles or lumps between two given size limits. The distribution of the size fractions in the total sample is known as the 'size distribution'.

Size stability The ability of a coal to withstand breakage during handling and shipping. In one test, size stability is determined by twice dropping a 22 kilogram sample of coal from a height of 2 metres on to a steel plate. From the size distribution before and after the test, the size stability is reported as a percentage factor.

Slope mine A mine similar to a *drift mine*, but in which the seams are at a perceptible angle to the horizontal line of entry.

Slug An engineering unit of mass. One slug = 32.174 pounds.

Slurry pipeline A pipeline in which solids are carried in their natural state) this is achieved by using a fluid medium, usually water, in which the solid is either ground down and transported as a slurry or reduced to granular form. A number of coal pipelines now exist; it is a particularly suitable method of fuel transport to furnaces at power stations where the pulverisation of coal before burning is required.

Slurry synthesis process A variant of the *Fischer–Tropsch synthesis process* in which the catalyst is maintained in suspension in an oil.

Smalls All coal down to dust that is smaller than a certain size. For example, 3 cm smalls consist of all the coal which passes through a screen with 3 cm holes.

Smithells separator A classical device for demonstrating the principle and nature of two-stage combustion. The jet of fuel gas entrains a fraction of its combustion air and burns sub-stoichiometrically at the top of the bunsen tube; the hot products contain carbon monoxide, hydrogen and other intermediate gases which burn as a diffusion flame at the top of the quartz tube in the ambient air. *See Figure S.3; stoichiometric.*

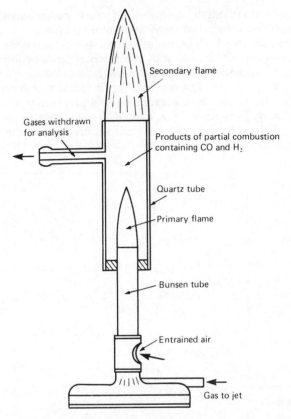

Figure S.3 Smithells separator

Smog Originally a term applied solely to a mixture of smoke and fog; it was first used by the late Dr H A Des Voeux, founder-president of the National Smoke Abatement Society, in 1905. Today the term is applied to any objectionable mixture of air pollutants, as in the case of **Los Angeles smog**, which is of photochemical origin. Other terms in use are 'smaze', a mixture of smoke and haze, and 'smust, a mixture of smoke and dust.

Smoke The visible product of incomplete combustion, consisting of minute carbonaceous particles mainly less than 1 μm in size. Meetham, in his *Atmospheric Pollution* (Pergamon Press, London and New York, 1956), stated that after examination of 100 000 smoke particles it was found that half the individual smoke particles were smaller than about 0.075 μm, but half the weight of smoke was in particles larger than about 0.51 μm, and there was a definite tendency for smoke particles to stick together in chains

315

perhaps 1 μm in length. *See Figure S.4*; *black smoke; brown smoke; clean air legislation; dark smoke; Ringelmann Chart*.

Smoke control area   An area containing domestic, commercial and industrial premises declared by order to be a 'smoke control area' by a local authority exercising its powers under Section 11 of the Clean Air Act 1956. The general effect is to prohibit the emission of all smoke from chimneys in the area, subject to any exemptions in force. By December, 1978, over 5000 smoke control areas had been confirmed, covering over 8 million premises in the United Kingdom.

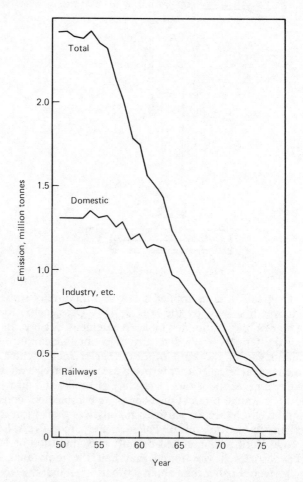

Figure S.4   Emission of smoke from coal combustion in the United Kingdom 1950–1977

Smoke density indicator   An instrument for measuring the opacity of smoke and particulate matter passing up a chimney. The principle is the projection of a beam of light of constant intensity across the interior of a stack. The beam of light, after crossing the stack, falls upon a ***photoelectric cell***, producing an electric current whose intensity depends upon the amount of light falling upon it. This is used to operate an electrical indicating or recording instrument. The amount of smoke in the path of the beam of light determines the amount of light which falls on the photoelectric cell and the intensity of the current produced; this is used to indicate the density of the smoke. The instrument may be fitted with an alarm and/or a continuous recorder. *See Figure S.5.*

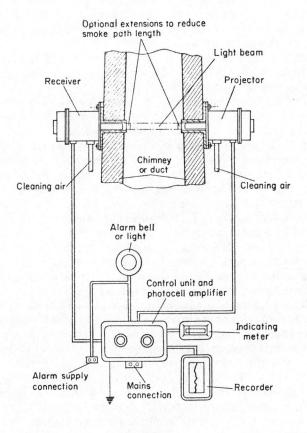

Figure S.5 Diagrammatic arrangement of smoke-density indicator with meter recorder and alarm. (Source: British Standards Institution)

Smoke eliminator door A special door for hand-fired natural draught Lancashire boilers developed by the British Fuel Research Station. It permits the introduction of additional volumes of secondary air after firing fresh fuel; it has openings for continuous air and for supplementary air.

Smokeless zone An area containing domestic, commercial and industrial premises declared by order to be a 'smokeless zone' by a local authority exercising its powers under local legislation. The general effect is to prohibit the emission of all smoke from chimneys in the area, subject to any exemptions in force. *See clean air legislation*.

Smoke point test A test for the burning quality of kerosine. It indicates the height of flame which can be obtained without the formation of smoke, and thus provides a measure of the illumination likely to be obtained from a particular kerosine used in a particular type of wick-fed lamp.

Smokestack America stocks In the United States a term descriptive of basic industrial stocks and shares, compared with glamour stocks, which may have shorter lives.

Smuts Small aggregates of soot, tar, unburnt solids of fly ash which may be emitted from chimneys as a result of soot-blowing or through inefficient combustion. *See acid soot*.

Snowy Mountains hydro-electric scheme An Australian hydroelectric development with a generating capacity of 3740 MW, designed to transfer waters from the Snowy River and its tributaries to an inland system, so that the water may be used for irrigation and power generation. The scheme, constructed by the Snowy Mountains Hydro-Electric Authority, is situated geographically about mid-way between the principal load centres of Sydney and Melbourne. It supplies each centre with valuable quantities of peak and mid-load electrical energy. The scheme utilises about one-half of the theoretical potential energy available in the Snowy mountains; it was completed in 1974.

Soaking pit An oil- or gas-fired furnace in which steel ingots are placed to provide an environment in which a uniform temperature throughout the ingot may be obtained; a uniform temperature ensures that the metal has the same degree of plasticity throughout the whole of the mass prior to rolling into sections or slabs.

Social benefits The value of gains accruing to the community as a consequence of the establishment of an industry, factory or facility, although the new development may not have taken place with that aim or purpose. Benefits may include increased opportunities for employment, steadier employment, improved roads, improved shops and amenities, elimination of a swamp or removal of unsightly existing premises.

318

Social capital Assets belonging to the community as a whole, rather than to private persons or industries, e.g. schools, hospitals, roads, etc.

Social costs Costs which do not appear in the accounts of a company, e.g. the costs to the public caused by air pollution, water pollution, noise and congestion, or the provision and maintenance of roads by the public for private distribution purposes. If a company is required by legislation to prevent or minimise its contribution to, say, air pollution, then the cost of doing this will appear in the accounts of the company. In this instance a general and difficult to measure social cost will have been converted into a specific and accurately measured private cost.

Social forestry The growing of trees for fuel as well as raw materials. In 1979 the *World Bank* financed a social forestry project in the Uttar Pradesh state of India; the project will make possible an increase of at least 20 per cent in the fuelwood supply for some 6 million people. Most of the tree plantings will be along roads, railways and canals, eliminating the need to walk long distances for fuel. A greater abundance of fodder will result in greater milk production; organic wastes now used for fuel will be freed for agricultural use. Forest-based cottage industries will be provided with more raw materials.

Social indicator An attempt to measure by means of an index the degree of human welfare or quality of life in a given area, national or regional. The need for such an index has arisen from widespread discontent with the use of the concept of gross national product as a measure of human well-being. An integrated social indicator reflects a set of social and environmental indices, as well as conventional economic indices. It is influenced by such factors as real income; employment, housing, educational and cultural characteristics; ease of access to social, community and transport services; recreational opportunities; depletion of non-renewable resources; air, water and noise pollution. In regional terms, a social indicator becomes a basis of comparison, and an important element in the development of urban and regional policies and the determining of priorities within and between programmes. Subjective social indicators deal with the degree of satisfaction felt by people with various aspects of their lives. Work on social indicators, both objective and subjective, has been conducted in several countries. The subject of social indicators is reviewed in the publication *Social Trends No. 4* (HMSO, London, 1974).

Sociosphere The area of study of the social scientist; analogous to the hydrosphere as the area of study of the oceanographer, the biosphere as the area of study of the biologist, and the lithosphere as the area of study of the geologist. The sociosphere embraces people, their roles and patterns of behaviour, their organisations and groups, and social interactions.

Soda-base grease   A grease with a high melting temperature; it is used in high-speed bearings of an anti-friction type.

Sodium   Na. **A metallic** *element*; at no. 11, at. wt. 22.99. It has a moderately low neutron-capture cross-section and can be used as a *coolant* in a *nuclear reactor*.

Sodium line reversal method   *See line reversal method*.

Sodium potassium alloy   NaK. A low-melting-point alloy consisting of sodium and potassium; it may be used as a *coolant* in certain types of *nuclear reactor*.

Sodium–sulphur battery   One of a group of batteries generally referred to as 'molten salt' batteries; potentially they offer both high energy density and high power density. Most of the high-energy battery development in the United Kingdom is now concentrated on the sodium–sulphur battery. The electrodes are liquids, while the electrolyte is solid. Liquid sodium is separated from liquid sulphur by a beta alumina ceramic which functions as both an electrolyte and a separator. This solid electrolyte is an ion filter, so that sodium passes through it to react with sulphur to form sodium sulphide, when an external current flows. Thus, the chemical energy of the reaction is converted directly into electrical energy; the cell is recharged by the regeneration of sodium and sulphur. The operating temperature of the battery is about 300° C. *See Figure S.6*.

Sodium Sulphite Na_2SO_3. An oxygen-removing agent suitable for treating boiler feed-water; it reacts with oxygen to give sodium sulphate. The reaction is somewhat slow and is often accelerated by adding a catalyst of cobalt salts. It adds appreciably to the dissolved solids in the boiler. *See hydrazine*.

Soft coal   *Coal* which is relatively friable and subject to degradation on handling. *See hard coal*.

Softening plant   Plant in which water is treated to remove *hardness*. *See ion exchange process; lime soda process*.

Solar energy   The energy of the sun which reaches the Earth in the form of short-wave radiation, visible light and near ultra-violet light. After penetrating the atmosphere, part of the energy heats the surface of the Earth and part of it is re-radiated back in the form of long-wave radiation and absorbed by water vapour and carbon dioxide in the atmosphere. The latter radiates again about half of the captured energy back towards the Earth's surface. The warming caused by the trapped energy has been called a *greenhouse effect*. The possible utilisation of solar energy for the generation of electricity is receiving an increasing amount of attention. Solar energy provides a continuous energy supply which, over Australia, averages about 5×10^{22} joules per annum, corresponding to a power averaged over the year of 200 watts per square metre. *See photovoltaic conversion.*

Solar Energy Research Institute (US)   A research institute established by the US Energy Research and Development Administration (ERDA), with the purpose of co-ordinating all solar energy research in the United States. The institute was established at Golden, Colorado, in 1977.

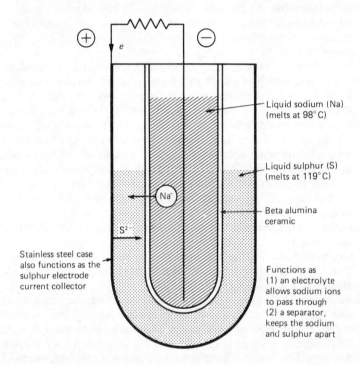

Figure S.6   Sodium sulphur cell (After Potter, D. E. *'Advanced Battery Design'*, conference on electric cars, Canberra, Australia, 1975.

Solar Energy Research Institute (WA)   A body established in 1979 by the Western Australian Government, under the control of a board of directors, to promote developments in the application of solar energy in the state. Associated with the Institute is a Solar Energy Advisory Committee which is responsible for the consideration of requests for funds.

Solar ethanol   *See ethanol.*

Solar oil   A description of liquid hydrocarbon fuels produced from crops on a renewable basis.

321

Solar One The United States' first solar-powered electricity generating plant, constructed near Barstow, California, for the *Department of Energy*, from a design developed by McDonnell Douglas. Computer-controlled mirrors follow the sun across the sky, focusing its rays on a tower-mounted boiler. The steam produced drives turbo-generators. Excess heat is channelled to an underground oil-rock 'storage battery' to keep the plant operating after the sun has set. Solar One generates enough electricity to meet the needs of a community of up to 10 000 people.

Solid lubricants Substances such as graphite or molybdenum disulphide which function as lubricants by reason of their crystal structure.

Solute A dissolved substance; hence, solute + solvent = solution.

Solvent extraction processes Processes in which solvents are used to dissolve out undesirable constituents, e.g. the removal of aromatics from kerosine by extraction with liquid sulphur dioxide.

Solvent-refined coal (SRC) process A coal hydrogenation process developed by the Pittsburgh and Midway Coal Mining Company, a pilot plant being completed in 1974. Liquid by-product yields have varied between 7 and 14 per cent, with gas yields of between 3 and 8 per cent; sulphur recovery is also carried out. Yields of solvent-refined coal vary between 60 and 67 per cent, based on moisture-free feed coal.

Solvents Liquid organic chemicals used in industry for dissolving oils, fats, waxes, resins and other similar materials. The most commonly used are the chlorinated solvents, namely trichlorethylene, 1,1,1-trichlorethane, carbon tetrachloride and perchlorethylene.

Soot blower A device used for cleaning boiler surfaces, using steam or air as the blowing medium. There are two main types of blower: (a) the single-jet or gun type, which is usually retractable; (b) the multi-jet type, which is usually non-retractable. Where deposits are likely to become hard, single-nozzle blowers are essential. Multi-jet blowers are more suitable for softer deposits.

Soot blowing A method of cleaning the external tube surfaces of a boiler while on load. Mechanical soot blowers are employed; these blowers may be operated by direct or remote manual control, or by a fully automatic electric or compressed air sequential system. Either steam or compressed air may be used as the blowing medium. The use of suitable equipment may raise the efficiency of a shell boiler by 5–7 per cent for a steam consumption of less than 0.5 per cent of the total boiler output. Soot blowing should be carried out at least once a shift.

Sour Having an unpleasant smell; positive to the *doctor test*. Sourness indicates the presence in gasolines, naphthas and refined

322

oils of hydrogen sulphide and/or mercaptans. The term 'sour gas' is applied to **natural gas**, containing an excessive amount of **hydrogen sulphide**.

Source control The elimination, before or during ultimate consumption, of potential air contaminants contained in raw materials, thus preventing the emission of contaminants to the atmosphere.

Source, point, line and area A point source is the plume from a single chimney; a line source is the emission from a row of industrial chimneys; an area source is a cluster of smoking chimneys in an urban residential or industrial neighbourhood.

Spark arrester A screen-like device situated on the top of a stack or chimney, or furnace exit, to reduce the amount of incandescent material expelled to the atmosphere.

Spark-ignition engine *See gasoline engine.*

Sparklers White-hot particles in a furnace; in a pulverised fuel-fired furnace this is an indication that oversized particles are passing the classifier.

Specific gravity (1) In respect of a substance, the density of that substance in relation to the density of water. Thus, the specific gravity of water is taken as unity. The specific gravities of some common substances are:

Aluminium	2.6	Gas/Diesel oil	0.84
Lead	13.6	Crude oil	0.8 to 0.97
Bitumen	1.00 to 1.10	Coal	1.2 to 1.7
Naphtha	0.67	Pyrites	4.0 to 4.9
Kerosine	0.79	Gypsum	2.3

(2) Respecting a gas, the ratio of the density of that gas to the density of air. Thus, the specific gravity of air is taken as unity. The specific gravities of some common gases are:

Steam (at 100° C)	0.469
Hydrogen	0.0606
Oxygen	1.105

Specific heat The amount of heat required to raise a unit mass of a substance through 1 degree, expressed as a ratio of the amount of heat required to raise an equal mass of water through the same range. If expressed on the Fahrenheit scale, the heat required is expressed in Btu; if on the Centigrade scale, the the heat required is expressed in Chu. Water has a specific heat of 1.0 at 15.5° C (60° F). Other substances have specific heats as follows: aluminium, 0.22; copper, 0.09; ice, 0.49; iron, 0.11; lead, 0.32; steam, 0.5–0.64. The specific heat of gases may be measured at constant volume or at constant pressure. For air the specific heat at constant volume is 0.17; at constant pressure, 0.24.

Specific impulse In rocket systems, the pounds force of thrust developed per unit mass of propellant per second.

Spectra Phenomena observed when all the electromagnetic radiations characteristic of a particular source of radiant energy are separated into an array of constituent colours, wavelengths and frequencies; the wavelengths may be separated by refraction in a transparent prism, by diffraction in crystalline solids or by diffraction from a ruled grating. *See spectrophotometry; spectroscopic units*.

Spectrophotometry A technique of chemical analysis based on the absorption or attentuation by matter of electromagnetic radiation of a specified wavelength or frequency; the region of the electromagnetic spectrum most useful for chemical analysis is that falling between 2000 Å and 300 μm, i.e. in the infra-red, visible and ultra-violet regions. A simple spectrophotometer consists of a source of radiation, such as a hydrogen or tungsten lamp; a monochromator containing a prism or grating which disperses the light so that only a limited wavelength or frequency range is allowed to irradiate the sample; and a detector, such as a photocell, which measures the amount of light transmitted by the sample. The absorbance of the sample is read directly from the measuring circuit of the spectrophotometer; when the absorbance of a sample is measured and plotted as a function of wavelength, an 'absorption spectrum' is obtained. Calibration or standard curves are prepared by measuring the absorption of known amounts of the absorbing material at the wavelength at which it absorbs most strongly. *See chromatography; spectra; spectroscopic units*.

Spectroscopic units Units of length used in stating the wavelengths of spectral lines in different parts of the spectrum; these include:

$$\text{micron (μm)} = 10^{-4} \text{ cm}$$
$$\text{millimicron (mμm)} = 10^{-7} \text{ cm}$$
$$\text{ångström (Å)} = 10^{-8} \text{ cm}$$
$$\text{X-unit (XU)} = 10^{-11} \text{ cm}$$

See spectra; spectrophotometry.

Spill-type burner A *wide-range pressure-jet burner*; the velocity of oil through the tangential ports is maintained at high discharge rates by recirculating a controllable amount of the oil which enters the swirl chamber back to the suction of the pump. *See Figure S.7*.

Spindle oil A term originally used to describe a stable low-viscosity oil used in the lubrication of textile spindles; the term now includes any low-viscosity mineral lubricating oil.

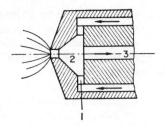

Figure S.7 Spill type burner showing (1) tangential ports, (2) swirl chamber and (3) return flow to pump

Spinning cup burner   *See rotary burner.*

Spinning reserve   The presence in an electricity supply system of generating units which are immediately ready to take on additional load, as distinct from other reserve or back-up units which would require some time to be brought to readiness.

Splint coal   Coal consisting predominantly of *attritus*.

Spontaneous combustion   Combustion which occurs of itself as a result of the combination of fuel with oxygen; this is liable to occur when the ventilation of a coal stack or heap is not sufficient to carry away the heat liberated, or when the stack is not sealed off well enough to prevent air from entering it.

Spot purchase   The purchase of commodities or supplies immediately available, as distinct from forward or future transactions. The spot price is the price paid for immediate delivery, as distinct from a forward price for delivery at some future time. Spot purchases occur in the international oil trade.

Spreader or sprinkler firing   A method of firing a boiler or furnace with solid fuel. In this method the coal is spread evenly and thinly over the entire fuel bed, but with rather less at the very back. The principle of this system is imitated by the automatic sprinkling stoker. It is claimed that this method produces better results and more steam than any other, but it is the most liable to create smoke. When hand-firing, the tendency to create smoke can be partly ameliorated by adopting the well-known principles 'little and often; level and bright'. The frequency of firing depends on the load, but it should be regular, say every 10 minutes. The volatiles are thus given a reasonable chance of being burnt. It is only when the boiler is working at moderate load, however, that smoke-free operation is likely to be achieved. *See firing by hand; sprinkler stoker.*

Spring-loaded safety valve   A type of *safety valve* in which the valve is kept in place by a spring; it has superseded the dead-weight valve on high pressure boilers. These valves may be single or double, with cast-iron body for the lower pressures, and cast or forged steel, or bronze, for higher pressures. An easing lever is fitted so that the valve can be lifted off its seat to make sure it is not sticking. In modern types a cap is fitted over the adjusting screw and secured to the valve spindle by a cotter and padlock to prevent unauthorised interference.

Springfields, UK   The Springfields Works of British Nuclear Fuels Ltd, a company formed from the United Kingdom Atomic Energy Authority in 1971 for the production of nuclear fuel elements. Uranium ore concentrates are processed through six stages, finishing with the manufacture of enriched uranium dioxide pellets and the sealing of the pellets in stainless steel or zircalloy cans.

Sprinkler stoker   A *mechanical stoker* which throws or sprinkles the fuel on to the grate, giving a thin and uniform fuel bed. As the volatiles are released from the top of the fuel bed, the mode of combustion is described as 'over-feed'. The thin fuel bed tends to lessen the effects of caking properties, and in this respect the stoker is flexible as regards type of fuel. It also greatly facilitates the combustion of the volatile matter given off by the coal. The chimney of a well-operated stoker should show only a slight haze, but the *grit carry-forward* is likely to be high when the fuel contains a considerable proportion of fines, particularly at high combustion rates. Coal is fed from a hopper, being pushed forward on to a feeder plate and projected into the furnace by either a rotary sprinkler or a spring-loaded shovel. *See Figure S.8.*

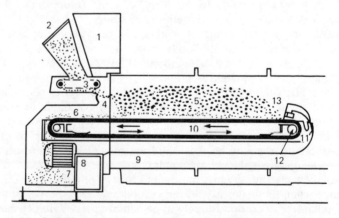

Figure S.8   Sprinkler or spreader stoker (Source: Edwin Danks & Co (Oldbury) Ltd)

1. Smoke box	8. Forced draught fan
2. Fuel hopper	9. Secondary air
3. Fuel feed regulator	10. Primary under grate air
4. Fuel distributor	11. Chain grate
5. Fuel	12. Needle bearing
6. Ash	13. Secondary rear air discharge
7. Ash collection	

Spudding Making the initial drilling for an oil well.

Spun glass An insulating material, but not suitable for surfaces above about 500° C.

Squirrel-cage induction motor The most popular type of electric motor using alternating current; this type of motor is the most economical and reliable when required to operate at nearly constant speed. A squirrel-cage motor may be designed to run at several speeds by providing alternative numbers of poles. Other alternating current motors are the slip-ring, synchronous and commutator.

Squish A method of obtaining turbulence in the combustion chamber of an internal combustion engine by using an irregular shape between the top of the piston and the cylinder head.

Stabiliser A petroleum distillation plant in which 'wild' or low-boiling hydrocarbons are removed under pressure from distillate or gasoline. The stabilised gasoline, with a boiling point between 30 and 70° C, may be used in blending low- and medium-grade motor spirits and aviation turbine gasolines.

Staffordshire kiln A top-fired transverse arch or chamber kiln used in the brick-making industry; it is fired by coal or oil.

Standard An established unit of measurement, or reference instrument or component suitable for use in the calibration of other instruments. Basic standards are those possessed or laid down by national laboratories or institutes. Examples are:

National Bureau of Standards (NBS)
American Specifications Institute (ASI)
American Society for Testing and Materials (ASTM)
American Public Health Association (APHA)
British Standards Institution (BSI)
National Physical Laboratory (NPL)
Australian Standards Institution (ASI)

Standard atmosphere A unit of pressure, defined as that exerted by a column of mercury 760 mm high at 0° C; equivalent to 101 325 newtons per square metre.

Standard atmosphere (ICAO) A standard composition for the general atmosphere adopted by the International Civil Aviation Organisation (ICAO), taken at 15° C or 59° F. Essentially, it comprises by volume, as percentages:

Nitrogen	78·084
Oxygen	20.948
Argon	0.934

plus traces of many other specified gases.

Standard temperature and pressure (STP) A conventional reference standard in relation to gases. The metric standard is 0° C, or more commonly 15° C, at 760 mmHg (approx. 1.013 25 bar) dry. In the United States of America (and the UK gas industry) the standard is 60° F at 30 inHg (approx. 1.015 92 bar) saturated. The international standard reference used by the United Nations is 15° C at 760 mmHg (approx. 1.013 25 bar) dry.

Stanton number A dimensionless coefficient, related to the ratio of convective to conductive heat transfer. For the same geometry in a nuclear reactor, an increase in Stanton number corresponds to improved heat transfer.

Star and delta connections Methods of interconnecting the windings in a three-phase generator or motor. Star–delta starting is a method of reducing the starting current for a three-phase squirrel-cage induction motor. To start the motor, the stator winding is connected to the electricity supply in a star connection; when the motor has run up to a sufficient speed, the connections are changed so that the windings are in a delta form. As a result, the starting current is only one-third of what it would be without the star starting connection.

State of emergency Proclaimed by the British Government from midnight, 13 November 1973, until mid-February 1974, in order to deal with the problems caused by a worsening oil supply situation and the reduction in coal supplies arising from an overtime ban in the mining industry. The main effect was the introduction of a 3 day working week for much of industry and commerce. The use of electricity for heating was prohibited in offices, shops, places of entertainment, bars and other public places, banks, petrol stations and film and television studios; and for advertising, display and floodlighting of outdoor arenas. In January 1974 restrictions on electricity consumption by large continuous process users were introduced; supplies of electricity to most of industry and commerce were confined to 3 days a week. Restrictions were also imposed on lighting levels in offices and some public buildings. Oil and petrol deliveries to customers were restricted, and arrangements were made for petrol rationing. A Department of Energy was created in January 1974.

States of matter The states in which matter can exist as a solid, liquid or gas. Solids are substances which tend to retain their shape and size indefinitely without deformation. Liquids are substances which flow to take up the shape of the vessel containing them. Gases also flow to take up the shape of the vessel containing them, but also expand to occupy all the available space. The state of a substance can be changed by a sufficient change in its heat content or its pressure.

Static pressure   The pressure of a fluid in motion, or at rest, exerted perpendicularly to the direction of actual or possible flow, as distinct from *velocity pressure*. *See total pressure*.

Static rectifier   A transformer-rectifier for converting the alternating current of normal electricity supply into high-tension direct current electricity; this unit has found wide use in electrostatic precipitators. The use of selenium elements has superseded the original copper-oxide type; silicon types have been developed but are more expensive. The main advantages of the static rectifier are: (a) silent operation; (b) total enclosure in oil without any moving parts; (c) requires no radio and television suppression devices; (d) simple to operate from a remote position; (e) no nitrous oxide is generated. For performance under erratic or adverse process conditions, they are not considered so adaptable as mechanical rectifiers. *See electrostatic precipitator; mechanical rectifier; rectifier.*

Static stability   A fundamental concept in meteorology, it refers to what happens to a parcel of air after it has been given an initial vertical displacement, either upwards or downwards. If, after an upward displacement, the parcel is found to be warmer (less dense) than its surroundings, its buoyancy will make it move farther from its original position; it will continue to move until its temperature (and density) becomes equal to that of its surroundings. In this instance, in which a parcel moves farther from its original level, the surrounding atmosphere is said to be statically unstable (or just 'unstable'). If after displacement, however, the parcel is found to be colder (more dense) than its surroundings, its buoyancy will tend to return the parcel to its original level; the surrounding atmosphere is then said to be statically stable (or just 'stable'). In the intermediate stage, when the vertical motion is neither encouraged nor opposed, the atmosphere is said to be neutrally stable. The concept of static stability is of considerable importance in the study of the dispersal of air pollutants in the atmosphere. *See lapse rate.*

Statoil   The Norwegian state-owned oil company created by Parliament in 1973 to engage in oil exploration, production, refining and the selling of petrol at service stations. It has a 50 per cent interest in the huge Statfjord field in the North Sea.

Stator   The stationary part of an *alternator*.

Steam   Generated from water, the most widely used heat transport fluid; it is cheap and non-toxic, offers ease of control and distribution, and has a high heat transport capacity. It contains about 25 times as much heat as the same weight of air at the same temperature.

Steam blast burner   A type of *oil burner* consisting of a double concentric tube which allows steam at pressure to impinge on the oil feed and atomise it. *See* **atomisation**.

Steam conditions The pressure and temperature of steam specified usually at the boiler main stop-valve.

Steam cracking An oil refinery process, the primary object of which is the production of ethylene, although the by-product is a valuable gasoline blending component. Feedstocks for this process are derived from straight-run gasoline, being generally referred to as 'light distillate feedstock'. On heating with steam at low pressure and temperatures in the range 700–820° C, the hydrocarbons are thermally cracked to yield a mixture of gases and liquids. The gases are removed for the production of chemicals, while the by-product is distilled to yield a fraction boiling in the gasoline range. Steam cracker gasoline is highly unstable and needs treatment in a hydrogen atmosphere over a catalyst to achieve acceptable stability.

Steam curtain A barrier of steam sprays placed between a potential source of ignition, such as a furnace, and possible points of leakage of flammable gases.

Steam-electric plant Electricity generating plant in which the prime movers connected to the generators are driven by steam, in contrast to hydroelectric plant, in which the prime mover is driven by water.

Steam engine A *heat engine* in which the expansion of steam is utilised to effect the movement of a piston in a cylinder; in its simplest form the steam expands from the initial pressure to the exhaust pressure in a single stage. In a multiple-expansion engine the expansion of steam is divided into two or more stages which are performed successively at falling pressures in cylinders of increasing size. In the compound engine high- and low-pressure cylinders are used in series. In a triple-expansion engine the steam expands successively in a high-pressure, intermediate pressure and low-pressure cylinder, working on the same crankshaft.

Steam flow/air flow ratio A relationship which has been employed as a means of combustion control in solid-fuel-fired steam boilers. *See* **combustion meter**.

Steam generating heavy water reactor (SGHWR) A British nuclear reactor design using slightly enriched uranium, a heavy water moderator and a light water coolant; it produces steam within the reactor core for direct use in the turbo-generator. A 100 MW pilot plant has been constructed in Britain; refuelling is carried out off-load about twice a year during a 4–6 day shut-down period. It was regarded by the British government as a strong contender for the role of successor to the *Advanced Gas-Cooled Reactor*. However, in 1977 the government decided not to proceed further with this reactor, preferring to build more Advanced Gas-cooled Reactors, and possibly one US light water reactor.

Steam reforming The treatment of hydrocarbons with steam in the presence of a catalyst to produce methane, carbon monoxide and hydrogen. *See Figure S.9.*

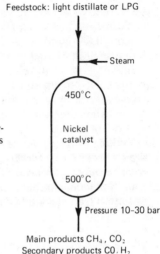

Feedstock: light distillate or LPG

Steam

450°C

Nickel catalyst

500°C

Pressure 10–30 bar

Main products CH_4, CO_2
Secondary products CO, H_2

Figure S.9 Steam reforming of light hydrocarbons: UK Gas Council catalytic rich gas process

Steam soaking A method of softening the hard deposits on the external surfaces of boiler tubes, by admitting low-pressure steam to the furnace when it is cold.

Steam tables Published tables giving experimentally tested data for total heat, volume and entropy of steam at various temperatures and pressures. The Callendar Steam Tables are an example, being available in both Fahrenheit and Centigrade temperature scales, and in complete and abridged forms.

Steam trap A device fitted at the lowest point of a steam pipework system to provide automatic *condensate* recovery. Steam traps may be classified into three main groups: (a) mechanical group which, through the action of a float, open to condensate and close to steam; (b) thermostatic group, which open or close according to the temperature; (c) thermodynamic group, which work on the difference in velocity between steam and condensate flowing across a simple valve disc. Other types are also available.

Steam turbine A *heat engine* in which jets of steam impinge upon blades attached to one or more discs or wheels mounted on a shaft supported in bearings. The shaft thus rotates, and may be used to drive an electric generator or other machinery. The successful development of the steam turbine was largely due to Sir Charles Parsons (1854–1931). *See back-pressure turbine; compound turbine; condensing turbine; gas turbine; impulse-type turbine; pass-out turbine; turbine cylinder; turbine rotor; turbine steam conditions; turbine steam rate; turbine thermodynamic efficiency; turbo-alternator.*

Steel An alloy of iron and carbon. A typical analysis of mild steel, constituents being expressed in percentages, is: carbon, 0.2; silicon, 0.04; sulphur, 0.05; phosphorous, 0.05; manganese, 0.5; balance, iron. In alloy steels some of the iron is replaced by elements such as nickel, chromium and tungsten. Steel is used for boiler shells, steam and water drums, economisers, air preheaters, conveyors, hopper, bunkers, tubes, pipes, ducts and all structural work. Alloy steels are used to meet extreme conditions of heat and/or corrosion.

Steelmaking furnace *See Ajax furnace; Bessemer converter; electric arc furnace; Kaldo furnace; Linz–Donawitz converter; Oberhausen rotor furnace; open-hearth furnace.*

Stefan–Boltzmann formula A formula for calculating the values for radiation from a solid body:

$$W = \delta \varepsilon T^4$$

where W = radiant energy per unit area per unit time;
δ = Stefan–Boltzmann constant;
ε = emissivity of the surface, a dimensionless number between 0 and 1;
T = absolute temperature.

Steradian Unit of solid angle; the solid angle which, having its vertex in the centre of a sphere, cuts off an area of the surface of the sphere equal to that of a square having sides of length equal to the radius of the sphere.

Stichting CONCAWE An organisation formed by the oil industry to deal with pollution problems; CONCAWE is an acronym for 'Conservation of Clean Air and Water, Western Europe'. Through the activities of its working groups, the organisation aims to develop a common view of the problems of air and water and noise pollution, and their abatement. In this way technical information is made available to the oil industry and to others. The offices of the organisation are in The Hague.

Still A closed chamber, usually cylindrical, in which heat is applied to a substance to change it into vapour.

Stoichiometric The calculated combining weights of chemical reactions or processes, based on the laws of conservation of mass and energy and the chemical laws of combining elements.

Stoichiometric air   *See theoretical air.*

Stokes   The unit of *kinematic viscosity* in the *metric system*. *See centistokes*.

Stokes' law   An equation by means of which the free-falling velocity attained by a particle under viscous flow conditions may be calculated:

$$v = \frac{d^2 g(\sigma - p)}{18\eta}$$

where v = free-falling velocity, cm/s;
  σ = density of particle, g/ml;
  p = density of fluid, g/ml;
  g = gravitational acceleration, 981 cm/s^2;
  η = absolute viscosity, poise;
  d = Stokes' diameter of particle, cm.

Stokes' diameter is the 'equivalent free-falling diameter' within the range of validity of Stokes' law, i.e. for which the *Reynolds number* is less than 0.2. The 'equivalent free-falling diameter' is the velocity of fall of a particle through a still fluid at which the effective weight of the particle is balanced by the drag exerted by the fluid on the particle. (Reference British Standard 2955; 1958.)

Stomata   The small apertures or pores in the surface or epidermis of a leaf or young stem; they allow the passage of gases and vapours into and out of leaves. *See photosynthesis*.

Storage heater   A specially constructed block which can store heat, providing space heating at a low cost. It is heated by electricity at 'off-peak' times, giving up the heat during the day.

Straight-run distillation   Continuous distillation which separates the products of petroleum in the order of their boiling points without cracking. Hence 'straight-run gasoline', or gasolines as supplied by primary distillation without further treatment.

Straight-run products   Products produced by straight-run distillation.

Strake   A cheap and simple method of preventing tall chimney stacks from oscillating in high winds, developed by the National Physical Laboratory: the method consists of applying metal strips, called strakes, to the top section of a stack, arranged in the form of a helix. Each strake consists of a plate attached edgewise to the surface of the cylinder. It is wound round the length of the cylinder in a rising spiral. The number of such strakes required is not critical, but experiments carried out to determine the optimum arrangement of strakes led to the recommendation of a configuration with strakes of 0.09 diameter high, wound with a helix pitch of five diameters. Structures or structural members of bluff section tend to oscillate in wind because of the vortices which are produced alternatively at the sides of the body and shed into the wake behind the body. This vortex formation gives rise to an alternating force in the cross-wind direction, and consequently the oscillatory displacements are produced in that direction. The strakes upset the correlation (i.e. the phasing) of the vortices shed from different parts along the length of the cylinder, and in this way a build-up of the forces tending to set up oscillations is prevented.

Strategic Petroleum Reserve   A scheme to reduce the vulnerability of the United States of America in its dependence on foreign oil; the first shipment of crude oil was delivered to special storage facilities in a Louisiana salt-dome in July 1977. It was intended to achieve an emergency reserve of one million barrels or more, which would reduce the effects of an interruption of oil supplies such as occurred as a result of the Yom Kippur War in 1973–74.

Stratosphere   The upper layer of the atmosphere, lying above the troposphere, i.e. above about 11 kilometres altitude, in which the temperature remains constant with height. Concern about the environmental consequences of supersonic aircraft, other than the effects of noise, derives from the fact that these aircraft cruise in the stratosphere. There is little vertical mixing in the stratosphere and washing-out by rain does not occur; particles from volcanic eruptions or atomic explosions tend to remain in layers for long periods, spreading horizontally from one hemisphere to the other but with little diffusion upwards or downwards. The residence time for particles may be between 1 and 2 years, in contrast to a few days in the troposphere. Apprehension has been expressed that supersonic aircraft will add to the pollution at this level, and furthermore that the exhaust gases will destroy the ozone which occurs naturally at these levels protecting the Earth from biologically dangerous ultra-violet radiation.

Stream days   In respect of plant, *time efficiency*, expressed as operating days per year.

Stringer   A series of fuel elements connected together in a channel of a nuclear reactor, so that they can be removed as one unit.

Strip mining   Or open-cut mining; a technique of mining employed when the coal is not more than about 30 metres below the surface; the overlying earth and rock are mechanically stripped to expose the coal, which is then removed with or without blasting. In August 1977 the United States established the first uniform federal controls over strip mining, which had previously been a state matter. The legislation requires the restoration of a stripped land so that it will serve the same function as it did prior to mining; it gives farmers and ranchers the right to veto the mining of their land, even if they do not own the mineral rights; and it sets up a $4.1 billion fund, financed by a federal tax on all coal from both strip and underground mines, to restore previously stripped land wherever desirable.

Sub-bituminous coal Coal intermediate between *lignite* and *bituminous coal* in appearance and properties.

Subcritical The condition of a *nuclear reactor* in which the rate of fissioning is not sufficient to sustain a chain reaction.

Subeconomic resources Accumulations or deposits of minerals that are not currently economically recoverable under current prices and costs using current technology.

Sub-hydrous coal *Coal* containing less *hydrogen* than is normal for the type species. *See orthohydrous coal; per-hydrous coal.*

Submarginal resources Resources that would require a substantially higher price or a major cost-reducing advance in technology to become economically recoverable.

Submerged production system (SPS) For offshore oil production, a system completely independent of platforms and surface connections; the submerged pipelines and equipment are maintained by divers or by remotely controlled underwater vehicles.

Subscribed demand tariff A *tariff* for the supply of electricity, consisting of a unit charge and a charge related to the demand for which the consumer wishes to subscribe.

Subsidence inversion An *inversion* formed usually well above the earth's surface as a result of the slow descent of air which has become warmer through *adiabatic* compression.

Subsistence economy An economy in which most production and consumption is local and immediate; a few items may enter from outside, such as salt and essential tools, and some goods and services may be exported to pay for these, but essentially the bulk of economic activities are carried on locally for local consumption. Characterised by the 'self-contained' village, and tending to be traditional and static.

Substation Any premises or enclosure containing apparatus for transforming or converting electricity from one voltage to another, e.g. (a) stepping up to a high voltage for transmission, or (b) stepping down to a low voltage for distribution.

Substitute natural gas (SNG) A description for methane which can be produced from coal as a product of pyrolysis, by methanation of synthesis gas produced by the steam–oxygen gasification of coal, and by direct reaction between carbon and hydrogen (hydrogasification). Under suitable conditions substantial methane yields are obtained during *pyrolysis*.

Substitution *See demand.*

Suction pyrometer A device for measuring temperatures when errors due to radiation from neighbouring surfaces are likely to occur; it comprises an open-ended tube containing a bare thermocouple surrounded by radiation shields. This assembly is inserted into a duct or flue and a sample of gas is drawn continuously through it. In a sonic suction pyrometer the gases are sucked through a nozzle at the speed of sound. In some cases suction pyrometers are water-cooled. *See pyrometer; thermoelectric pyrometer; thermometer*.

Sulphate Salts of sulphuric acid, comprising the SO_4 group of atoms. Sea-spray is a substantial source of airborne sulphate particles. Sulphur dioxide emitted during the combustion of sulphur-containing fuels slowly oxidises in the atmosphere to sulphur trioxide; in the presence of moisture this becomes sulphuric acid. Combined with basic materials such as ammonia or metals and their oxides, particulate sulphates are formed.

Sulphur A non-metallic *element*; relative at. wt. 32.064. It occurs in nature in the free state, and combined as sulphides and sulphates. It is present in coal and oil, being derived from the substances from which they were formed. The sulphur may be either chemically or physically mixed with the fuel. The sulphur content of British coals ranges from 0.5 to 3.5 per cent, with an average of 1.6 per cent. Sulphur is present in coal in three forms: (a) *pyrites*; (b) organic sulphur compounds, cleaning processes having virtually no effect on this type of sulphur; (c) sulphates, present only in very small quantities and rarely exceeding 0.03 per cent. The ratio of pyritic to organic sulphur varies considerably. High-*rank* coals tend to be low in sulphur, medium-rank coals high in sulphur; low-rank coals appear to be fairly average. During the combustion of coal, most of the sulphur is released to the atmosphere, mainly as *sulphur dioxide* together with a small amount (3–5 per cent) of *sulphur trioxide*. With coal and coke burned in domestic heating appliances, about 20 per cent of the sulphur is retained in the ash or clinker. In industrial boilers and furnaces only about 10 per cent of the sulphur is retained in the ash. The crude oils of the Middle East have an average sulphur content of about 2.5 per cent; most of this sulphur appears in the final products, being less than 1 per cent in diesel and gas oils, and up to 4 per cent or more in residual oils.

Sulphur content, statutory limitations on Legal limitations on the amount of sulphur allowed in fuels which are to be burned in certain areas; the purpose is to restrict the amount of *sulphur dioxide* emitted to the atmosphere by source control. In the 1960s the City of New York introduced restrictions on the sulphur content of fuels, progressively leading to an upper limit of 1 per cent. The City of London has also decreed that the sulphur content of fuel oil shall be restricted to a maximum of 1 per cent. Many cities have now introduced restrictions. The restrictions may be imposed by specific legislation, or by limitations set down in licences issued under environment protection legislation. *The European Economic Community* has issued a directive requiring all member states to impose limits on the sulphur in gas oils: Type A, being for general use and having an upper limit of 0.3 per cent; and Type B, for use in zones where sulphur dioxide levels are low, having an upper limit of 0.5 per cent by weight. In the United Kingdom this requirement was implemented by the Motor Fuel and Oil Fuel (Sulphur Content of Gas Oils) Regulations 1976, made under Part IV of the Control of Pollution Act 1974.

Sulphur cycle The circulation of sulphur atoms brought about mainly by living things. The decomposition of proteins containing sulphur (e.g. egg albumin) under aerobic conditions results in the formation of sulphates which are odourless and relatively harmless. However, when such proteins undergo decomposition by anaerobic bacteria, the foul-smelling gas *hydrogen sulphide* is produced. Other malodorous sulphur compounds may also be produced under certain conditions, e.g. methyl mercaptan (CH_3SH) will be produced when a river or other body of water is devoid of both dissolved oxygen and nitrate. Sulphur circulates globally between air, land and sea. A large part of the sulphur in the global atmosphere (as distinct from the atmosphere of towns) is emitted originally as hydrogen sulphide from natural sources; much of this hydrogen sulphide is converted later to *sulphur dioxide*. Sulphur, in the form of either sulphur dioxide, *sulphur trioxide*, hydrogen sulphide or sulphate salts, is removed from the atmosphere in rain, drizzle and fog, and by gaseous absorption in the oceans. Only about one-third of the discharge to atmosphere is due to combustion processes, the balance being attributable mainly to release from sea-spray and biological decay.

Sulphur dioxide SO_2. A colourless, pungent gas formed when sulphur burns in air. It is considered to be one of the most important air pollutants; most of the sulphur dioxide in the general atmosphere comes from the combustion of the sulphur present in most fuels. The following are the average percentages of sulphur in fuels in common use in the United Kingdom: coal, 1.6; coke, 1.3; domestic fuel oil, 0.1; gas and diesel oil, 0.3–1.5; industrial fuel oil, 1.0–4.0; kerosine (paraffin), 0.03; coal tar fuels, 0.5–1.0; gas 0.02. All the sulphur in oil, and from 80 to 90 per cent of that coal and coke, is emitted from the chimney as sulphur dioxide, the remainder being retained in the ash. *See Table S. 2* for an estimate of SO_2 emissions to atmosphere in North America during 1977–78.

Sulphur trioxide SO_3. A constituent of flue gases from *sulphur*, bearing fuels, frequently to the extent of 3–5 per cent of the *sulphur dioxide* present. Several mechanisms appear to contribute to its formation; in every case a supply of oxygen is necessary. Thus, the reduction of excess air tends to inhibit the formation of this corrosion-promoting gas. *See acid soot; dew point*.

Table S.2 SULPHUR DIOXIDE EMISSIONS: NORTH-EASTERN NORTH AMERICA

Source	SO_2 emissions $\times 10^3$ tons/year	
	Urban	Utilities
Canada		
Ontario	1 741.0	191.0
United States		
Ohio	578.4	2 338.3
Indiana	433.5	1 666.4
Kentucky	74.2	1 387.7
Illinois	332.4	1 255.6
Michigan	189.4	1 158.8
Pennsylvania	535.5	1 119.6
West Virginia	149.7	1 020.9
Tennessee		721.4
Missouri	226.4	499.0
District of Columbia		216.5
New York	636.8	209.5
Wisconsin		192.1
Maryland	129.5	
Massachusetts	139.9	
Virginia	141.6	
US Total	3 567.3	11 785.8

Source: Ontario Ministry of the Environment, 1979

'Sunbrite' A *hard coke*, carefully prepared and selected for domestic heating appliances such as room heaters, boilers, cookers, fires with under-floor primary air for combustion and fan-assisted open fires. It is manufactured by the *National Coal Board*, the steel companies and other hard-coke producers. *See authorised fuels; smoke control area*.

Sun day A day designated by the *United Nations Environment Program* to help promote the use of renewable energy resources; in 1978, 3 May was designated.

Superdiabetic lapse rate A *lapse rate* greater than the dry adiabatic lapse rate, i.e. greater than 3° C (5.4° F) per thousand feet. The dry adiabatic lapse rate is often exceeded by a factor of several times near a land surface which is strongly heated by solar radiation. Turbulence in the atmosphere is strong, and the dilution of waste industrial gases is more rapid than in average or neutral conditions. *See inversion.*

Supercritical gas solvent extraction (SGSE)   A process for treating coal which yields a hydrogen-rich extract suitable for use as an aromatic chemical feedstock, and for refining into liquid transport fuels. Extraction is performed at temperatures of 350–450° C and pressures of 100–200 bars, using a light organic solvent (typically toluene) under supercritical conditions. The product is thought to be particularly suitable for use in the manufacture of plastics, resins, rubbers, paints and artificial fibres. The char is usable as a solid fuel and as a gasification feedstock. The process is being developed in both the United States of America and the United Kingdom.

Supercritical once-through boiler   Known also as the Benson boiler, after its British inventor, a water-tube forced circulation boiler in which the feed-water is heated, evaporated and superheated in a single passage, through a number of tubes in parallel. The original design was based on the principle that water at the critical pressure of about 218 bars needs no latent heat for conversion into steam. All the water flashes into steam when the necessary temperature is reached, so that no steam release surface and no steam drums are required. The use of small-bore tubes and the absence of drums reduces the weight of the pressure parts and of the supporting structures. Britain's first commercial supercritical O.T. boiler was commissioned at the Margam 'B' power station of the Steel Company of Wales Ltd by Messrs Simon Carves Ltd.

Supercritical steam pressure   Steam pressure at or above the critical pressure of 218 bars (3200 lb/in² abs.), at which pressure no latent heat is required to convert water into steam.

Supereconomic boiler   A three-pass *economic boiler* in which the second and third passes of fire-tubes are situated below the main furnace tubes; this system ensures maximum heat transfer and imparts a vigorous circulation in what in other designs is a 'dead water' area.

Super-grid   A high voltage electricity transmission system operated by the *Central Electricity Generating Board*. It comprises a 275 kV (275 000 V) transmission network now being supplemented by a 400 kV (400 000 V) transmission system. *See grid.*

Superheated steam   Steam subjected to additional heating after leaving the boiler; the heat added to the steam is *sensible heat*. A sufficient amount of superheat prevents harmful and wasteful condensation in steam turbines and steam engine cylinders. *See superheater.*

Superheater   A heat exchanger used in boilers to raise the temperature of the steam above that at which it leaves the boiler.

The installation of a superheater leads to a reduction in steam requirements wherever it can be usefully employed, e.g. in supplying superheated steam to steam engines and turbines. Superheaters are constructed of small-bore carbon and alloy steel tubes. In design, a superheater may consist of a series of U-shaped tubes connected to headers, or a number of 'hair pin' bends. In shell boilers the superheater is installed at the back end of the main furnace tubes, where gas temperatures may range from 500 to 850° C. Superheaters exposed to the radiant heat of the furnaces are described as 'radiant superheaters'; those which derive heat solely from the hot flue gases are known as 'convector superheaters'. In the *water-tube boiler* their description varies with position: if in the space over the water-tubes, they are known as 'over-deck'; if between the water-tubes as 'inter-tube', and if between banks of water-tubes, as 'inter-bank'. *See superheated steam.*

Super Phénix   A commercial-size fast breeder reactor, constructed by France at Creys-Malville; it was based on the Phénix prototype. *See fast breeder reactor*.

Surface combustion   *Combustion* in the immediate vicinity of a hot surface; it has been found that hot surfaces have the property of increasing the rates of combustion of gases in air, temperatures of up to 1900° C being attainable.

Surface condenser   A *condenser* in which the cooling water flows through a large number of tubes, the steam condensing on the outer surfaces of the tubes.

Surface ignition   In respect of an internal combustion engine, hot deposits in the cylinders which may ignite the cylinder charge at the wrong time.

Surface moisture   Another name for *free moisture*.

Surface tension   A characteristic of liquid surfaces whereby, owing to unbalanced molecular cohesive forces near the surface, they appear to be covered by a thin elastic membrane in a state of tension. Surface tension is measured by the force acting across unit length in the surface, the force being expressed in dyn/cm. At 10° C water has a surface tension of 74.22 dyn/cm and benzene a surface tension of 30.22 dyn/cm.

Suspension firing   The firing of *pulverised fuel*, which burns in suspension. The common method of firing are vertical, horizontal, tangential, cyclonic and opposed inclined.

Suspension gasifier   A gasifier in which powdered fuel is held in suspension by the gasification medium, e.g. air.

Suspensoid   suspended matter.

Sustained yield conservation   That aspect of *conservation* which seeks wise use and continuing productivity of renewable natural resources; originating with forestry, the principle has been extended to other fields.

Sweet Having a pleasant smell; negative to the *doctor test*. Sweetness indicates the abscence in gasolines, naphthas and refined oils of hydrogen sulphide and mercaptans.

Sweetening The process by which petroleum products are improved in odour and colour by oxidizing the sulphur products and unsaturated compounds.

Swelling number In respect of coal, the number of the standard profile most nearly corresponding to the coke button obtained under test, taking the average of five determinations; the property of swelling is determined by the crucible swelling test described in BS 1016; Part 12: 1959. The standard numbered profiles are from 1 to 9 half-units. *See Gray–King assay*.

Swidden agriculture Also known as 'slash and burn', or shifting cultivation, or shifting agriculture; a type of agriculture involving the clearing of land by cutting and burning the vegetation to open the soil for planting. It is a major agricultural technique used by the people of Central and South America, Asia, Africa and other parts of the world. After limited use the land is abandoned and a fresh clearing made elsewhere; thus, only a fraction of land area is under cultivation at any one time. Most agriculturists consider swidden agriculture to be wasteful and inefficient, while large amounts of land are required to operate it.

Swimming pool reactor A *nuclear reactor* using water as *moderator* and *coolant*, in the form of a tank of water with the fuel elements suspended well below the surface so that the water also acts as a shield. Such reactors are often used for the study of shielding problems.

Swivel-type damper A damper or draught regulator used, for example with economisers. A level is attached to the swivel, and works in a quadrant with two holes in it, so that the damper may be held firmly in position by a steel pin.

Sydney Metropolitan Waste Disposal Authority Established in 1971, a statutory body of the New South Wales Government, Australia; it is responsible for the disposal of waste within a region covering about 3900 square kilometres with a population of over 3 million. While mainly relying on landfill disposal of solid waste, the Authority is continually examining various methods of waste treatment, including pulverisation, incineration, pyrolysis, compaction, composting and baling, as well as recycling and resource recovery. The Authority anticipates that resource recovery will be a significant factor in waste management and natural resource conservation by the mid-1980s.

Sydney Oxidant study A study of photochemical smog in Sydney,

New South Wales, Australia, launched by the State Pollution Control Commission in 1974, assisted by Sydney and Macquarie Universities and the Commonwealth Scientific and Industrial Research Organisation. The study took about 5 years to complete and cost over $A1 million. Photochemical smog in Sydney is exceeded only in Los Angeles and Tokyo, occurring principally during the summer months.

Sydney pollution index A daily indicator of general atmospheric pollution released by the State Pollution Control Commission of New South Wales, Australia. The index has three ranges of figures: low (0–30), medium (31–45) and high (above 45). On the basis of this index, some 45 days in 1977 were rated as high in the Sydney metropolitan area. The index figure is based on the highest hourly average of both suspended matter and ozone occurring before 4.00 p.m. on each day.

Symbiosis A phenomenon characterised by a mutually beneficial relationship between two organisms, factors or elements.

Synchronous motor A motor designed for alternating current, the speed of the motor being directly proportional to the frequency of the supply.

Synchroton An accelerator of the cyclotron type, in which the magnetic field is modulated but the electric field is maintained at a constant frequency.

Syncrude Or synthetic crude oil, being the product of coal liquefaction processes.

Synergistic effect The tendency of chemicals and processes to react together to form possibly unforeseen combinations which may have a wholly new or markedly more powerful effect than the substances or processes taken separately. Synergistic effects occur in the formation of photochemical smogs, and have been suspected to have occurred also in the London smog episodes of 1952 and 1962. A synergistic effect may link different elements in the total biosphere, utilising, for example, organic substances and the characteristics of light. The formation of highly toxic methyl *mercury*, through the interaction of organic matter and mercury, is another example of a synergistic effect.

Synfuels Or synthetic fuels, a term used loosely for fuels derived from other substances, e.g. oil or gaseous fuel from coal, or oil from shale rock. The term is misleading, since it implies something artificial or manufactured, rather than the transformation of basic fuels such as coal and oil into more usable forms.

Synoptic Affording an overall view. In meteorology, a synoptic chart displays the state of the atmosphere over a large area at one time.

Synroc An Australian technique for handling radioactive waste substances, the waste being embodied in a synthetic material.

Synthesis A reaction of simple substances to produce more complex ones, e.g. the production of hydrocarbons from carbon monoxide and hydrogen.

Synthesis gas A mixture of gases made specifically for use in a synthesis process, e.g. gas used in the manufacture of ammonia and other chemicals such as methanol. *See ammonia synthesis*.

Synthoil process A process being developed in the United States in which a slurry of coal in a recycle carrier oil is propelled by a highly turbulent flow of hydrogen through a bed of catalyst, the result being a high rate of conversion of coal to oil. The process appears most efficient in the production of substitute heavy fuel oil, while being significantly less efficient in the production of gasoline.

Système International D'Unités *See International system of Units*.

System fuel savings Fuel cost savings arising from the reduced operation of the existing power stations in an electricity supply system, consequent upon the commissioning of new and more efficient power stations. The new power stations supply more cheaply some of the energy previously supplied by the older stations which have moved down the order of merit. *See merit order*.

System load factor *See load factor, system*.

System planning The formulating evaluating and choosing between the various courses of action suggested in the light of system objectives, and the need to meet those objectives at minimum cost.

T

Tail gas Residual gas left after recovering the desired product from the gas stream.

Tanbark A bark residue remaining after bark has been used in tanning operations. It contains a high percentage of water (60–70 per cent) and has a low calorific value of between 6 and 7 kJ/g.

Tangential firing *See corner-fired furnace*.

Tank farm Land on which a number of storage tanks are located.

Tank gauge A device to indicate the level of a liquid fuel in a storage tank. Tank gauges can be divided, in general, into two types: (a) floating indicator type, where a float is coupled to an external indicator which is marked to give the oil in the tank in litres; (b) diving-bell type, in which the oil in the tank exerts a pressure on the air in the bell, causing an indicating fluid to stand at a level equivalent to that of the oil in the tank.

Tapping Drawing off molten metal from a furnace, e.g. an *open-hearth furnace* or a foundry *cupola*.

Tar A black viscous liquid resulting from the distillation of solid material such as coal or wood; the term is sometimes used to describe heavy liquid residues derived from petroleum processes.

Tariff Method of charging for services, e.g. supplies of gas or electricity. *See block tariff; bulk supply tariff; flat rate tariff; installed load tariff; load/rate tariff; maximum demand tariff; restricted hour tariff; seasonal tariff; subscribed demand tariff; time of day tariff; two-part tariff.*

Tar distillation The fractional distillation of tar by the application of heat to produce *coal tar fuels* and other products. Two types of still are in use: (a) batch still, perhaps hand-fired with coke for intermittent production; (b) pipe still, for continuous distillation. In one example crude tar is heated in a pipe still in two stages—to effect dehydration and subsequent fractionation into light oil, carbolic oil, naphthalene oil, creosote oil, anthracene oil and pitch of medium hardness.

Tar oil A product obtained by the distillation of *coal tar*, comprising 'creosote oil', 'anthracene oil' and other substances. It has a calorific value of 35–38 kJ/g. *See coal tar fuels.*

Tar sands Geological deposits of sand and clay, impregnated with heavy asphaltic oil. Tar sand has to be mined and treated to separate the bitumen from the sand. Large reserves exist, e.g. the Athabasca tar sand deposits of Canada. Generally, commercial extraction has not been attractive, but may become so with increasing oil shortages.

Technological effects Or 'real' effects; effects which alter the total production possibilities or the total welfare opportunities for consumers in the economy. The effects are described as economies when they are favourable and as diseconomies when they are unfavourable. Classical examples of external real diseconomies are air and water pollution; the effects are borne by persons and firms other than those who cause the pollution, and in consequence production and consumption opportunities elsewhere in the economy are reduced.

Teeming The pouring of steel into ingot moulds, which yields an *ingot* after the mould is stripped. The ingot is then 'soaked' with heat in a *soaking pit*.

TEL *See tetra-ethyl lead.*

Telechiric mining Mining and transport of coal by conventional machines, erected, operated and maintained by telechirs; telechirs are remote hands and eyes, carried on a suitable vehicle, so that a miner on the surface can drive them to any place in the mine and perform the same work as though he was down the mine at the coal-face. The **National Coal Board** has carried out feasibility studies into telechiric mining.

Telemetering   The transmission of signals from a measuring instrument by radio or telephone line to a remote point where the results are recorded. By this method the levels of pollution at a number of locations can be recorded and displayed at a single centre.

Temco kiln   A thin-walled, steel-encased circular kiln, in use in the ceramic industry mainly for firing steel works refractories from 1100 to 1150° C with either mechanical stokers or oil. The crown and walls, some 22–27 cm thick, are constructed of refractory insulation bricks. A kiln of similar type, fired by natural gas, is in use in the United States for firing ceramic glazed and unglazed sewer pipes and other building products.

Temperature   The degree of hotness or coldness of a substance measured in terms of a temperature scale. *See temperature scales.*

Temperature entropy diagram   A graph used when consideration is being given to the optimum efficiencies obtainable with power station steam cycles of various temperatures and pressures.

Temperature scales   The scales in terms of which temperature may be expressed. The four in common use are: degrees Centigrade (Celsius) (° C); degrees Fahrenheit (° F); Kelvins (K) and degrees Rankine (° R). The last two are *absolute temperature* scales. In 1948 the General Conference of Weights and Measures adopted the Celsius in place of the Centigrade scale; the scales are identical, although the term 'Centigrade' remains in common use. *See Centigrade (Celsius) scale; Fahrenheit scale; International Temperature scale; Kelvin scale; Rankine scale.*

Tempering   The adding of water to coal, usually as it reaches a bunker, to reduce the resistance of fines in the fuel bed to the passage of combustion air. Sufficient water is added to make the coal 'ball' in the hand when squeezed, without wetting the hand. Some coals are difficult to wet with water alone, and for these exhaust steam is admitted at the bottom of the coal hopper or chute to complete the wetting.

Tennessee Valley Authority (TVA)   An agency of the United States federal government, created by Act of Congress in 1933, with full responsibility for meeting all the electricity requirements within a large area of operation. With over 18 000 MW of electricity generating capacity, the TVA is the largest electricity supply system in single ownership in the United States.

Terajoule (TJ)   A unit comprising 1000 gigajoules, 10^{12} J.

Teratogenicity   The ability of a substance to induce malformation in a fetus.

Terawatt (TW)   A unit comprising 1000 gigawatts.

Terminal settling velocity   The maximum velocity reached by a falling particle when the gravitational force upon it is balanced by the viscous drag of the fluid through which it is falling.

Tertiary air   A third supply of air sometimes introduced along the path of a flame after the *secondary air* inlets, to assist with the complete combustion of long-flaming coals.

Tertiary industry   The provision of a wide range of services required by the primary and secondary industries and by individual consumers, e.g. the provision of transport and communication; commercial and financial services such as insurance and banking; the provision of health, legal and educational services; and the wholesale and retail trade.

Tesla   Unit of magnetic flux density; the tesla is equal to 1 *weber* per square metre of circuit area.

Tetraethyl lead   *See gasoline additives; lead susceptibility.*

Tetrahydrothiophene   *See odoriser.*

Theoretical air   Or stoichiometric air; the amount of air required in theory to burn completely a given amount of fuel. It may be calculated from the chemical composition of the fuel. Theoretical air requirements per 10 000 kJ gross for different fuels are on average: anthracite and coke, 3.37 kg; bituminous coals, 3.23 kg; and petroleum oils, 3.21 kg. The weight of air required per kilogram of fuel may be calculated roughly from the formula:

$$W_a = 25.3\,C + 75.9 \left(H - \frac{O}{8} \right) + 9.5\,S$$

where W_a = theoretical amount of air, kg;
   C = carbon content of fuel, per cent;
   H = hydrogen content of fuel, per cent;
   O = oxygen content of fuel, per cent;
   S = sulphur content of fuel, per cent.

The values for C, H, O and S are obtained directly from the *ultimate analysis* for solid and liquid fuels; for gaseous fuels a conversion from the volumetric analysis may be carried out. In practice, additional or *excess air* is required to ensure complete combustion.

Therm   A unit of heat containing 100 000 British thermal units. *See British thermal unit.*

Thermal cracker   An oil refinery unit for the cracking of *hydrocarbons*, utilising heat and pressure only; heavier straight-run products such as gas oils and residual stocks can be thermally cracked into lighter fractions by heating to about 510° C at pressure within the range 20–40 bars. This type of cracking was the first developed, but the *gasoline* produced by this method is inferior to that obtained when cracking takes place in the presence of a catalyst. *See fluid catalytic cracking unit (FCCU)*.

Thermal efficiency   The calorific content of the total useful heat or energy produced by a plant, expressed as a percentage of the calorific content of the total fuel consumed, or simply:

$$\frac{\text{heat output}}{\text{heat input}} \times 100$$

In respect of a boiler, 'heat output' relates to the energy value of the useful hot water or steam produced. In considering electricity generation, however, the overall thermal efficiency is the total energy value of the electricity 'sent out', expressed as a percentage of the energy value (gross as fired) of the fuel used. This overall concept takes account of both boiler and steam turbine losses. A modern power station achieves about 35 per cent thermal efficiency.

Thermal fission   Fission produced by *thermal neutrons*.

Thermal liquid   A liquid used as a heat-transporting medium in the field of process heating and cooling. Thermal liquids include hot water, mercury, diphenyl-diphenyloxide (Dowtherm 'A'), O-dichlorobenzene (Dowtherm 'E'), molten salt mixtures and mineral oils. Dowtherm 'A' and 'E' are products of the Dow Chemical Company. Of the mineral oils, Mobiltherm 600 and Mobiltherm Light, produced by the Socony Mobil Oil Company, are aromatic mineral oils of lower viscosity than conventional mineral oils.

Thermal neutrons   Neutrons in thermal equilibrium with the material in which they are moving.

Thermal plant   Electricity generating plant which uses heat to produce electricity; such plants may burn coal, oil or gas, or use nuclear fuel, in contrast to water-driven hydro-plant.

Thermal pollution   The transfer of heat from industrial processes to bodies of water or air in such quantities as to be detrimental to the environment. Effluents having a significant effect on the temperature of a river may affect the oxygen content of the water and accentuate other adverse conditions. At sufficiently high temperatures, fish may die; fish populations could be destroyed also by the lethal effects of increased temperature either on another animal on which they feed or on their own spawn. Heat will accelerate the decomposition processes taking place in water; in relation to oxygen-sag curves, the effect will be to lower the curve and also to depress the saturation line. In the presence of sufficient oxygen, it appears that heat alone will shorten the stretch of river affected by the pollution load, accelerating purification. In a river with a 'light' pollution load this would be beneficial; however, in a river with a 'heavy' pollution load the effect is likely to be harmful. The harmful effects may include a reduction of the oxygen concentration in g/m^3 below the level critical to the organisms living in the water. A sufficient loss of oxygen will result in anaerobic conditions. Sewage fungus, found frequently in polluted rivers, grows rapidly in warmed water.

Power stations use large quantities of water in condensers for the cooling and condensing of exhaust steam from the turbines.

Thermal reactor A *nuclear reactor* in which the chain reaction is sustained primarily by fission brought about by thermal neutrons.

Thermal reformer An oil refinery unit for upgrading heavy gasoline, or naphtha, obtained as a fraction from the crude oil units. The naphtha is charged to a furnace where, at a temperature of about $538°$ C and a pressure of 100 atmospheres, the structures of the hydrocarbons are changed to yield a reformate having anti-knock properties much superior to those of the original naphtha. The reformate is chemically treated and, with light straight-run gasoline, then forms the basis of regular-grade gasoline.

Thermal shield A metallic shield placed around a *nuclear reactor* between the core and the concrete *biological shield*. It is usually made of steel or other high-density material, and is intended to absorb some of the thermal neutrons, gamma radiation, etc., emitted by the core, thus reducing the energy dissipated in the concrete shield and protecting it from thermal damage.

Thermal storage boiler Or Ritchie boiler; a boiler in which provision is made for meeting steam peaks or valleys by means of a substantial and controlled rise and fall of water level. The boiler shell has a diameter very much larger than in a normal boiler of similar heating surface. It operates at a pressure higher than that required in the factory, the process pressure being held constant by a reducing valve. A master firing gauge indicates the need for increasing or decreasing the rate of firing. The water level is allowed to rise during periods of low steam demand and to fall when meeting peak steam requirements. It is claimed that the thermal storage boiler enables peaks of 25–40 per cent above average steam demand, and valleys of similar size, to be met without any change in process steam pressure and without any change in firing rate.

Thermionic converter A low-voltage high-current device; a typical 100 W thermionic cell produces 100 amperes at 1 volt. Units must be arranged in series and in parallel to produce the desired output in terms of voltage and current. The converter consists of evacuated or plasma-filled cells in which electrons are 'boiled out' of a hot anode and are collected at the cold anode. The theoretical limit of efficiency for thermionic devices is about 40 per cent.

Thermo-compressor A compressor using high-pressure steam to compress low-pressure vapour.

Thermocouple. *See thermoelectric pyrometer.*

Thermodynamics, laws of Fundamental statements regarding heat, energy and work. The First Law states that:

heat supplied = work done + increase of internal energy

The Second Law states that it is impossible to cause heat to pass from one body to another at a higher temperature without the aid of some external supply of energy.

Thermoelectric generator A *heat engine* with no moving parts, based on the principle that a temperature difference between the functions of two dissimilar conductors forming a loop will generate an electric current (the Seebeck effect). If a load resistance is placed in the circuit, there is direct thermoelectric conversion into electrical energy of part of the heat energy used to set up and maintain the temperature difference. Thermoelectric generators which employ bottled gas heaters are in common use for the supply of small amounts of electrical energy.

Thermoelectric pyrometer A device for measuring temperature. It consists of two wires of dissimilar conductors joined together at one end—the 'hot junction'—and enclosed in a protective sheath of steel, nickel–chromium alloy or refractory materials. The two wires form a thermocouple. When the junction is heated, an electromotive force is generated, the magnitude of which depends on the temperature difference between the 'hot junction' and the cold ends of the conductors—the 'cold junction'—as well as on the type of conductors used. The 'cold junction' is connected to a measuring instrument. Base metal thermocouples are suitable for the temperature range $-200°$ C to $1100°$ C; rare metal thermocouples from $0°$ C to $1450°$ C generally. Thermocouples of the rhodium–platinum type can be used up towards the melting point of platinum, $1770°$ C; some are capable of even higher temperatures. *See Figure T.1; pyrometer; radiation pyrometer; suction pyrometer; thermometer.*

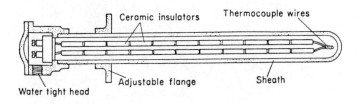

Figure T.1 Construction of a thermocouple

Thermometer   A device for measuring temperatures. *See **bimetallic thermometer; constant-volume gas thermometer; liquid-in-glass thermometer; liquid-in-steel thermometer; resistance thermometer; vapour pressure thermometer.** See also **line reversal method; pyrometer.***

Thermometer pocket   A metal or glass sheath to protect a thermometer from damage when it is inserted into a gas or liquid flow stream.

Thermopile, flux-measuring   A device consisting of a number of thermocouples connected in series with alternate hot junctions coated with a neutron absorber such as boron; the current flowing is a measure of a neutron flux. This device has been used to investigate the variation of flux across a ***nuclear reactor.***

Thermostat   A device designed to respond to temperature changes in a plant, operating controls either directly or indirectly should the temperature rise or fall above or below predetermined levels. Thermostats may be of the: (a) rod-and-tube type; (b) bimetallic strip type; (c) liquid-expansion type; (d) thermocouple type. *See **pneumatic controller.***

Thoron   A radioactive gas which is released from the soil into the atmosphere. Thoron gas decays before it has diffused more than a short distance above the ground.

Three-phase system   In respect of an electric alternator or motor, a system of supplying or using simultaneously three separate alternating currents of the same voltage and frequency, but with phases differing by one-third of a period or cycle. A three-phase alternator gives a larger output than a single-phase machine of the same size; three-phase motors run more smoothly and efficiently than single-phase motors. *Figure T.2* shows the three voltages, V_1 V_2 and V_3, produced by a three-phase alternator; there is a phase difference of 120° between them. *See **alternating current.***

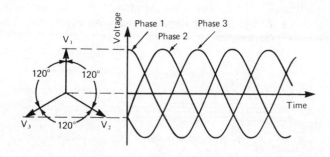

Figure T.2   Three-phase supply

Three Ts, the Assuming that a sufficient supply of oxygen has been directed to a fuel bed, the three essentials for efficient combustion—time, temperature and turbulence.

Threshold limit value The concentration of any of a number of pollutants regarded as a maximum to which healthy adult workers may be exposed for 8 hours a day, 5 days a week, without adverse effects. The list is compiled by the American Conference of Governmental Industrial Hygienists and used extensively in many countries.

Throttling calorimeter Instrument for determining the *enthalpy* and the quality of wet steam. Steam taken from a header in a sampling tube passes through a needle valve, where it is throttled to a lower pressure; the temperature and pressure are taken in the low-pressure calorimeter chamber. With the necessary readings, a *Mollier chart* is used to determine steam quality.

Throughput The quantity per unit of material undergoing a particular process or treatment.

Thylox process A wet scrubbing process for the removal of *hydrogen sulphide* from refinery and petroleum oil gas streams. The gas is scrubbed with a solution of ammonium thioarsenate:

$$(NH_4)_3AsO_2S_2 + H_2S \rightarrow (NH_4)_3AsOS_3 + H_2O$$

See hydrogen sulphide removal.

Tidal energy The use of the rise and fall of the tides to generate electrical energy; there are, however, few examples to be found. The only significant commercial plant in the western world is a 240 MW power plant in the La Rance Estuary in France; it also functions as a dam and bridge. The Soviet Union and Canada are considering tidal power plants. The north-west coast of Australia has an exceptional tidal range, but the considerable expense in harnessing and conveying energy from this location has discouraged development. The size, shape and orientation of the Severn Estuary in Britain lends itself to tidal generation; a Severn Barrage Committee has been set up to consider the advantages and disadvantages of possible schemes, and the implications of any scheme for the economy, the coastal regime, the transport system and environment generally. The possible Severn Estuary Barrage is discussed in the British Government's White Paper, 'The Development of Alternative Sources of Energy' (Cmnd 7236). Department of Environment studies have indicated that changes in tidal height, periodicity, salinity, pollutant dispersion and other factors could cause major alterations to the general ecology of the estuary and coastal habitats, with consequent effects on sea fisheries and wild life in general.

Time of day tariff A *tariff* for the supply of electricity which, in its simplest form, has a high rate in the daytime and a low rate at night; but a number of variations are possible.

Time efficiency The fraction of the total time available during which a plant is in productive operation.

Tokamak magnetic field configuration Pioneered in the Soviet Union, a magnetic configuration which appears to have the best chance of reaching all the main plasma conditions required in a nuclear fusion reactor. Tokamaks are expected to produce simultaneously the correct plasma density, temperature and confinement time in a relatively efficient magnetic field system. The system has two basic components: a set of field coils placed around a toroidal vacuum vessel creating a strong toroidal field, and a poloidal field produced by passing a large electric current through the plasma itself. The resultant field has a helical structure around the torus. The plasma current not only creates the poloidal field, but also heats the plasma. *See nuclear fusion*.

Tokyo economic summit meeting A meeting of the representatives of seven major industrial nations held in Tokyo in June 1979 to debate the energy crisis and devise means for restricting the growth in demand for oil, principally by the control of imports and by conservation measures. The summit meeting was a response to growing oil shortages and the pricing policies of the *Organisation of Petroleum Exporting Countries (OPEC)*. The countries involved were the United States of America, West Germany, Britain, France, Japan, Italy and Canada. The seven countries represented 15 per cent of the world's population, 55 per cent of the world's gross national product and 60 per cent of the world's oil consumption. The meeting set specific import targets for the seven nations individually up to 1985. The United Kingdom, France, West Germany and Italy were to freeze oil imports at the 1978 level. The United States adopted a goal of keeping oil imports below the 1977 level of 8.5 million barrels a day. On the other hand, Japan, because of an almost complete dependence on imported energy, was to be allowed to increase its oil imports by 26–38 per cent by 1985 over the 1978 level of 5 million barrels a day, i.e. to 6.3–6.9 million barrels a day. Canada was to be allowed to triple its net oil imports from 200 000 to 600 000 barrels a day by 1985, on the grounds that its major oil-producing province, Alberta, would shortly witness a dramatic decline in output. The Tokyo communique stated that the guiding principle was 'to obtain fair supplies of oil products for all countries, taking into account the differing patterns of supply, the efforts made to limit oil imports, the economic situation in each country, the quantities of oil available, and the potential of each country for energy conservation'. In addition, the meeting urged rapid development in the use of coal without damaging the environment; elimination of national policies that result in the subsidisation or underpricing of oil; and speedy development of nuclear power and alternative energy sources. The meeting criticised the Organisation of Petroleum Exporting Countries for price increases, indicating that it will cause more world-wide inflation, unemployment, balance of payments difficulties and reduced economic growth. The oil prices would press most seriously upon the developing countries.

Ton   The unit of measurement for the rate of refigeration. A ton of refrigeration represents the rate at which heat is extracted to produce 1 ton of ice in 24 hours; this represents 12 MJ/h. The ton of ice refers to a short (US) ton of 2000lb.

Tonne   The metric ton, equal to 1000 kilograms, or 2204.62lb. In Britain, the unit of mass mostly used when dealing with *nuclear reactor* fuels.

Top-hat kiln   An electric ceramic kiln in which the cover (sides and crown) is lowered over a base on which the ware is set for firing.

Topography   The characteristics of an area in terms of height, with reference to both natural features and buildings.

Topped crude   Crude oil from which some of the lighter constituents have been removed by distillation.

Topping plant   Distillation equipment for the removal of the more volatile fractions of an oil.

Torbernite   A uranium-bearing mineral, being hydrated copper–uranium phosphate.

Toroidal combustion   A technique of combustion which uses the principle of the vortex to achieve continuous mixing of fuel and air. The action of the toroid arrangement is to increase the residence time of an atomised oil particle once it has left the burner. A toroidal system may be set up by a system of jets through which combustion air enters the combustion chamber. Toroidal combustion permits stoichiometric conditions to be maintained. *See Figure T.3; combustion; stoichiometric.*

Torrey Canyon incident   A major oil spill which occurred in March 1967, when the tanker *Torrey Canyon* went aground on the Seven Stones reef about 24 kilometres north-east of the Scilly Isles, near Land's End, UK. Bound for Milford Haven, Wales, with a cargo of 120 000 tonnes of Kuwait crude oil when wrecked, *Torrey Canyon* was subjected to bombing by the Royal Air Force. A great deal of the cargo finally landed on the Cornish beaches. Some drifted across the English Channel, necessitating prompt action by the French Government to safeguard the beaches and oyster fisheries of Brittany. *See Amoco Cadiz incident.*

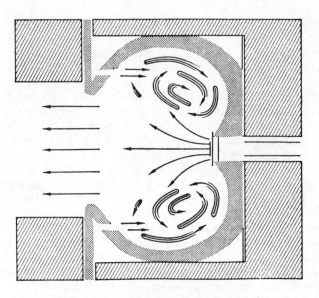

Figure T.3   A single toroidal combustion chamber

Total energy system   A system which meets all the energy requirements of a site by on-site generation of electricity, waste heat recovery from the prime movers, and the provision of supplementary heat as required. The term is also applied more generally to combined heat and power systems where on-site generation and heat recovery optimise the use of fuel within the site, usually to meet the bulk of the energy requirements. *See Figure T.4.*

Total radiation pyrometer   A device for measuring temperatures, by measuring the intensity of all the wavelengths of the radiation emitted from a hot body. The heat rays radiated are focused by a mirror on to a sensitive thermocouple. The instrument is suitable for temperatures from 500° C upwards. *See radiation pyrometer.*

Town gas   Gas produced by a public utility for the general use of domestic, commercial and industrial consumers. In Britain much of this gas has been produced, until recently, almost entirely by the *carbonisation* of coal in gas retorts, the gas as supplied to consumers having an *energy value* of 20 MJ/m³. Town gas has been manufactured in the *horizontal retort*, the *intermittent vertical retort* and the *continuous vertical retort*. In recent years an increasing amount of town gas has come from new processes, such as the *Lurgi process* and gas-reforming processes using heavy and light oils and refinery *tail gases*. Supplies of gas have been further augmented by the importation of liquefied *natural gas* from the

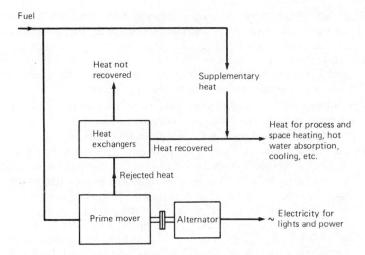

Fuel

Heat not
recovered

Supplementary
heat

Heat
exchangers

Heat recovered

Rejected heat

Prime mover

Alternator

Heat for process and
space heating, hot
water absorption,
cooling, etc.

~ Electricity for
lights and power

Figure T.4 Diagrammatic representation of a total energy scheme

Sahara. The current development of North Sea natural gas resources has resulted in a rapid displacement of conventional gasification processes.

Trace heating The heating of oil in a pipe by means of steam or electric heating elements.

Tracer technique A method of studying the path of a pollutant by the injection or release of a readily identifiable substance. A dye may be introduced to trace the destination of polluted underground water, or to identify the outlet from a particular drain. Smoke may be released to simulate the path of gaseous pollutants.

Tractor vaporising oil (TVO) Or vaporising oil; a blended kerosine-type petroleum distillate suitable for certain types of spark-ignition engine, including those used for agricultural purposes.

Transformer Apparatus for converting electrical energy received at one voltage to electrical energy sent out at a different voltage; in its simplest form it consists of two coils on a common soft iron or mild steel core. Alternating current in the 'primary' winding sets up an alternating flux in the core, threading the 'secondary' winding and generating an induced electromotive force across it. In a 'step-down' transformer the output volts are less than the input volts; in a 'step-up' transformer the secondary volts are higher than the primary volts. Three types of power transformer are used for the voltage changes necessary in electricity transmission systems. Generator transformers are installed at power stations to step up the voltage developed at the generator terminals to that required by the transmission system to which they are connected. Transmission transformers provide interconnection between transmission networks at different voltages. Bulk supply transformers provide electricity at sub-transmission voltages for local distribution networks. Transformers may be cooled by oil or water heat exchangers, or by air-blast cooling.

Transformer oil A well-refined pale petroleum distillate of low viscosity, resistant to oxidation under conditions of use; it is used in transformers for cooling and for electrical insulation.

Transmission lines A system component in electricity supply systems for conveying large amounts of energy from the sources of generation to bulk sub-stations; from there energy passes into local distribution networks at reduced voltages. Transmission lines, usually suspended between lattice towers, have increased in capacity over the years; lines handling 275 kV, 330 kV, 400 kV and 500 kV are quite common. Transmission involves some energy losses, while transmission costs and constraints modify strict merit order operation. *See* **grid; merit order**.

Travelling grate A *mechanical stoker* which differs from a *chain grate stoker* only in relatively minor respects. The drive is independent of the grate, so that the expansion of the grate surface can be most easily allowed for, and, in addition, individual grids of sections of the grate can be replaced without disturbing the drive. In a chain grate stoker the drive is effected through the grate.

Trickle feed stoker A *mechanical stoker* to serve top-fired continuous brick kilns. It consists essentially of a hopper holding about 25kg of small coal, at the bottom of which a rotating table discharges the coal in very small quantities at regular intervals through a feedspout to the kiln firehole. One stoker is required for each firehole, the mechanisms of each being operated by an electro-mechanical driving gear.

Triple point of water The point where water, ice and water vapour are in equilibrium; it is 273.16 K.

Tritium A radioactive isotope of hydrogen, with a nucleus containing one proton and two neutrons. It occurs naturally in minute quantities, and is a by-product of *nuclear fission*. It is produced, in relatively large quantities, by *nuclear fusion*.

Trommel A cylindrical screen with a centre shaft mounted nearly horizontally; the screen has round or square holes, and as coal or other material passes through it, the smalls pass through the holes.

Tropena *See Bessemer converter*.

Tropopause The boundary between the *troposphere* and the *stratosphere*.

356

Troposphere The lower layer of the atmosphere, extending up to about 11 kilometres above the surface of the earth, and in which temperature normally falls with increasing height.

True specific gravity The ratio of the weight of a given volume of a sample of dried coal or coke, ground to pass through a 72 mesh BS test sieve, to the weight of an equal volume of water at the same atmospheric temperature. *See apparent specific gravity; porosity; specific gravity; voidage.*

Tubular precipitator An *electrostatic precipitator* consisting of vertical hexagonal tubes, arranged side by side, through which the gas passes upward. The tubes may be up to some 25 cm in diameter and 4 m in length, and each contains an axial discharge electrode. While gas cannot by-pass the treatment zone, uniformity of gas distribution between the tubes is difficult to obtain; the re-entrainment of dust is high, since it has to fall into the collecting hoppers through the incoming gas stream.

Tungsten–halogen lamp Or quartz–halogen lamp; a tungsten-filament lamp in which the filament is enclosed in a small quartz envelope containing the vapour of a halogen, e.g. iodine or bromine. The filament can operate at a much higher temperature than in a conventional lamp, giving much greater brilliance and longer life. *See electronic Harlarc.*

Tunnel mixing burner A *gas burner* in which the processes of mixing and combustion take place together in a refractory-lined quarl or tunnel. It provides a stable flame both for high and low flame speed gases, e.g. town and natural gas, over a wide range of conditions. Several variants of tunnel mixing burner designs are available.

Turbidity A muddy condition of water from some natural sources, e.g. a river. If used in a boiler without prior treatment, the suspended solids would quickly accumulate and produce a foam with the steam bubbles. All suspended solids should be filtered out before any other water treatment; if the solids are fine, a coagulation process may be needed as well as filtration.

Turbine *See gas turbine; steam turbine.*

Turbine cylinder The casing assembly of a steam turbine which houses the fixed blades and the rotor. *See steam turbine.*

Turbine rotor The rotating part within a *turbine cylinder* to which the moving blades are attached. Rotors may be of the drum, disc, solid-forged or welded type.

Turbine steam conditions The temperature and pressure of the steam at the turbine stop-valve. *See steam turbine.*

Turbine steam rate The amount of steam, in kilograms, consumed per kilowatt-hour. *See steam turbine.*

Turbine thermodynamic efficiency The ratio between the heat energy in the steam entering a turbine and the heat converted by the turbine into mechanical energy. *See steam turbine.*

Turbo-alternator The combination of a *steam turbine* and an electricity generating unit. In power stations, the turbine shaft is directly coupled to an *alternator*, which, when driven by the turbine, produces electrical energy. The world's most powerful generating unit (1000 MW) is at Ravenswood power station, New York; it is a cross-compound unit, i.e. with two lines of shafts. It was connected to the grid in June 1965. The largest single-line shafts are to be found in the United Kingdom serving new 2000 MW power stations; these units have a capacity of 500 MW or 660 MW.

Turbo-generator Synonymous with *turbo-alternator*.

Turbo-jet engine A *gas turbine* used for aircraft propulsion. Air is first compressed, fuel being injected into it and burnt; the burnt gases at high pressure and temperature provide the energy source. In the turbo-jet engine sufficient energy is taken from the exhaust gas to drive the compressor; the remainder provides the propulsive thrust for the aircraft.

Turbo-prop A *gas turbine* used for aircraft propulsion, in which the energy of the hot gases is absorbed by a turbine to drive both the compressor and a propeller. *See turbo-jet engine.*

Turbulence The random movements of the air which are superimposed upon the mean wind speed. An individual movement is called a 'turbulent eddy'; it may have almost any size, and may move in any direction and at any speed. Turbulence is the most important factor controlling the dispersal of pollution. The intensity of turbulence depends on the following: wind strength at a standard height; profile of wind variation with height; variation of wind direction with height; profile of temperature variations with height; rate at which heat is exchanged between the atmosphere and the ground and outer space; nature of the earth's surface; cloud formation, rain, etc. Only in exceptional circumstances will high pollution levels be found when the wind is strong and possessing considerable turbulence. With low wind strengths and reduced turbulence, pollution levels frequently increase; the most severe pollution occurs in non-turbulent conditions during an *inversion*, when air temperature, instead of decreasing, increases with height.

Turn-down ratio The ratio between full output and minimum output of an oil burner.

Tuyères Air ports which distribute air to a fuel bed.

Twelve Principles for Energy Policy The body of principles adopted by the countries participating in the *International Energy Agency* at a meeting in Paris in October 1977; the principles are intended

358

to guide IEA countries in implementing national energy policy measures. *See Table T.1.*

Table T.1 THE TWELVE PRINCIPLES FOR ENERGY POLICY ADOPTED BY
THE INTERNATIONAL ENERGY AGENCY, OCTOBER 1977

1. Reduce oil imports by conservation, supply expansion and oil substitution.
2. Reduce conflicts between environmental concerns and energy requirements.
3. Allow domestic energy prices sufficient to bring about conservation and supply creation.
4. Slow energy demand growth relative to economic growth by conservation and substitution.
5. Replace oil in electricity generating and industry.
6. Promote international trade in coal.
7. Reserve natural gas to premium users.
8. Steadily expand nuclear energy generating capacity.
9. Emphasise R & D, increasing international collaborative projects.
10. Establish a favourable investment climate; establish priority for exploration.
11. Plan alternative programmes should conservation and supply goals not be fully attained.
12. Co-operate in evaluating world energy situation, R & D and technical requirements with developing countries.

Twin fluid atomiser Another name for the ***blast atomiser***, the term 'fluid' being applied to both oil and air.

Two-colour pyrometer An instrument which measures the colour temperature of the individual particles in a gas stream. If the spectral emissivity of the coal particles does not vary substantially over the range of wavelengths viewed by the instrument, the colour temperature is little different from the true temperature. For carbon at 2000 K the colour temperature is known to be only 16 K less than the true temperature. The colour temperature is determined by measuring the total amount of energy in each of two different narrow wavelength bands. The principle of the method is shown in *Figure T.5*. Images of particles moving in the furnace are focused on to a 300 micron aperture. The light from these images is then divided by a semi-silvered mirror so as to fall on two photomultipliers of different spectral response characteristics. The output pulses from the two photomultipliers are displayed on a double-beam oscilloscope. The ratio of the pulse heights gives the colour temperatures of the individual particles in the gas stream. The range of the instrument is from 950–2000° C with an accuracy estimated to be $\pm 40°$ C at 2000° C. *See colour/temperature scale; pyrometer*.

Two-part tariff A composite charge for gas or electricity, comprising a fixed charge per accounting period and a charge per therm or unit of gas or electricity consumed. The underlying principle is that the fixed minimum charge should cover roughly the fixed or

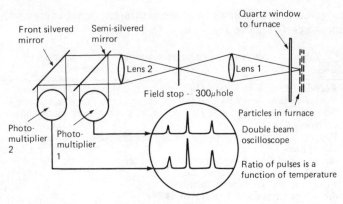

Figure T.5 Optical system for two-colour pyrometer measuring colour temperature.
(Source: BCURA)

overhead costs of production and distribution, and the charge varying with consumption should cover the variable costs of production. *See tariff.*

Two-out-of-three system A system used in the control of nuclear reactors in which certain units, such as amplifiers, are installed in triplicate; the reactor is not shut down unless two units indicate an excessive power level simultaneously. This system avoids a reactor being shut down unnecessarily when an electronic fault occurs in one of the units.

U

Ullage The empty space above the liquid in a tank or similar container; it is estimated by measuring the distance from the top of the container to the surface of the liquid held in the container.

Ultimate analysis A form of analysis which divides up a fuel into the percentages of carbon, hydrogen, oxygen, nitrogen, sulphur and ash which it contains; as a reminder of these constitutents, students sometimes use the mnemonic 'no cash'. *See proximate analysis.*

Ultrasonic agglomerator A device to agglomerate particles in suspension, thus facilitating their subsequent removal from a waste gas stream. Gases enter an agglomerating tower and are subjected to high frequency sonic vibrations; the fine dust, converted into larger aggregates, passes to a *cyclone*, or other device for collection.

Underfeed combustion. *Combustion* in which *ignition* takes place from the top of the fuel bed, the *ignition place* travelling downwards

360

against the air flow. The principle is illustrated in *Figure U.1*. This form of combustion is virtually smoke-free; it does not tend to form large coke masses and is relatively insensitive to coal **rank**. It occurs in the **underfeed stoker** and in the **chain grate stoker**. *See overfeed combustion*.

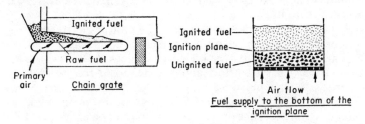

Figure U.1   Underfeed combustion as illustrated in the chain grate stoker.

Underfeed stoker.   A *mechanical stoker* in which fuel is supplied by an Archimedean screw, or by a reciprocating ram, into the bottom of a retort below the fire and is gradually forced up into the combustion zone. Combustion air is injected through ports or tuyères into the fuel bed just below the combustion level. As the coal rises in the retort, the volatiles released pass upwards through the burning fuel, mixing with the incoming combustion air, and burning at the top of the fire with a short flame. In correct conditions, when the fuel reaches the surface of the fuel bed only coke remains. This coke burns as it moves outwards towards the perimeter of the fire; combustion completed, ash accumulates around the retort and must be removed by hand at intervals. All the air required for combustion is delivered through the tuyères; it is not normally necessary to admit secondary air through the fire-door. The underfeed stoker is fitted to horizontal, vertical, sectional and central heating boilers. Where there is intermittent operation smoke may be emitted at times of stopping and starting, as the correct combustion conditions tend to be disturbed; this characteristic has been largely overcome by the development of overfire air jets and other techniques to maintain efficient combustion at all times. *See black centre burning*.

Underground gasification.   The gasification of coal while still in the seam.

Underground storage.   The storage of gas in underground reservoirs and natural strata, instead of in gasholders above ground.

Unitary pricing system   In respect of oil supplies, the price system adopted by the *Organisation of Petroleum Exporting Countries (OPEC)*, scrapping the nomenclature of posted prices, equity and participation crude. Saudi Arabian light crude oil is still used as

361

a standard 'marker', however, variations within the system being related to it. In 1979 a two-tier price arrangement was announced, with Saudi Arabian crude being the cheapest at $18 a barrel, while the ceiling for any crude was set at $23.50 a barrel. Despite the diversity of prices which has re-emerged, the system is still unitary in so far as the agreed prices are handed down by the OPEC cartel.

Unit of electricity or kilowatt-hour   The unit of electrical energy, being the amount of energy corresponding to a power of 1 kilowatt (1000 watts) sustained for 1 hour. One kilowatt-hour = 3600 kJ.

Unit system   A system for supplying *pulverised fuel*, to boilers or furnaces. One pulveriser supplies fuel direct to one or two furnaces. Unlike the *bin and feeder system*, this method does not require storage space or bins. It is the most popular system of P.F. firing.

United Nations Conference on the Human Environment, 1972   Held in Stockholm, from 5 to 16 June 1972, the first United Nations Conference on the Human Environment; the main purpose of the Conference was defined as being: 'to serve as a practical means to encourage, and to provide guidelines for, action by Government and international organisations designed to protect and improve the human environment and to remedy and prevent its impairment, by means of international co-operation, bearing in mind the particular importance of enabling developing countries to forestall occurrence of such problems.' The Conference was attended by 113 delegations. The principal achievements of the Conference were the agreements reached on:

(a) A Declaration on the Human Environment.

(b) An extensive programme of international action (the Action Plan).

(c) A permanent environment secretariat (now based in Nairobi, Kenya).

(d) An environment fund of $100m for expenditure in the first 5 years to support new environmental initiatives.

The major disappointment of the Conference was the failure of the Soviet Union and countries from eastern Europe, with the exception of Yugoslavia and Rumania, to attend. However, the Conference was attended by China.

United Nations Environment Programme (UNEP)   A programme set up by the United Nations in consequence of the *United Nations Conference on the Human Environment*, held in Stockholm in 1972. The programme is managed by a Governing Council of 54 members; activities are financed from an Environment Fund. The programme secretariat is based in Nairobi, Kenya. The

362

Governing Council has named energy as one of the top six priority areas relating to the environment.

Up-draught kiln Or bottle kiln; a ceramic kiln in which the gases pass through openings in the crown of the kiln. The kiln is surmounted by a superstructure in the shape of a bottle.

Uraninite and pitchblende Uranium-bearing minerals, being uranium oxides.

Uranium U. The heaviest element occurring in significant quantities in nature; at. no. 92, at. wt. 238.03. Natural uranium contains 1 part of U^{235} to 139 parts of U^{238}. Uranium has a typically metallic appearance, being bright, hard and heavy. It is half as heavy again as lead. *See uranium enrichment*; *Table W.1* (page 379).

Uranium enrichment Processes to raise the content of the fissile isotope, uranium-235, from the 0.71 per cent normally found in nature to a higher level; low-enriched uranium, containing 2–4 per cent of uranium-235, is used as fuel in many types of nuclear reactor; while high-enriched uranium, containing perhaps more than 90 per cent of uranium-235, is used as fuel in some reactors and to make nuclear weapons. *See gas centrifuge process; gas diffusion process*.

Urquhart oil gasifier A high-duty gasification unit in which fuel oil is caused to react with less than the stoichiometric quantity of air to produce a gas with little or no carbon deposition; the reactions can be controlled to provide a range of gases varying in composition and temperature. *See stoichiometric*.

U-tube manometer *See draught gauge*.

U-tube viscometer An instrument for measuring the *kinematic viscosity* of fluids.

V

Vacuum distillation Distillation under reduced pressure. The effect is to reduce boiling temperature sufficiently to prevent decomposition or cracking of the material being distilled.

Vacuum jets Steam ejectors for removing air and non-condensable gases from barometric condensers on distillation equipment.

Valency In respect of an *element*, the number of atoms of hydrogen that one atom of the element can combine with or replace.

Valency band *See energy bands*.

Vanadium V. A metallic *element*; at. no. 23, at. wt. 50.942. It occurs as an impurity in fuel oil to the extent of about 0.003 per cent.

Vane control The control of fan output by a system of adjustable vanes at the air inlet.

Vaporising burners Or pot burners; oil burners used in conjunction with domestic central heating boilers. The oil is vaporised by heat, the vapour mixing with the correct amount of air to achieve efficient combustion. There are two types of burner: (a) natural-draught, utilising a chimney of sufficient height to obtain the necessary draught; (b) assisted-draught, in which air is supplied by a fan. Vaporising burners are designed for light grades of fuel oil.

Vaporising oil A fuel used in tractor engines of low compression ratio, i.e. about 4.5 to 1. The boiling range is similar to that of kerosine and it has an *octane number* of about 50. The fuel lacks readily volatile constituents, and it is necessary to start tractor engines with motor gasoline before switching over to vaporising oil. Also known as power kerosine.

Vapour A gas at a temperature below its critical temperature; the latter being that above which the gas cannot be liquefied. Below the critical temperature a vapour can be liquefied by a sufficient increase in pressure, and cooling is unnecessary. Otherwise a gas must be cooled below its critical temperature and sufficiently compressed before it will liquefy.

Vapour lock A term applied to the condition when fuel starvation occurs in an engine, due either to the low fuel *volatility tolerance* of the engine fuel system or to the excess volatility of the fuel itself. Vapour lock manifests itself either by hesitations during acceleration, or by stalling of the engine.

Vapour pressure The pressure exerted by a vapour, either by itself or in a mixture of gases; saturated vapour pressure means the pressure exerted by a vapour in contact with its liquid form. Saturated vapour pressure increases with rise of temperature.

Vapour pressure thermometer A device for measuring temperature. It comprises a bulb partly filled with a volatile liquid such as methyl chloride or toluene, and a pressure measuring element, the two being connected by a capillary tube. A change in temperature causes a change in pressure of the saturated vapour above the liquid. The instrument is suitable for a temperature range $-20°$ C to $+350°$ C. *See pyrometer; thermometer.*

Variable speed electric motor Electric motor the speed of which can be regulated; there are three types in common use, the AC slip-ring motor, the AC commutator motor and the DC motor. They have either shunt regulation or variable voltage control.

Vel A unit of free centimetre falling speed. A particle has a free falling speed of 1 vel if it falls 1 centimetre per second in still air at $20°$ C. *Table V.1* gives the relationship between vels and particle sizes in microns, assuming a particle density of 2 grams per cubic centimetre.

Table V.1

Vel grading	Microns (m × 10⁻⁶) (Density 2 g/c³)
40	89
20	59
10	41
6	31
4	25
2	18
1	13
0.6	10
0.4	8

Velocity pressure   Pressure caused by and related to the velocity of the flow of fluid, as distinct from *static pressure*; a measure of the kinetic energy of the fluid. *See **total pressure***.

Venturi mixing gas burner   A type of *gas burner* which uses the momentum of the gas to entrain the air. A single venturi has a limited turn-down ratio, and it is necessary to have a number of venturis in parallel if a high turn-down ratio is required. To extend the turn-down of venturi burners, a variable jet system has been introduced.

Venturi pneumatic pyrometer   A *pyrometer* based on the principle that if a quantity of hot gas is drawn through a venturi restriction, then cooled and passed through a similar restriction, the ratio of the differential pressures produced is inversely proportional to the ratio of the densities of the gas at the two points. If the gas obeys the *gas laws*, then the ratio is proportional to the ratio of the temperatures of the gas, measured from *absolute zero*. If the temperature of the gas is measured at the 'cold' venturi, the temperature of the hot gas may be calculated:

$$T_{\mathrm{h}} = K \; \frac{\triangle P_{\mathrm{h}}}{\triangle P_{\mathrm{c}}} \, T_{\mathrm{c}}$$

where T_{h} = absolute temperature of gas at 'hot' venturi, K;
   T_{c} = absolute temperature of gas at 'cold' venturi, K;
 $\triangle P_{\mathrm{h}}$ = pressure at 'hot' venturi;
 $\triangle P_{\mathrm{c}}$ = pressure at 'cold' venturi;
   K = a constant.

The instrument may be used for gas temperatures above about 1400° C; it responds to changes in less than 2 seconds.

Venturi scrubber   A device for scrubbing industrial waste gases to remove dust before they pass to atmosphere. In a typical example the gas to be cleaned flows into a venturi nozzle placed just before a point at which water is injected. The velocity of the gas may be raised to 100 m/s or higher, this high gas velocity atomising the water. The dust in the gas, now thoroughly wet, enters a second section, where more water is added; the dust is removed from the gas by cyclonic action. This technique is highly efficient, figures as high as 99.8 per cent being quoted. The pressure drop is relatively high; power consumption is also high.

Venturi tube A tube which narrows to a throat and then gradually increases to the original diameter. It may be used as a device for measuring the flow of a gas or liquid, or with spray injectors as a means of scrubbing gases. *See venturi scrubber.*

Vertical Boiler A steam-raising plant which in its simplest form consists of a cylindrical vertical shell surrounding a fire box (combustion chamber) in the bottom of which is a grate on which the fuel is burnt. The fire box may contain cross-tubes to assist circulation. Some types incorporate one or two banks of 'smoke' tubes which increase the effective heating surface of the boiler. The working pressure is normally up to 10 bars. *See boiler.*

Vertical firing *See down-fired furnace.*

Vibrating screen A screen used for removing the smaller sizes, say < 1 cm, from washed coal before sending it to market. The screen is electrically vibrated at a rate ranging up to 3000 vibrations per minute.

Victorian Brown Coal Development Council A statutory authority created by the Victorian Government, Australia, in 1978 to promote the orderly use of Victoria's huge brown coal resources. The Council replaced the Victorian Brown Coal Research and Development Committee, while assuming wider responsibilities and powers.

Virgin stock Oil derived directly from crude oil; 'straight-run' stock. *See straight-run distillation.*

Viscometer An instrument for measuring the *viscosity* of a liquid. *See Ostwald viscometer; Redwood viscometer; Saybolt system; U-tube viscometer.*

Viscosity The ease with which a liquid will flow. It is determined by recording the time required for a measured volume of liquid at a specified temperature to flow through an orifice of prescribed dimensions. Different methods are employed in various countries, those most commonly used being Redwood No. 1 and No. 2 (British), the results being expressed in 'Redwood seconds'; Engler (Continental European); Saybolt Universal and Saybolt Furol (American). Results are frequently expressed in centistokes. *See dynamic viscosity; kinematic viscosity; viscometer.*

Viscosity breaking The lowering or 'breaking' of the viscosity of residuum by thermal cracking; this technique may be necessary to prepare a *residuum* for fuel oil blending.

Viscosity index An index which compares the change in the viscosity of a fluid obtained over the temperature range 40–100°C with the change obtained over the same temperature range with a Pennsylvanian oil (of high viscosity index, good viscosity temperature characteristic) and a Gulf Coast oil (of low viscosity index, poor viscosity temperature characteristic).

Viscosity index improver A lubricating oil additive which is intended to minimise changes of viscosity with temperature; typical compounds are polymeric hydrocarbons (isobutene) or esters (fumarate, methacrylate).

Viscous oil filter An *air filter* consisting of closely spaced corrugated metal plates which are 'wetted' with a special oil. Impurities in the air adhere to the oil and are removed by the continuous flushing of the plates with fresh oil. The resulting sludge is removed from the oil automatically. Collecting efficiencies of 85–90 per cent are claimed for 5 µm particles.

Vitrinite A substance in coal derived from partly decayed bark and woody tissue completely impregnated by initially liquid decomposition products; a major constituent of *clarain*, in which it occurs in uniformly brilliant black layers.

VLCC Very large crude carrier. See *Oil Tankers.*

VLN steelmaking process *See Bessemer converter.*

V-notch meter A device for measuring the supply of feed water to a boiler. It can only be employed on the suction side of the feed pump. One type consists of a tank divided into three parts by means of a vertical baffle plate and a water-tight partition which carries a gun-metal V-notch plate or weir. The flow through the V-notch is automatically controlled in accordance with the demands of the feed pump by means of an equilibrium valve actuated by lever and float in the catch-box or hotwell end of the tank. The instrument is operated by a float which rises and falls with the level of the water in the V-notch, and may be of the indicating, recording or counting type, or a combination of all three. *See feed-water meter*.

Volatile matter The percentage loss in weight resulting from the heating of 1 gram of coal under test conditions in a crucible from which air is excluded; moisture is excluded from this calculation. Volatile matter consists of a complex mixture of tarry vapours and combustible gases such as hydrogen, methane, ethane, benzene, etc. The volatile matter content of various solid fuels, expressed as a percentage of total weight, is as follows: (a) coke, 1–2; (b) anthracite, 4–8; (c) Welsh steam coal, 8–16; (d) caking coal, 20–

34; (d) free burning coal, 34–40. Practically the whole of a liquid fuel consists of volatile matter. *See volatility.*

Volatility The ease with which a liquid fuel vaporises. Tests on gasoline show the volume of vapour formed per unit volume of liquid at a series of different temperatures and indicate the tendency of a fuel to vapour locking. *See vapour lock; volatility tolerance.*

Volatility tolerance The ability of an engine to cope with excess volatility in a fuel; the volatility tolerance of an engine is interpreted in terms of the ambient temperature at which *vapour lock* would occur on a standard motor fuel. Major factors which determine the volatility tolerance of a vehicle are the temperatures of the fuel system components, the most critical of which is usually the fuel pump, and the vapour-handling capacity of the system.

Volt The unit of potential difference and electromotive force; it is the difference of electric potential between two points of a conducting wire carrying a constant current of 1 *ampere*, when the power dissipated between these points is equal to 1 *watt*. Named after the famour Italian scientist, Alessandro Volta (1745–1827).

Voltage The potential, potential difference or electromotive force of a supply of electricity, measured in volts. *See volt.*

Volumetric SO$_2$ apparatus An instrument for measuring sulphur dioxide concentrations in the general atmosphere; air is bubbled through dilute hydrogen peroxide, the sulphur dioxide being oxidised to sulphuric acid. Measurement is by titration with alkali. The instrument usually incorporates a smoke filter. Readings are usually taken every 24 hours. *See lead peroxide candle; National Survey of Air Pollution, British.*

W

Wagon Tippler A cradle in which a wagon of coal is secured; the cradle and wagon are rotated to empty the coal into a conveyor.

Wall-fired furnace A *water-tube boiler* in which an array of pulverised-fuel burners is situated low on one wall; a typical array may consist of three rows each of six burners. The burners may be short-flame turbulent burners producing parallel coaxial annular swirling jets. The inner annulus feeds primary air and coal, while the outer annulus feeds secondary air. Through the very centre of the burner, tertiary air may be admitted. The outlet from the furnace is at the roof; a projection from the rear wall of widely spaced water-tubes, termed a nose, assists in protecting the superheater tubes by partially chilling and partially deflecting the hot gases. The burners, instead of being all on one wall, are sometimes arranged on opposite walls. The opposing jets produce a greater degree of turbulence in the furnace, compared with the one-wall arrangement. Front-wall-fired furnaces are used with finely ground high-volatile coals. *See Figure W.1.*

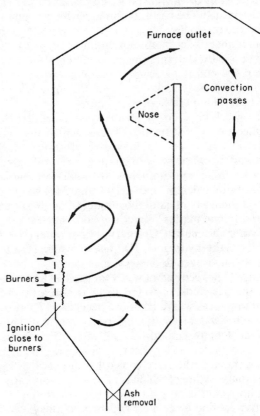

Figure W.1 General shape of chamber and flow pattern in wall-fired water-tube boiler.

Warm front The sharp boundary between two extensive air masses, warm air displacing cold air.

Washability curve A curve obtained by plotting the results of a *float and sink test*, from which the theoretical yield of floats and sinks may be read off.

Washed coal *Coal* that has undergone a wet cleaning process to improve its quality.

Washery A *coal* preparation plant using wet cleaning processes. *See coal cleaning*.

Washout Or rainout; the removal of gases and particles from the atmosphere by their solution in, or attachment to, moisture falling as rain.

Waste-heat boiler A boiler possessing a multi-fire tube system depending mainly or entirely on waste gases as a source of heat.

Waste management The current term describing a comprehensive, integrated and rational systems approach towards the achievement and maintenance of acceptable environmental quality. It involves preparing policies; determining environmental standards; fixing emission rates; enforcing regulations; monitoring air, water and soil quality, and noise emissions; and offering advice to government, industry, land developers and public.

Water The working fluid of steam systems, and a heating or cooling medium. Constituents of water which may give rise to difficulties in boilers are: (a) corrosive substances such as acid solutions and *dissolved gases*, including carbon dioxide, oxygen, hydrogen sulphide and ammonia; (b) scale-forming substances such as salts of calcium and magnesium; (c) foam producing substances such as oil and the products of decomposition of sewage and humic matter. Water is used in nuclear reactors both as a *moderator* and *coolant*. Ordinary or light water (H_2O) contains ordinary or light hydrogen (H); heavy water (D_2O) contains mainly heavy hydrogen or deuterium.

Water cooling The recovery of water for reuse in a process. Besides reducing the basic cost of water, there is also a saving in water treatment processes where these are necessary. There are several ways of cooling water so that it may be recirculated. These include (a) blast coolers resembling a car radiator with mechanical draught; (b) water–water heat exchangers in which tubes containing the hot water are cooled by an external flow of cold water, the temperature of which is raised; (c) evaporative cooling, in which hot water is brought into contact with the atmosphere. Evaporative cooling is the most widely used today.

Water equivalent In relation to a body, the weight of water which would require the same amount of heat to raise the temperature by 1 degree:

water equivalent = weight of body × specific heat of body

Water gas Gas produced by the action of steam on red-hot carbon. The reaction is an *endothermic reaction*:

$$C + H_2O \text{ (steam)} \rightarrow CO + H_2 - 10 \text{ kJ/g/carbon}$$

The calorific value of the gas averages $12MJ/m^3$. It burns with a non-luminous bluish flame; often called 'blue water gas'. The manufacturing process is cyclic. Coke in a large cylindrical generator is blown with air to raise its temperature; this is followed by a 'run' with steam. 'Blow' gases are discharged to atmosphere. A typical analysis, the constituents being expressed in percentages, is as follows: hydrogen, 48; carbon monoxide, 42; carbon dioxide, 4.5; nitrogen, 5.0; methane, 0.5. *See carburetted water gas.*

Water gas shift reaction A reaction in which carbon monoxide and steam are partially converted to carbon dioxide and hydrogen. The reaction is:

$$CO + H_2O \rightarrow CO_2 + H_2$$

which proceeds to equilibrium only and not to completion. This reaction occurs in gasifiers at temperatures above about 700° C, influencing the composition of the final gas. It is also carried out over a catalyst at about 400° C to increase the H_2/CO ratio in gas, e.g. in the manufacture of hydrogen from water gas.

Water gauge A device to indicate the level of water in a boiler. All boilers should have two water gauges fitted at drum level, suitably protected by means of a shield of specially toughened glass. In large boilers, operating at pressures exceeding 60 bar, the water gauges may be replaced by two independent remote-level indicators of the compensated manometric type. Water gauges must be so arranged that the lowest visible part of the gauge is not lower than the lowest safe working level of the water in the boiler. The gauges should be fitted with isolation cocks, together with a drain cock or valve with a discharge pipe. Water gauges should be tested regularly, a set procedure being followed by the stoker at the beginning of each shift. With large water-tube boilers, the water gauges are situated far above the firing floor. Illuminated indicators are often fitted so as to enable the gauges to be read from the firing floor.

Water lancing A process complementary to *soot blowing*, the use of a water jet to remove deposits from boiler surfaces. Often deposits with the greatest resistance to soot blowing are those most readily removed by water lancing.

Water pollution Substances, bacteria or viruses present in such concentrations or numbers as to impair the quality of the water, rendering it less suitable or unsuitable for its intended use and presenting a hazard to man or to his environment. Pollution may be caused by:

(a) Bacteria, viruses and other organisms that can cause disease, e.g. cholera, typhoid fever and dysentery,

(b) Inorganic salts that cannot be removed by any simple conventional treatment process, making the water less suitable for drinking, for irrigation and for many industries.

371

(c) Plant nutrients such as potash, phosphates and nitrates which, while largely inorganic salt, have the added effect of increasing weed growth, promoting algal blooms and producing, by photosynthesis, organic matter which may settle to the bottom of a lake.

(d) Oily materials that may be inimical to fish life, cause unsightliness, screen the river surface from the air, thus reducing re-oxygenation, accumulate in troublesome quantities, or have a high oxygen demand.

(e) Specific toxic agents, ranging from metal salts to complex synthetic chemicals.

(f) Waste heat that may render the river less suitable for certain purposes.

(g) Silt that may enter a river in large quantities, causing changes in the character of the river bed.

(h) Radioactive substances.

Water smoking Or steaming, the release of water vapour during the period when ceramic ware is being dried by passing large volumes of air, at temperatures up to 250° C through the setting.

Water treatment Measures to condition water so that it is suitable for use in boilers; the main aims of water treatment are: (a) to 'soften' the water, i.e. to remove or neutralise the scale-forming salts; (b) to achieve the correct alkaline condition in the boiler water; (c) the removal of excess oxygen and carbon dioxide from the water. By adopting the correct methods to achieve these aims, scale and deposits, corrosion and embrittlement can be largely avoided, and the steam supply protected against impurities. *See carboxylic resin; caustic embrittlement; chelating agents; hydrazine; ion exchange process; lime-soda process; sodium sulphite.*

Water-tube boiler A steam generator consisting of a large number of closely spaced water-tubes connected to one or more drums which act as water pockets and steam separators. Draught fans, firing equipment, superheaters, economisers and air preheaters complete the installation. The decade 1960–70 saw the introduction of 500 MW units which are now in general use in North America and the United Kingdom, and in Europe generally. Water-tube boilers are universally used in power stations, most of them being fired by pulverised coal; the remainder are oil- and natural-gas-fired, with a few using other fuels. Thermal efficiencies of 90 per cent are attained. *See boiler; La Mont boiler; supercritical once-through boiler; wall-fired furnace.*

Watt The smaller unit of electric power; it is the power which gives rise to the production of energy at the rate of 1 *joule* per second. With *direct current*, volts × amperes = watts. One kilo-watt = 1000 watts. Named after the famous Scottish engineer James Watt. *See kilowatt.*

Watt Committee on Energy Formed in the United Kingdom in 1976 on the initiative of The Institution of Mechanical Engineers, a body which provides a cohesive national forum on energy matters. The Committee brings together representatives from nine institutions embracing the disciplines of engineering, science, building, accountancy and economics.

Wave energy The utlisation of the energy of waves, as distinct from tides, for the generation of electricity. The energy of waves is very diffuse, and the harnessing of this energy requires large and robust structures to achieve even low conversion rates. The cost of generating electricity appears to be between 10 and 20 times the cost of generation by conventional methods. Most proposals for using wave energy comprise a system of floats allowing the movement of waves to compress or elevate a fluid which drives turbo-generators. In the United kingdom the availability of wave energy is high and coincides with seasonal demands for electricity; the subject has become one of interest, although friction losses means that the theoretical efficiency of most systems will be low.

Weakly caking coal *Coal* which does not become sufficiently plastic during carbonisation to form a mechanically strong coke.

Weather The condition of the atmosphere at a certain time or over a certain short period, as described by various meteorological phenomena such as atmospheric pressure, temperature, humidity, rainfall, cloudiness, and wind speed and direction.

Weathered coal *Coal* the character of which has changed owing to the combined effects of chemical and physical action on exposure to weathering processes.

Weathered crude Crude petroleum which, during storage and handling, has lost an appreciable quantity of the more volatile components owing to natural causes.

Weber A unit of magnetic flux; it is the magnetic flux which, linking a circuit of one turn, produces in it an electromotive force of 1 *volt* as it is reduced to zero at a uniform rate in 1 second.

Welsh dry steam coal A high-*rank* coal containing between 9 and 19.5 per cent *volatile matter*; it is therefore somewhat more reactive than *anthracite*. As the name implies, it is a product of the South Wales coalfields.

Wet-back economic boiler *See economic boiler.*

Wet-bottom furnace A furnace in which the ash is deposited and leaves the furnace in a molten condition. From 30 to 50 per cent of the ash leaves through the ash hopper. Also known as a slag-tap furnace. *See dry-bottom furnace.*

Wet gas A petroleum gas containing a relatively high proportion of hydrocarbons recoverable as liquids.

Wet gas holder An inverted metal dome floating in a water tank with inlet and outlet gas connectors; the pressure on the gas is controlled by a system of balancing weights. *See piston-type gas holder.*

Wet gas meter A gas meter consisting of a casing in which a hollow measuring drum rotates in water. The drum consists of several gas compartments of equal volume; gas enters one compartment, causing the drum to revolve, this gas being discharged while the next compartment is filling. The drum is connected to dials indicating accurately the volume of gas passed through. *See dry gas meter.*

Wet saturated steam Steam containing entrained water droplets. *See dryness fraction; dry saturated steam.*

Wet washer Or scrubber, a device for removing particulate matter from gas streams. Wet washers or scrubbers fall into two main categories: (a) water film types, in which the gases are brought into contact with water-covered surfaces so that the dust particles are retained and discharged in the water; (b) water spray types, in which the gases are sprayed with water in a finely divided state. With high-energy scrubbers having pressure drops in excess of 500 mm water gauge, collecting efficiencies of up to 98 per cent can be achieved, but power consumption is high. *See dust arrester; venturi scrubber.*

Whale oil substitute *See jojoba bean.*

Whipstock A long cylindrical steel billet with a tapering face cut at the desired angle which is inserted into a borehole when it is necessary to deflect the drilling bit and change the direction of the borehole.

White cast-iron Hard *cast-iron* which presents a silvery lustre at a fractured surface; carbon is present as carbide 'cementite', Fe_3C. Being both hard and brittle, it has poor machining properties.

White oils Oils which are substantially colourless and without bloom, being made from light lubricating oils by a drastic process of refining. They are used for medicinal purposes and in the manufacture of toilet preparations.

White oil ships Oil tankers which carry light-coloured petroleum products up to gas oil.

White products A term applied to the more volatile petroleum products such as gasoline, petroleum spirits, white spirit and kerosine.

White spirit A refined distillate intermediate in distillation range between gasoline and kerosine, i.e. with a distillation range of about 150–200° C. It is used as a paint thinner and for dry cleaning. The terms 'mineral turpentine' and 'turpentine substitute' are sometimes used for white spirit. In the United States of America the term 'petroleum spirits' is used for white spirit.

Wick-char burning test A test for the carbon formation of kerosine. *See IP burning test.*

Wide-cut gasoline Light hydrocarbons intended for use in aviation gas-turbine power units.

Wide range pressure jet burner An *oil burner* with a *turndown ratio* of up to 10:1. *See spill-type burner.*

w.i.f. boiler A sectional header *water-tube boiler*, it comprises straight solid drawn steel tubes inclined over the furnace at an angle of 15° to the horizontal, connecting the uptake header with the downtake header, each header being connected to the same water and steam drum. The wrought iron front (w.i.f.) today consists of steel. Designed and constructed by Babcock and Wilcox Ltd, the type has tended to be superseded by more modern bi-drum designs by the same makers. *See boiler.*

Wigner effect The phenomenon in which an atom is displaced from its normal lattice position by neutron bombardment; the atom displaced may then come to rest in an interstitial position. The effect occurs in a graphite moderator and causes a change in shape of the graphite blocks. Moreoever, if the atoms return subsequently to their normal positions, the energy stored is given off in the form of heat.

Wigner energy The energy stored by the atoms of a moderator which have been displaced from their normal lattice positions by the *Wigner effect*. If the *moderator* normally operates at a sufficiently high temperature, the thermal agitation tends to cause the displaced atoms to return to their normal positions continuously; this process is known as 'annealing'. With nuclear reactors which operate at lower temperatures, however, it is customary to carry out an annealing or Wigner energy release procedure periodically by raising the temperature above that used in normal operation.

Wild cat A well which has been drilled without a complete geological exploration of the locality.

Wild gasoline A light petroleum spirit containing appreciable quantities of material which are normally gaseous at atmospheric temperatures and pressures, i.e. an unstabilised gasoline.

Wind energy Solar-derived energy; wind energy is proportional to wind speed, which in turn varies with geographic location, height above ground, time of day, season of year, topography, turbulence and diurnal heating. The relationships are complex; for a given mean wind speed over a period of time, there will be more energy in a more variable wind. Sailing and water-pumping and corn-grinding systems are early examples of man's use of wind energy. In 1941 a windmill generating some 1.25 kW of electricity was built in Vermont, USA. During the 1950s and 1960s, wind machines of up to 100 kW were built in the United Kingdom, Germany, France, Denmark and the Soviet Union. In 1976 the construction of a 100 kW experimental wind turbine was completed in Ohio by the NASA-Lewis Research Centre. Two upgraded versions of 200 kW have now been completed. Wind energy conversion has become part of the US Department of Energy development programme; a number of wind energy projects are now supported by federal funds. Savonius–Darrieus windmill combinations are favoured for electricity generation because of the high speeds attainable.

Wind rose A diagram indicating the frequency and strength of winds in a definite locality for a given period of years. It is conventional to consider the wind direction as the direction from which the wind blows, e.g. a north-east wind will carry pollutants to the southwest of the source.

Windscale chemical processing plant A UK Atomic Energy Authority plant located at Windscale, Cumbria, England, for the processing of irradiation fuel elements, i.e. burned-up nuclear reactor fuel elements. The elements are stored to allow the radioactive iodine to decay. After solution in acid, the irradiated uranium passes through a complex chemical separation process which first separates the radioactive fission products. After the removal of the bulk of the fission products, the *plutonium* is separated from the *uranium*. *See uranium enrichment.*

Windscale Inquiry An inquiry held in 1977 into a proposal to expand substantially facilities at Windscale for reprocessing irradiated oxide fuels from both British and non-British nuclear power stations; the inquiry lasted 100 days. The outcome was a recommendation that the expansion could safely proceed, subject to a range of safeguards. The inquiry was conducted by Sir Roger Parker. The British Government accepted the recommendation; immediately after the decision, a major contract with Japan was signed.

Windscale nuclear reactor incident A major accident which occurred at the Windscale establishment of the UK Atomic Energy Authority on 10 October, 1957; a fire occurred inside the No. 1 reactor and volatile fission products were released to the atmosphere. The accident occurred during a routine Wigner release, i.e. the heating of graphite above its normal operating temperature to get ride of *Wigner energy*. Local overheating of the uranium fuel elements occurred, canning failed and exposed uranium oxidised. A filter in the ventilation stack arrested most of the strontium and caesium, but radioactive fission products escaped to atmosphere, including iodine-131 and tellurium-132. *See Figure W.2; **Harrisburg nuclear power plant incident***.

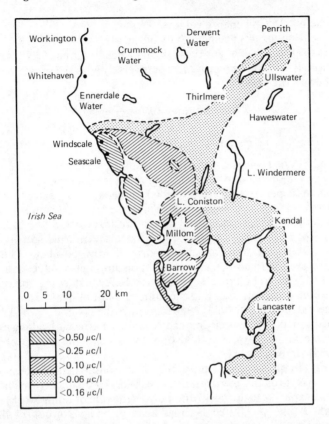

Figure W.2 Map of the Windscale area showing contours of radio-iodine contamination in milk on 13th October, 1957 (Source: Medical Research Council, *The Hazards to Man of Nuclear and Allied Radiation*, Cmnd. 1225, London: HMSO, 1960)

Wind shear The change in speed and direction of wind, usually with height.

Winkler system The gasification of coal utilizing a fluidised bed in a brick-lined chamber and an air–steam blast. Finely divided fuel is fed into the grate by a screw; ash is removed from the grate by a plough. This system has been widely used in Germany on lignites.

Wobbe number For any gas passing through a given orifice:

$$\text{Wobbe number} = \frac{\text{e.v.}}{\sqrt{\text{sp. gr.}}}$$

where   e.v. = energy value of the gas;
   sp. gr. = specific gravity of the gas.
As the discharge of gas through an orifice in terms of kJ/h varies directly with the *energy value*, but inversely with the square root of the specific gravity:

$$\text{kJ rate} = K \frac{\text{e.v.}}{\sqrt{\text{sp. gr.}}}$$

where K is a constant dependent on the orifice.

Wood A *fuel* consisting mainly of cellulose, a substance containing nearly 45 per cent of oxygen. The moisture content varies, being about 25 per cent for freshly felled timber and 15 per cent for air-dried timber. The *energy value* varies with the moisture content and the different amounts of oils and resins which different kinds of wood contain. A typical value for an oven-dried softwood is about 20 kJ/g, and for an oven-dried hardwood about 18 kJ/g. However, with normal amounts of moisture energy values of about 14 kJ/g are to be expected. Wood is a bulky fuel, being about half as dense as coal; to obtain the same amount of heat as any given weight of coal will yield, some four times the volume of wood must be burned. Wood is clean, readily ignites, burns with a long clean flame under correct combustion conditions, and leaves only a small amount of ash.

Work In physics, descriptive only of an event when a force moves a mass through a vertical distance. Work is distinct from energy (the capacity to do work) and power (the rate of doing work).

World Bank A United Nations agency, being a group of three institutions: the International Bank for Reconstruction and Development, the International Development Association and the International Finance Corporation. The common objective of these institutions is to help raise the standards of living in developing countries by channelling financial resources from developed countries. The Bank was established in 1945. During 1976 the Bank made lending and investment commitments totalling $6877 million. In recent years about 10 per cent of Bank assistance has been in relation to energy projects. For example, in 1979 the International Development Association announced a $US 175 million credit for a rural electrification project in India. This project was expected to benefit some 2.5 million households in 15 000 villages; in addition, about 500 000 pumps were electrified. The credit was for 50 years, with 10 years' grace on repayment; an interest rate of less than 1 per cent per annum was imposed to cover administrative expenses.

World Energy Conference   Founded in 1924, an association of the representatives of some 70 nations, widely differing in their degree of industrialisation yet concerned with all forms of commercial energy, primary and secondary, from production through utilisation. The Conference is concerned with the best and most efficient use of energy; it advocates using energy in such a way as to protect the environment while enabling people to enjoy the maximum benefit so made available to them. The Conference is held annually at important centres around the world; the actual work is undertaken by national committees. In its 1974 survey of coal resources the Conference adopted the twofold classification 'known reserves' and 'additional reserves'.

Table W.1   ESTIMATED WORLD CONSUMPTION, RECOVERABLE RESERVES AND RESOURCES OF COAL, OIL, NATURAL GAS, ETC. (1975)

	Consumption in 1975 (energy units: 10^{18} joules)	Reserves		Ultimately recoverable resources (physical units)
		(Energy units: 10^{18} joules)	(Physical units)	
Coal (black and brown)	73	15 080	665×10^9 tonnes	$5400-7300 \times 10^9$ tonnes
Oil	112	4 360	110×10^9 m^3	$210-300 \times 10^9$ m^3
Natural gas	45	2 500	65×10^{12} m^3	$80-170 \times 10^{12}$ m^3
Oil shale	..	..	80×10^9 m^3	$180-255 \times 10^9$ m^3
Bitumen rocks	..	..	56×10^9 m^3	$160-400 \times 10^9$ m^3
Uranium (thermal reactors)	3	1 130[a]	2700×10^3 tonnes	$3800-5000 \times 10^3$ tonnes
Thorium	..	..	320×10^3 tonnes	$2000-2800 \times 10^3$ tonnes

Sources: 1975 consumption—information provided by IAEA; reserves and resources—U.S. ERDA, *Creating Energy Choices for the Future* (1976).
[a] Using conversion factor referred to in the text. Note that 1 tonne of uranium is equivalent to 1.30 short tons of U_3O_8.

World energy resources   Resources encompassing coal (black and brown), oil, natural gas, oil shale, bitumen rocks, uranium and thorium considered as recoverable reserves and resources. *Table W.1* indicates the size of these reserves and resources.

World Environment Day   The fifth of June each year, as adopted by the United Nations Conference on the Human Environment 1972 to mark the beginning of the first conference and as a means of focusing attention on national and world environmental problems.

World Health Organisation A United Nations organisation established in 1948 which has as its basic objective 'the attainment by all peoples of the highest possible levels of health'. Its work is guided by the World Health Assembly, composed of the representatives of all member states. The organisation has taken a strong interest in the nature and significance of air pollution, and the long-term objectives that comprise a satisfactory air quality. It is directly concerned also with nuclear radiation hazards.

World parity pricing *See import parity pricing.*

X

Xenon Xe. An inert gas; at. no. 54, at. wt. 131.30; produced as a fission product in nuclear reactors. Xenon-135 has a very large neutron-capture cross-section and is one of the most important poisons in a reactor.

X-ray diffraction An accurate method of determining the details of the internal atomic structure of a substance; the planes of the atoms of crystals act as a diffraction grating to *X-rays* which are scattered by them.

X-rays Electromagnetic radiation, with a wavelength of less than 100 Å; produced by bombarding a metal target with fast electrons in an evacuated X-ray tube.

X-unit The unit of wavelength for electromagnetic waves; equal to approximately 10^{-3} Ångström, or 10^{-11} centimetre. *See spectroscopic units.*

Xylene A colourless liquid, $C_6H_4(CH_3)_2$, of the aromatic group of hydrocarbons, made by the catalytic reforming of certain naphthenic petroleum fractions. It is used as a high-octane motor and aviation spirit blender and as a solvent. Its isomers are metaxylene, orthoxylene and paraxylene.

Y

Yellow cake A concentrate resulting from the leaching of uranium from its ore; this concentrate contains about 75–85 per cent uranium oxide. In a natural uranium cycle this concentrate is purified and transferred to a fuel element fabrication plant. In an enriched uranium cycle it is converted to uranium hexafluoride and fed to a gaseous diffusion or centrifuge enrichment plant.

Yorkshire boiler An early type of shell boiler, similar in design to the *Lancashire boiler* save that the two internal furnace tubes were tilted upwards from front to rear, and the diameter of the tubes increased gradually from front to back. The purpose of this arrangement was to obtain uniform heat transmission.

Z

Zero energy growth A proposed objective in respect of the energy consumption of affluent societies, aimed at avoiding shortages and undue pressure on supplies through the gradual tapering off of the growth of demand for energy; a concurrent objective is the more efficient use of energy so that more goods and services are provided for a given input of energy. Proponents believe that while this objective should be adopted by all affluent countries, demand in developing countries should be allowed to grow with improving living standards.

Zero energy reactor An experimental *nuclear reactor* operating at very low neutron flux and power level, so that no forced cooling is required, and the fission product activity is so small that the fuel elements can be handled after use.

'Zig-zag' filter An *air filter* in which the filtering medium of paper, fabric, glass fibre or plastics, is formed into a zig-zag shape over a wire framework. This type of filter incorporates a large filtration area in a comparatively small space. Collecting efficiencies claimed depend on the filter medium, and vary from 45 to 95 per cent in the 0.1 to 5 micron particle size range.

Zirconium Zr. A metallic *element* at. no. 40, at. wt. 91.22; with a low neutron-capture cross-section. It has been considered as a possible canning material for fuel elements in nuclear reactors.

Zwentendorf nuclear power station Austria's first nuclear power station; on 5 November 1978, in a national referendum, Austrians voted against opening the station, which had cost $250 million to construct.

APPENDIX I: CONVERSION FACTORS

	Multiply number of	by	to obtain equivalent number of	Multiply number of	by	to obtain equivalent number of
Length	Inches (in)	25.4	millimetres (mm)	Millimetres	0.039 37	inches
		2.54	centimetres (cm)	Centimetres	0.393 7	inches
	Feet (ft)	30.48	centimetres (cm)		39.370 1	inches
		0.304 8	metres (m)	Metres	3.280 8	feet
	Yards (yd)	0.914 4	metres		1.093 6	yards
	Fathoms (6ft)	1.828 8			0.546 81	fathoms
	Miles (land: 5 280 ft)	1.609 344	kilometres (km)	Kilometres	0.621 37	miles (land)
	Miles (UK sea: 6 080 ft)	1.853 184			0.539 61	miles (UK sea)
	Miles, international nautical	1.852	kilometres		0.539 96	miles, international nautical
Area	Sq. inches (in²)	645.16	sq. millimetres (mm²)	Sq. millimetres	0.001 55	sq. inches
		6.451 6	sq. centimetres (cm²)	Sq. centimetres	0.155 0	sq. inches
	Sq. feet (ft²)	929.030 4	sq. centimetres (cm²)	Sq. metres	10.763 9	sq. feet
		0.092 903	sq. metres (m²)		1.195 99	sq. yards
	Sq. yards (yd²)	0.836 127	sq. metres	Hectares	2.471 05	acres
	Acres	4 046.86	sq. metres	Sq. kilometres	247.105	acres
		0.404 686	hectares (ha)		0.386 1	sq. miles
		0.004 047	sq. kilometres (km²)			
	Sq. miles	2.589 99	sq. kilometres			
Volume and capacity	Cu. inches (in³)	16.387 064	cu. centimetres (cm³)	Cu. centimetres	0.061 02	cu. inches
	UK pints	34.677 4	cu. inches	Litres	61.024	cu. inches
		0.568 3	litres (l)		0.035 3	cu. feet
	UK gallons	4.546 09	litres		0.264 2	US gallons
	US gallons	3.785	litres		0.219 97	UK gallons
	Cu. feet (ft³)	28.317	litres	Hectolitres	26.417	US gallons
	Cu. feet	0.028 317	cu. metres (m³)		21.997	UK gallons
	UK bushels	0.363 7	hectolitres (hl)		2.838	US bushels
	US bushels	0.352 4	hectolitres		2.750	UK bushels
	UK gallons	1.200 95	US gallons	Cu. metres	35.314 7	cu. feet
	US gallons	0.832 674	UK gallons		1.307 95	cu. yards
	UK bulk barrels	36	UK gallons		264.172	US gallons
		43.234 2	US gallons		219.969	UK gallons
		0.163 7	cu. metres		6.110 26	UK bulk barrels

(weight)				Grams	0.035 27	ounces, avoirdupois
	Ounces, avoirdupois (oz)	28.349 5	grams (g)		0.032 15	ounces, troy
	Ounces, troy (oz tr)	31.103 5	ounces, troy	Kilograms	35.274	ounces, avoirdupois
	Ounces, avoirdupois	0.9115	grams		32.151	ounces, troy
	Pounds avoirdupois(lb)	453.592 37	kilograms (kg)		2.204 62	pounds, avoirdupois
		0.453 59		Metric quintals	220.462	pounds, avoirdupois
	Hundredweights (cwt) (112 lb)	0.05	metric quintals (q)	Tonnes	2 204.62	short tons
	Short tons 2 000 lb	0.508 023	long tons		1.102 31	long tons
		0.892 857	tonnes (t)		0.984 207	
	Long tons (2.240 lb)	0.907 185	short tons			
		1.12				
		1.016 05	tonnes			
Velocity	Miles/hour	1.609 344	kilometres/hour	Kilometres/hour	0.621 37	miles/hour
		0.868 976	international knots		0.539 96	international knots
		1.466 7	feet/second			
Fuel consumption	Miles/UK gallon	0.354 01	kilometres/litre	Kilometres/litre	2.824 81	miles /UK gallon
	Miles/US gallon	0.425 14	kilometres/litre	Litres/100 kilometres	2.352 15	miles/US gallon
	UK gallons/mile	282.481	litres/100 kilometres		0.003 54	UK gallons/mile
	US gallons/mile	235.215			0.004 25	US gallons/mile
Energy units	British thermal units (Btu)	1.055 06	kilojoules (kj)	Watt-hour (Wh)	3.412 14	British thermal units (Btu)
		0.293 071	watt-hours (Wh)		3.6	kilojoules (kj)
		0.251 996	Kilocalories ($kcal_{IT}$)		0.859 845	kilocalories ($kcal_{IT}$)
	Kilojoule (kj)	0.947 817	British thermal units (Btu)	Kilocalorie ($kcal_{IT}$)	3.968 32	British thermal units (Btu)
		0.277 778	Watt-hours (Wh)		4.185 5	kilojoules (kj)
		0.238 846	kilocalories ($kcal_{IT}$)		1.162 64	watt-hours (Wh)
Heat energy content	Btu/lb	0.002 326	kJ/g	US gallon = 231 cubic inches 0.003 8 cubic metre 3.785 litres 0.832 7 UK gallons 0.023 8 barrel US barrel = 42 US gallons 0.159 0 cubic metre 158.99 litres 34.97 UK gallons		
	kJ/g	429.923	Btu/lb			
	Btu/ft³	0.037 258 9	MJ/m³			
	MJ/m³	26.839 2	Btu/ft³			
Other energy equivalents	UK gallon = 277.42 cubic inches 0.004 5 cubic metre 4.546 litres 1.201 US gallons 0.028 6 barrel					

APPENDIX 2: TEMPERATURE CONVERSION TABLE

To convert any temperature in Centigrade or Fahrenheit degrees to the other, select the figure to be converted in the centre column, and if converting from Centigrade to Fahrenheit, read off to the right-hand column; if converting from Fahrenheit to Centigrade, read off to the left-hand column.

Examples: $1400°$ C $= 2552°$ F
$390°$ F $= 199°$ C

C		F	C		F	C		F
−17.8	0	32	5.56	42	107.6	28.9	84	183.2
−17.2	1	33.8	6.11	43	109.4	29.4	85	185.0
−16.7	2	35.6	6.67	44	111.2	30.0	86	186.8
−16.1	3	37.4	7.22	45	113.0	30.6	87	188.6
−15.6	4	39.2	7.78	46	114.8	31.1	88	190.4
−15.0	5	41.0	8.33	47	116.6	31.7	89	192.2
−14.4	6	42.8	8.89	48	118.4	32.2	90	194.0
−13.9	7	44.6	9.44	49	120.2	32.8	91	195.8
−13.3	8	46.4	10.0	50	122.0	33.3	92	197.6
−12.8	9	48.2	10.6	51	123.8	33.9	93	199.4
−12.2	10	50.0	11.1	52	125.6	34.4	94	201.2
−11.7	11	51.8	11.7	53	127.4	35.0	95	203.0
−11.1	12	53.6	12.2	54	129.2	35.6	96	204.8
−10.6	13	55.4	12.8	55	131.0	36.1	97	206.6
−10.0	14	57.2	13.3	56	132.8	36.7	98	208.4
−9.44	15	59.0	13.9	57	134.6	37.2	99	210.2
−8.89	16	60.8	14.4	58	136.4	38	100	212
−8.33	17	62.6	15.0	59	138.2	43	110	230
−7.78	18	64.4	15.6	60	140.0	49	120	248
−7.22	19	66.2	16.1	61	141.8	54	130	266
−6.67	20	68.0	16.7	62	143.6	60	140	284
−6.11	21	69.8	17.2	63	145.4	66	150	302
−5.56	22	71.6	17.8	64	147.2	71	160	320
−5.00	23	73.4	18.3	65	149.0	77	170	338
−4.44	24	75.2	18.9	66	150.8	82	180	356
−3.89	25	77.0	19.4	67	152.6	88	190	374
−3.33	26	78.8	20.0	68	154.4	93	200	392
−2.78	27	80.6	20.6	69	156.2	99	210	410
−2.22	28	82.4	21.1	70	158.0	100	212	414
−1.67	29	84.2	21.7	71	159.8	104	220	428
−1.11	30	86.0	22.2	72	161.6	110	230	446
−0.56	31	87.8	22.8	73	163.4	116	240	464
0	32	89.6	23.3	74	165.2	121	250	482
0.56	33	91.4	23.9	75	167.0	127	260	500
1.11	34	93.2	24.4	76	168.8	132	270	518
1.67	35	95.0	25.0	77	170.6	138	280	536
2.22	36	96.8	25.6	78	172.4	143	290	554
2.78	37	98.6	26.1	79	174.2	149	300	572
3.33	38	100.4	26.7	80	176.0	154	310	590
3.89	39	102.2	27.2	81	177.8	160	320	608
4.44	40	104.0	27.8	82	179.6	166	330	626
5.00	41	105.8	28.3	83	181.4	171	340	644

C	F		C	F		C	F	
177	350	662	466	870	1598	754	1390	2534
182	360	680	471	880	1616	760	1400	2552
188	370	698	477	890	1634	766	1410	2570
193	380	716	482	900	1652	771	1420	2588
199	390	734	488	910	1670	777	1430	2606
204	400	752	493	920	1688	782	1440	2624
210	410	770	499	930	1706	788	1450	2642
216	420	788	504	940	1724	793	1460	2660
221	430	806	510	950	1742	799	1470	2678
227	440	824	516	960	1760	804	1480	2696
232	450	842	521	970	1778	810	1490	2714
238	460	860	527	980	1796	816	1500	2732
243	470	878	532	990	1814	821	1510	2750
249	480	896	538	1000	1832	827	1520	2768
254	490	914	543	1010	1850	832	1530	2786
260	500	932	549	1020	1868	838	1540	2804
266	510	950	554	1030	1886	843	1550	2822
271	520	968	560	1040	1904	849	1560	2840
277	530	986	566	1050	1922	854	1570	2858
282	540	1004	571	1060	1940	860	1580	2876
288	550	1022	577	1070	1958	866	1590	2894
293	560	1040	582	1080	1976	871	1600	2912
299	570	1058	588	1090	1994	877	1610	2930
304	580	1076	593	1100	2012	882	1620	2948
310	590	1094	599	1110	2030	888	1630	2966
316	600	1112	604	1120	2048	893	1640	2984
321	610	1130	610	1130	2066	899	1650	3002
327	620	1148	616	1140	2084	904	1660	3020
332	630	1166	621	1150	2102	910	1670	3038
338	640	1184	627	1160	2120	916	1680	3056
343	650	1202	632	1170	2138	921	1690	3074
349	660	1220	638	1180	2156	927	1700	3092
354	670	1238	643	1190	2174	932	1710	3110
360	680	1256	649	1200	2192	938	1720	3128
366	690	1274	654	1210	2210	943	1730	3146
371	700	1292	660	1220	2228	949	1740	3164
377	710	1310	666	1230	2246	954	1750	3182
382	720	1328	671	1240	2264	960	1760	3200
388	730	1346	677	1250	2282	966	1770	3218
393	740	1364	682	1260	2300	971	1780	3236
399	750	1382	688	1270	2318	977	1790	3254
404	760	1400	693	1280	2336	982	1800	3272
410	770	1418	699	1290	2354	988	1810	3290
416	780	1436	704	1300	2372	993	1820	3308
421	790	1454	710	1310	2390	999	1830	3326
427	800	1472	716	1320	2408	1004	1840	3344
432	810	1490	721	1330	2426	1010	1850	3362
438	820	1508	727	1340	2444	1016	1860	3380
443	830	1526	732	1350	2462	1021	1870	3398
449	840	1544	738	1360	2480	1027	1880	3416
454	850	1562	743	1370	2498	1032	1890	3434
460	860	1580	749	1380	2516	1038	1900	3452

C		F	C		F	C		F
1043	**1910**	3470	1254	**2290**	4154	1466	**2670**	4838
1049	**1920**	3488	1260	**2300**	4172	1471	**2680**	4856
1054	**1930**	3506	1266	**2310**	4190	1477	**2690**	4874
1060	**1940**	3524	1271	**2320**	4208	1482	**2700**	4892
1066	**1950**	3542	1277	**2330**	4226	1488	**2710**	4910
1071	**1960**	3560	1282	**2340**	4244	1493	**2720**	4928
1077	**1970**	3578	1288	**2350**	4262	1499	**2730**	4946
1082	**1980**	3596	1293	**2360**	4280	1504	**2740**	4964
1088	**1990**	3614	1299	**2370**	4298	1510	**2750**	4982
1093	**2000**	3632	1304	**2380**	4316	1516	**2760**	5000
1099	**2010**	3650	1310	**2390**	4334	1521	**2770**	5018
1104	**2020**	3668	1316	**2400**	4352	1527	**2780**	5036
1110	**2030**	3686	1321	**2410**	4370	1532	**2790**	5054
1116	**2040**	3704	1327	**2420**	4388	1538	**2800**	5072
1121	**2050**	3722	1332	**2430**	4406	1543	**2810**	5090
1127	**2060**	3740	1338	**2440**	4424	1549	**2820**	5108
1132	**2070**	3758	1343	**2450**	4442	1554	**2830**	5126
1138	**2080**	3776	1349	**2460**	4460	1560	**2840**	5144
1143	**2090**	3794	1354	**2470**	4478	1566	**2850**	5162
1149	**2100**	3812	1360	**2480**	4496	1571	**2860**	5180
1154	**2110**	3830	1366	**2490**	4514	1577	**2870**	5198
1160	**2120**	3848	1371	**2500**	4532	1582	**2880**	5216
1166	**2130**	3866	1377	**2510**	4550	1588	**2890**	5234
1171	**2140**	3884	1382	**2520**	4568	1593	**2900**	5252
1177	**2150**	3902	1388	**2530**	4586	1599	**2910**	5270
1182	**2160**	3920	1393	**2540**	4604	1604	**2920**	5288
1188	**2170**	3938	1399	**2550**	4622	1610	**2930**	5306
1193	**2180**	3956	1404	**2560**	4640	1616	**2940**	5324
1199	**2190**	3974	1410	**2570**	4658	1621	**2950**	5342
1204	**2200**	3992	1416	**2580**	4667	1627	**2960**	5360
1210	**2210**	4010	1421	**2590**	4694	1632	**2970**	5378
1216	**2220**	4028	1427	**2600**	4712	1638	**2980**	5396
1221	**2230**	4046	1432	**2610**	4730	1643	**2990**	5414
1227	**2240**	4064	1438	**2620**	4748	1649	**3000**	5432
1232	**2250**	4082	1443	**2630**	4766	1705	**3100**	5619
1238	**2260**	4100	1449	**2640**	4784	1760	**3200**	5722
1243	**2270**	4118	1454	**2650**	4802	1816	**3300**	5972
1249	**2280**	4136	1460	**2660**	4820	1871	**3400**	6152

APPENDIX 3: SHORT BIBLIOGRAPHY

Adams, P. J. *The Origin and Evolution of Coal*. London: HMSO, 1960.

American Chemical Society. *Fuel and Energy from Renewable Resources*. Proceedings of a Symposium held in Chicago, August, 1977. New York: Academic Press, 1978.

Anderson, Larry L. and Tillman, David A. (Eds). *Fuels from Waste*. New York: Academic Press, 1977.

Armstead, H.C.H. *Geothermal Energy*. New York: Spon, 1979.

Association of the Coal Producers of the European Community. *Energy in Europe—The Importance of Coal*. London: National Coal Board, 1974.

Australian Institute of Energy. *Energy Resource and Technology*. A monthly publication. Sydney, Australia.

Babcock and Wilcox. *Steam: Its Generation and Use*. New York.

Barden, R. G. (Ed.). *Sound Pollution*. University of Queensland Press, 1976.

Berkowitz, David A. and Squira, Arthur M. (Eds). *Power Generation and Environmental Change*. Cambridge, Mass. Massachusetts Institute of Technology, 1971.

Bradbury, Kathleen. *Solid Fuel in the Home: A Study in Domestic Heat Services*. London: Women's Solid Fuel Council, 1973.

Bradley, J. H. *Flame and Combustion Phenomena*. London: Methuen, 1969.

Brame, J. S. S. and King, J. G. *Fuel: Solid, Liquid and Gaseous*. London: Edward Arnold, 1967.

British Petroleum Company. *Our Industry Petroleum*. London: BP Educational Service, 1979.

Central Electricity Generating Board. *Annual Reports*. London.

Central Electricity Generating Board. *Statistical Yearbook 1980 – 81*. London.

Chapman, P. *Fuel's Paradise: Energy Options for Britain*. Harmondsworth, Middlesex: Penguin, 1975.

Cheremisinoff, Paul N. and Regino, Thomas C. *Principles and Applications of Solar Energy*. New York: Wiley, 1979.

Coal Research Establishment, UK. *Annual Reports*. Stoke Orchard, Cheltenham.

Commoner, B. *The Poverty of Power: Energy and the Economic Crisis*. New York: Cape, 1976.

Commonwealth Scientific and Industrial Research Organisation. *Proceedings of Conference on Electrostatic Precipitation*. North Ryde, New South Wales, Australia, 1979.

Corbett, A. H. *Energy for Australia: Resources, Technology and the Environment.* Harmondsworth: Penguin, 1976.

Council on Environmental Quality, US. *Ninth Annual Report.* Washington, DC, 1978. Also: *Solar Energy: Progress and Promise,* 1978; *Analysis of Oil Spill Trends,* 1976; *Coal Surface Mining and Reclamation: An Environmental and Economic Assessment of Alternatives,* 1973.

Cowan, Edward. *Oil and Water: The Torrey Canyon Disaster.* Philadelphia: Lippincott, 1968.

Crenson, Mattew A. *The Un-Politics of Air Pollution: A Study of Non-Decisionmaking in the Cities.* Baltimore: Johns Hopkins, 1971.

Department of Energy, UK. *Annual Reports.* London: HMSO.

Department of Energy, UK. *Coal Industry Examination.* Interim Report, June 1974; Final Report, October 1974.

Department of the Environment UK. *Clean Air Today.* London: HMSO, 1974.

Digest of Energy Statistics. London: HMSO.

Dorf, R. C. *Energy, Resources and Policy.* London: Addison-Wesley, 1979.

Earth Resources Foundation. *Australia's Mineral Energy Resources: Assessment and Potential.* New South Wales, Australia; University of Sydney, 1978.

England, Glyn. *The Gas Turbine—A Successful Developing Technology.* London: Central Electricity Generating Board, 1978.

England, Glyn. *Coal and Coal Prices.* London: Central Electricity Generating Board, 1977.

Environment Protection Authority, Victoria. *Report on the Environmental Effects of Newport Power Station.* Victoria, Australia: Government Printer, 1973.

Environmental Council, N. Z. *Energy in the New Zealand Environment.* Wellington, New Zealand: Government Printer, 1977.

Environmental Protection Agency, US. *Air Pollution Engineering Manual,* 2nd Ed. North Carolina, 1973.

European Economic Community. *Continuation and Implementation of a European Community Policy and Action Programme on the Environment.* Brussels, 1976.

Ezra, Sir Derek. *Coal and Energy.* London: Ernest Benn, 1978.

Financial Times. Oil and Gas International Year Book. London.

Fisher, J. C. *Energy Crisis in Perspective.* London: Wiley, 1974.

Fox, R. W. *et al. Ranger Uranium Environment Inquiry:* First Report 1976; Second Report 1977. Canberra: Australian Government Publishing Service.

Francis, W. *Coal: Its Formation and Composition.* London: Arnold, 1954.

Francis, W. *Fuels and Fuel Technology.* London: Pergamon, 1965.

Gilpin, Alan. *The Prospects for Nuclear Power Generation in Australia.* Brisbane: Unpublished Ph. D. thesis, University of Queensland, 1974.

Gilpin, Alan. *Australia's Future Fuel Requirements: the Magnitude and Significance of Resultant Atmospheric Emissions.* Sydney.

Gilpin, Alan. *Control of Air Pollution.* London, England: Butterworths, 1963.

Gilpin, Alan. *Air Pollution.* Brisbane, Australia: University of Queensland Press, 1978.

Gilpin, Alan. *Dictionary of Economic Terms*, 4th edn. London England: Butterworths, 1977.

Gilpin, Alan. *Dictionary of Environmental Terms.* Brisbane, Australia: University of Queensland Press, 1976. London: Routledge and Kegan Paul, 1976.

Griffin, A. R. *Coal Mining.* London: Longmans, 1971.

Hawksley, P. G. W., Badzioch, S. and Blackett, J. H. *Measurement of Solids in Flue Gases.* Leatherhead: British Coal Utilisation Research Association, 1963.

Llitch, C. H. (Ed.). *Modelling Energy—Economy Interactions: Five Approaches.* Baltimore; Johns Hopkins, 1977.

Institute of Energy, UK. *The Efficient Use of Energy.* London: IPC Science and Technology Press, 1978.

Institute of Energy, UK. *Fluidised Combustion.* Proceedings of Institute of Energy Conferences, 1975, 1980.

Institute of Energy, UK. *Energy from Waste Burning*, 1979.

Institute of Energy, UK. *Energy World.* A monthly publication.

Institute of Energy, UK. *Fuel Abstracts and Current Titles* (FACTS). A regular publication.

Institute of Energy UK. *Energy for the Future: Report from the Working Party*, 1973.

Institute of Energy, UK. *Progress in the Incineration of Industrial and Domestic Waste*, 1979.

Institute of Energy, UK. *Incineration of Municipal and Industrial Waste*, 1969

Institute of Energy, UK. *Potential for Power*, 1979.

Institute of Energy, UK. *Proceedings of the Conference on Science in the Use of Coal*, 1958.

Institute of Energy, UK. *Proceedings of the Second Conference on Pulverised Fuel*, 1957.

Institute of Energy, UK. *Proceedings of the Third Conference on Liquid Fuels*, 1966.

Institute of Energy, UK. *Proceedings of a Symposium on the Inorganic Constituents of Fuel*, 1964.

Institute of Energy, UK. *The Nuclear Fuel Cycle*, 1979.

Institute of Petroleum. *Modern Petroleum Technology.* London.

Institute of Petroleum. *Energy: from Surplus to Scarcity?* Proceedings of the Institute of Petroleum Summer Meeting, June 1973. London: Applied Science Publishers, 1974.

International Atomic Energy Agency. *Environmental Contamination by Radioactive Materials.* Proceedings of a Seminar jointly Organised by IAEA, FAO and WHO. Vienna, 1969.

Inter-Governmental Maritime Consultative Organisation (IMCO). *Proceedings of the International Conference on Tanker Safety and Pollution Prevention, 1978.* London.

Krevelen, D. W. *Coal.* Amsterdam: Elsevier, 1961.

Krevelen, D. W. and Schuyer, J. *Coal Science: Aspects of Coal Constitution.* Amsterdam: Elsevier, 1957.

Joel Rayner. *Basic Engineering Thermodynamics.* London: Longmans Green, 1966.

Jensen, W. G. *Energy and the Economy of Nations.* London: Foulis, 1970.

Lovins, A. B. *World Energy Strategies: Facts, Issues and Options.* London: Friends of the Earth, 1973.

Loftness, R. L. *Energy Handbook.* Wokingham: Van Nostrand Reinhold, 1979.

Lyle, Sir Oliver. *The Efficient Use of Steam.* London: HMSO, 1947.

Medical Research Council. *The Hazards to Man of Nuclear and Allied Radiations* (First Report, Cmnd 9780, 1956; Second Report, Cmnd. 1225). London: HMSO, 1960.

Ministry of Power. *Fuel Policy.* Cmnd 3438. London: HMSO, 1967.

Nathan, R. A. (Ed.). *Fuels From Sugar Crops.* New York: US Department of Energy, 1979.

National Academy of Sciences, US. *Lead: Airborne Lead in Perspective.* Washington, DC, 1972.

National Coal Board, UK. *Annual Reports,* published since 1947. London

National Coal Board, UK. *Plan for Coal.* London, 1974.

National Coal Board, UK. *Black Diamonds—Silver Anniversary 1947—72.* London

National Coal Board, UK. *Coal: Its Origin and Occurrence.* London, 1970.

National Energy Advisory Committee. *Australia's Energy Resources: An Assessment.* Report No 2. Canberra: Department of National Development, 1978.

National Industrial Fuel Efficiency Service. *The New Stoker's Manual.* London.

Odell, Peter R. *Oil and World Power: Background to the Oil Crisis,* 4th edn. Harmondsworth: Penguin, 1975.

Organisation for Economic Co-operation and Development. *Photochemical Oxidant Air Pollution.* Paris, 1975.

390

Organisation for Economic Co-operation and Development. *Long Range Transport of Air Pollution*. Paris, 1977.

Payne, Gordon A. *The Energy Managers' Handbook*. Guildford: IPC, 1979.

Phillips, Owen. *The Last Chance Energy Book*. Baltimore: Johns Hopkins, 1979.

Political and Economic Planning. *The British Fuel and Power Industries; A Report*. London: PEP, 1947.

Posner, M. V. *Fuel Policy: A Study in Applied Economics*. London: MacMillan, 1973.

Raggatt, H. G. (Ed.). *Fuel and Power in Australia*. Melbourne: Cheshire, 1969.

Rasmussen, N. C. *An Assessment of Accident Risks in U. S. Commercial Nuclear Power Plants*. Washington, DC: U.S. Atomic Energy Commission, 1974.

Reid, G. L. *The Nationalised Industries*. London: Heinemann, 1973

Royal College of Physicians. *Air Pollution and Health*. London, 1970.

Royal Commission on Environmental Pollution. *Air Pollution Control: An Integrated Approach*. Fifth Report, Cmnd. 6371. London, England: HMSO, 1975.

Royal Commission on Environmental Pollution. *Nuclear Power and the Environment*. Sixth Report. London, England: HMSO, 1976.

Royal Commission on Petroleum. *Towards a National Refining Policy: Fifth Report*. Canberra: Australian Government Publishing Service, 1976.

Sach, J. S. (Ed.). *Coal Tar Fuels: their Derivation, Properties and Application*. London: Association of Tar Distillers.

Select Committee on Science and Technology, UK. *Report on Energy Conservation*, 1975; and *Energy Conservation: The Government's Reply to the Report from the Select Committee on Science and Technology*, Cmnd. 6575: London: HMSO, 1975.

Senate Select Committee on Science and the Environment. *Solar Energy*. Canberra: Australian Government Printing Office, 1977.

Severn, R. T., Dineley, D. L. and Hawker, L. G. (Eds). *Tidal Power and Estuary Management*. New York: John Wright, 1979.

Shacker, Sheldon R. *The Complete Book of Electric Vehicles*. Sydney: Australia and New Zealand Book Company, 1979.

Shell Briefing Service. *Improved Energy Efficiency*. London and Rotterdam: Royal Dutch/Shell Group, 1979.

Siddall, M. *The Science and Art of Coal Mining*. Robens Coal Science Lecture 9 October 1972. London: British Coal Utilisation Research Association, 1972.

Sinclair, J. *Geological Aspects of Mining*. London: Pitman, 1958.

Smith, Irene. *Carbon Dioxide and the Greenhouse Effect: An unresolved Problem*. London: International Energy Agency/Coal Research, 1978.

Spiers, H. M. (Ed.). *Technical Data on Fuel*. London: British National Committee, World Energy Conference.

Stern, A. C. (Ed.). *Air Pollution*. Vols I–V, 1968–80. New York and London: Academic Press, 1968.

Teller, Edward. *Energy from Heaven and Earth*. New York: Freeman, 1979.

Thring, M. W. *The Science of Flames and Furnaces*. London: Chapman and Hall, 1962.

Turvey, Ralph. *Optimal Pricing and Investment in Electricity Supply*. London: Allen and Unwin, 1968.

United Nations Environment Programme. *Review of the Impact of Production and Use of Energy on the Environment*. Nairobi, 1976.

Williams, J. M. *Boiler House Practice*. London: Allen and Unwin, 1960.

World Health Organisation. *Air Quality Criteria and Guides for Urban Air Pollutants*. Technical Report Series No 506. Geneva: 1972.

World Energy Conference. *Energy Resources: Availability and Rational Use*. A digest of the 10th World Energy Conference. Guildford: IPC, 1979.

WITHDRAWAL